Economic Geography

This book series serves as a broad platform for scientific contributions in the field of Economic Geography and its sub-disciplines. *Economic Geography* wants to explore theoretical approaches and new perspectives and developments in the field of contemporary economic geography. The series welcomes proposals on the geography of economic systems and spaces, geographies of transnational investments and trade, globalization, urban economic geography, development geography, climate and environmental economic geography and other forms of spatial organization and distribution of economic activities or assets.

Some topics covered by the series are:

- Geography of innovation, knowledge and learning
- Geographies of retailing and consumption spaces
- Geographies of finance and money
- Neoliberal transformation, urban poverty and labor geography
- Value chain and global production networks
- Agro-food systems and food geographies
- Globalization, crisis and regional inequalities
- Regional growth and competitiveness
- Social and human capital, regional entrepreneurship
- Local and regional economic development, practice and policy
- New service economy and changing economic structures of metropolitan city regions
- Industrial clustering and agglomeration economies in manufacturing industry
- Geography of resources and goods
- Leisure and tourism geography

Publishing a broad portfolio of peer-reviewed scientific books *Economic Geography* contains research monographs, edited volumes, advanced and undergraduate level textbooks, as well as conference proceedings. The books can range from theoretical approaches to empirical studies and contain interdisciplinary approaches, case studies and best-practice assessments. Comparative studies between regions of all spatial scales are also welcome in this series. Economic Geography appeals to scientists, practitioners and students in the field.

If you are interested in contributing to this book series, please contact the Publisher.

Tatiana López

Labour Control and Union Agency in Global Production Networks

A Case Study of the Bangalore
Export-garment Cluster

 Springer

Tatiana López
University of Würzburg (JMU)
Würzburg, Germany

Berlin Social Science Center (WZB)
Berlin, Germany

ISSN 2520-1417   ISSN 2520-1425 (electronic)
Economic Geography
ISBN 978-3-031-27389-6   ISBN 978-3-031-27387-2 (eBook)
https://doi.org/10.1007/978-3-031-27387-2

This Springer imprint is published by the registered company Springer Nature Switzerland AG
The registered company address is: Gewerbestrasse 11, 6330 Cham, Switzerland

Foreword

For a long time, labour and its conditions were neglected in economic geography and related disciplines. The dominant research interest focussed on companies in their competitive regional and international environments. In this context, workers and employees were portrayed in a subordinate role: Theoretical concepts framed the working population either as passive and controlled by capital actors or as a mere location factor in managers' decision-making processes. However, since the end of the twentieth century, international work in economic geography has started to deal with labour control questions and the active role of blue- and white-collar workers. In this context, female work and gender relations have increasingly come into focus as well.

Since the beginning of the twenty-first century, labour geography has developed into a highly dynamic field of research focussing on human labour in the context of capitalist production. Within the structural context of capitalist competition, translated into labour control within companies, work organisation is hardly self-determined and provides limited meaningful task designs for many workers. A Taylorist or neo-Taylorist division of labour processes into small tasks often prevents meaningful work. This tendency is further exaggerated due to technological development, such as digitisation. In the context of digital work restructuring, the interests of workers and employees and aspects of attractive task design are often not systematically considered.

How work takes place and how it is controlled differs according to the level of competencies of the workers and employees. In general, the work of professionals is often more demanding and characterised by a higher degree of responsibility and complexity. In contrast, simple work is often repetitive; this is the case, especially for work on the shop floor, such as in factories and logistic centres. The extent of high and low work requirements varies from sector to sector as well as along different stages of the same value chains. Moreover, the geographical location makes a difference: Working conditions differ in the Global North and the Global South (and, on a smaller scale, between regions of a country) and between metropolises and rural areas.

The focus within labour geography on the spatial conditions of labour sheds light on the international integration of local labour and employment relationships on the one hand. Still, it also highlights resistance, labour conflict, co-determination and negotiation, on the other hand. Local institutional settings, embedded in political dynamics (such as labour laws and co-determination laws), and networks of local actors (such as trade unions, works councils and the local state), play a central role here. At the same time, the analysis of local institutions and actors needs to be linked back to the intertwined world economy. Not only companies operate internationally, but also organisations that represent workers' interests.

The present book by Tatiana López addresses this topic through the lens of a case study of the garment industry in the Indian city of Bangalore and its surrounding areas. The study illustrates the interplay of dynamics of labour control with the agency strategies of worker organisations. The book introduces the reader to current debates at the intersection of labour geography and Global Production Network (GPN) analysis using a relational, practice-oriented research approach. The juxtaposition of 'decent work' approaches and critical political-economic perspectives on work control is central to the theoretical concept.

Furthermore, the book addresses a research gap in previous scalar analyses of labour control in GPNs. The relational approach developed by Tatiana López in this book not only reveals how labour control as a structural context is constructed through interrelated practices of different state and capital actors in Bangalore and beyond, but it also sheds light on spaces of intervention and change.

Based on a qualitative, single-embedded case study of Bangalore's garment export cluster, the book substantiates the constraints for creating decent work. These include, among other things, the recruitment of female surplus labour force from rural areas, tight workplace control by factory managements and 'pro-business' state practices. At the same time, the book also shows 'spaces of labour agency'. In this regard, the study reveals certain yet limited opportunities for improving working and social conditions for labourers in the industrial export sectors of the Global South through international regulation, such as supply chain laws and consumer campaigns. The book illustrates that to bring about sustained improvements of working conditions, building local union bargaining power in production countries remains central.

The book is, therefore, a highly valuable contribution worth reading for anyone interested in human and humane labour, its conditions and potential for improvements. In addition, the book provides valuable cause for thought for scholars involved in theory building in Labour Geography and GPN analysis. It is an important basis for practitioners working in trade unions and international organisations committed to improving working conditions in the global garment industry.

Cologne, Germany   Martina Fuchs
October 2022

Acknowledgements

This book is the outcome of my Ph.D. research project conducted between 2015 and 2021 at the Department for Economic and Social Geography at the University of Cologne. Here, I wish to express my deepest and most sincere gratitude to all the fantastic people I have met on this journey who have supported me in one way or another.

First of all, I would like to sincerely thank my Ph.D. supervisor Prof. Martina Fuchs, who has provided me with invaluable academic and practical support throughout all stages of my Ph.D. I would like to also extend my thanks to my second Ph.D. supervisor Prof. Matthias Pilz for reviewing the manuscript for this book and to my mentor Dr. Manuela Maschke. Their tips and support were crucial during the writing process. I am also grateful to my first postdoc supervisor Prof. Martin Krzywdzinski for his support in the publication phase.

Second, I thank my colleagues at the University of Cologne and the WZB Berlin Social Science Center who have been an important part of my journey and who have helped me in so many ways—from bouncing conceptual ideas around over lunch to developing strategies for coping with stress, insecurities and other challenges. In this line, I would like to also thank the student assistants who have helped me transcribe interviews and create maps for this book.

Third, I am grateful to all the fantastic people from the TIE Global union network I have had the luck to meet and work with over the past six years: Thank you for allowing me to become a part of this great network, for the political and theoretical discussions, for the inspiring and motivating meetings and for your friendship.

Fourth, I am deeply indebted to all the people and organisations who have enabled me to conduct the field research for this thesis: first and foremost, the union leaders and activists from GATWU, GLU, KGWU and the AITUC Bangalore chapter who have given me many hours of their valuable time. My sincere appreciation also extends to all other interviewees for sharing their knowledge and insights with me.

Fifth, I am thankful for the financial support from the Centre for Modern Indian Studies, the Global South Studies Centre Cologne and the Department of Social and Economic Geography at the University of Cologne, for my various field research trips. Moreover, I am grateful for funding from the WZB Berlin Social Science

Center, the Leibniz Open Access Monograph Publishing Fund and the Department of Social and Economic Geography at the University of Cologne, who made the open-access publication of this book possible.

Last but not least, I want to thank my family and friends for always being there for me through all the ups and downs. Special thanks go to my husband Bruce for his constant encouragement and support.

Without the input and support from every single individual or organisation mentioned here, this book would not exist.

Cologne   Tatiana López
October 2022

About This Book

This book explores the conditions that enable and constrain the capacity of local unions in garment producing countries to bring about lasting improvements for workers. It makes the central argument that even in globalised production, local worker power and agency remain central for promoting positive change in working conditions. Theoretically, the book integrates concepts and frameworks from labour geography with Global Production Networks (GPN) analysis and relational, practice-oriented research approaches in economic geography.

After the introductory chapter, which sets out the context and relevance of this book, Chap. 2 situates this book within broader debates on labour in global value chains (GVCs) or GPNs. First, the chapter delineates two contrasting analytical approaches to labour in GVCs/GPNs—the 'Decent Work' approach and the 'Marxist political economy approach'—and places this study within the latter approach. The chapter then proceeds to discuss major conceptual contributions from economic and labour geography with regard to explaining labour control and labour agency in GVCs/GPN. In this context, the chapter highlights the lack of attention by hitherto dominant scalar conceptualisations of labour control and labour agency in GPNs for the interrelations and interdependencies between processes at different levels—from the workplace to the international level of the global production network.

To tackle this research gap, Chap. 3 develops a practice-oriented, relational approach to studying labour control regimes and union agency in GPNs. It conceptualises labour control regimes at specific nodes of a GPN as emerging from place-specific articulations of six processual labour control relations stretching across various distances with localised labour processes. These labour control relations are, on the one hand, sourcing relations at the vertical dimension of the GPN and, on the other hand, territorially embedded workplace, wage, employment, industrial and labour market relations at the horizontal dimension of the GPN. Chapter 3 further presents a heuristic framework for studying labour's networked agency strategies in GPNs through the lens of three relational 'spaces of labour agency' constructed by unions through relationships with different actors: (1) spaces of organising constituted through unions' practices of building relationships with workers; (2) spaces of collaboration constituted through unions' relationships with other labour or civil society organisations; and (3) spaces of contestation constructed by unions around specific labour struggles through practices of targeting employers, lead firms and state actors, and through practices of 'drawing' allies into struggles.

Chapter 4 introduces this study's case study research design and discusses the different methods used for data collection and analysis. After that, Chap. 5 situates the empirical case of this study—the Bangalore export-garment cluster—within the garment GPN: The chapter first lays out the central characteristics of the garment GPN. It then describes the historical and geographical evolution of the Bangalore export-garment cluster.

Chapters 6 and 7 finally present the empirical analysis of the labour control regime and unions' agency strategies in the Bangalore export-garment cluster. Chapter 6 analyses how the labour control regime in the Bangalore export-garment industry emerges from the articulations of labour control dynamics in the workplace with labour control dynamics at the state and international levels. It illustrates how the capital accumulation regime in the garment GPN is reproduced through a complex mesh of labour control practices, including local managers' gendered worker exploitation practices, state authorities' 'pro-business' practices and international lead firms' predatory purchasing practices. Chapter 7 then scrutinises the different agency strategies of the three local garment unions active in the Bangalore export-garment cluster regarding their potential for bringing about sustained improvements for workers. It highlights local unions' practices of building relationships with actors at various levels as a central element of unions' networked agency strategies. These include community organising practices to evade the tight management control within factories and gain the trust of women workers, who constitute around 85% of the workforce in the Bangalore export-garment cluster. Moreover, unions build alliances with consumer and labour organisations from the Global North. However, the chapter exposes that collaborations with consumer organisations and NGOs from the Global North tend to have mixed or constraining effects on unions' capacities to build associational, organisational and workplace bargaining power.

The concluding chapter (Chap. 8) summarises the central findings in light of the posed research questions and discusses this book's empirical and theoretical contributions. In terms of empirical contributions, the book highlights local worker organisations' central role in improving garment industry working conditions. At the same time, it sheds light on the complex, networked labour control structures that constrain the terrain for labour agency in garment producing countries. Against this background, the book stresses the need for unions to develop networked agency strategies that employ coalitional and moral power resources from international consumer and labour organisations to open spaces for workplace organising and collective bargaining. At the same time, unions need to be wary of the risks of relying on Global North actors' power and financial resources. When power flows unilaterally from North to South, this can hamper internal union democracy and the space for workers and unions to develop strategic capacities, thereby further constraining unions' capacities to develop associational power resources. Regarding the conceptual contribution, the analytical approach developed and applied in this study reinvigorates a relational understanding of labour control and agency in GPNs as exercised through power-laden, networked relationships at the vertical and horizontal dimensions of the GPN. The study addresses a central gap in past scalar analyses of labour control and labour agency in GPNs, which have not sufficiently explored the links between network dynamics and territorial outcomes for labour at specific nodes of a GPN.

Contents

About the Author

Dr Tatiana López is an economic and labour geographer researching labour in global production and in the digital platform economy. She holds a Ph.D. in economic geography and a diploma in Latin American Studies from the University of Cologne, Germany. Currently, she is a postdoctoral research fellow at the University of Würzburg in the working group Economic Geography and a guest researcher at the Berlin Social Science Center in the research group Globalization, Work, Production.

Abbreviations

ADWU	Avery Dennison Workers Union
AEPC	Apparel Export Promotion Council
AITUC	All India Trade Union Confederation
ATDC	Apparel Training and Design Centre
BMS	Bharatiya Mazdoor Sangh
CCC	Clean Clothes Campaign
CITU	Centre of Indian Trade Unions
CLB	China Labour Bulletin
CoC	Codes of Conduct
CSR	Corporate Social Responsibility
CWM	Centre for Workers' Management
ETI	Ethical Trading Initiative
FoA	Freedom of Association
GATWU	Garment and Textiles Worker Union
GCC	Global Commodity Chain
GLU	Garment Labour Union
GPN	Global Production Network
GVC	Global Value Chain
HMS	Hindh Mazdoor Sabha
ILO	International Labour Organisation
INTUC	Indian Trade Union Congress
ISDS	Integrated Skill Development Scheme
KGWU	Karnataka Garment Workers Union
LCR	Labour Control Regime
LPT	Labour Process Theory
MoU	Memorandum of Understanding
MSI	Multi-Stakeholder-Initiative
NGO	Non-governmental Organisation
NOLA	Networks of Labour Activism

NTUI	New Trade Union Initiative
OEM	Original Equipment Manufacturer
WRC	Worker Rights Consortium

List of Figures

List of Tables

Part I
Introduction

Chapter 1
Introduction: Why We Need Stronger Unions in the Global Garment Industry

Abstract For newly industrialising countries, the global garment industry is considered a vehicle for economic and social development, especially for increasing women's participation in the labour market. At the same time, the garment industry has also been widely criticised for frequent labour rights violations, low wages and bad working conditions. Media and public discourses have focussed largely on private regulatory mechanisms and international labour standards as tools for promoting 'decent work' in the global garment industry. However, this chapter argues that lasting improvements for workers can only be achieved through the agency of strong local unions in garment producing countries. Against this background, this chapter introduces two central research questions that remain understudied in existing literature on labour in Global Value Chains (GVCs) and Global Production Networks (GPNs): (1) How do labour control regimes at specific nodes of the garment GPN shape and constrain the terrain for worker and union agency in garment producing countries? (2) Which relationships and interactions enable unionists and workers in garment producing countries to develop strategic capacities and power resources that allow them to shift the capital-labour power balance in favour of workers?

Keywords Bangalore · Garment industry · Global production networks · Labour power · Labour control regime · Union agency

1.1 Towards a Relational Analysis of the Enabling and Constraining Conditions for Local Union Agency in Garment Producing Countries

For newly industrialising countries worldwide, the global garment industry is considered a vehicle for economic and social development (Dicken 2015; Gereffi 1994, 1999). Particularly in Asian countries, the emergence of export-garment industries producing for US and European retailers has propelled industrialisation processes and created jobs for millions of workers (ILO 2015). In particular for low-skilled or unskilled workers, the garment industry can facilitate access to the formal labour market and—where state and industry actors invest in vocational education and

© The Author(s) 2023

T. López, *Labour Control and Union Agency in Global Production Networks*,
Economic Geography, https://doi.org/10.1007/978-3-031-27387-2_1

training—also important upskilling opportunities (Maurer 2011). Today, Asian countries account for eight out of the top ten global garment exporters (WTO 2020: 119). In these countries, on average, half of all manufacturing jobs are in the garment sector (ILO 2018: vii). The garment industry is also considered an important driver of female economic empowerment in the region, with women comprising the major share of the workforce in most Asian countries (ILO 2015). In India, for example, the textiles and apparel[1] sector provides direct employment for about 45 million people, of which about 70% are women (Indian Ministry of Textiles 2018; Make in India n.d.). Approximately 50% of the Indian textiles and apparel sector correspond to the ready-made garment industry with one quarter of produced apparels being sold on the global market (CareRatings 2019: 1).

Despite the significant contribution of the Asian export-garment industry to economic development and employment creation in the region, the industry has been widely criticised for frequent labour rights violations, low wages and bad working conditions (see, e.g., Hale and Wills 2005; Jenkins and Blyton 2017; Mezzadri 2017; Ruwanpura 2016). Anti-sweatshop movements and consumer organisations from the Global North have attributed these bad conditions to global fashion retailers' and brands' 'predatory purchasing practices' that 'squeeze' suppliers and workers (Anner 2019, 2020; Esbenshade 2004). Against this background, public and academic discourse has largely focussed on private regulatory mechanisms—such as Codes of Conduct or Multi-Stakeholder Initiatives—as well as on international labour standards as tools for promoting 'decent work' in the global garment industry (see, e.g., Bartley and Egels-Zandén 2015; Hess 2013; Hughes et al. 2008; Lindholm et al. 2016; Lund-Thomsen and Lindgreen 2018). In the same way, literature from economic geography and labour studies has predominantly highlighted the role of Northern actors—and particularly of transnational consumer and NGO networks—as main agents for change (Hauf 2017; see, e.g., Kühl 2006; Merk 2009). Most recently, legislative projects obligating companies in Global North countries to implement due diligence obligations along their supply chains—such as the French 'Loi de Vigilance'[2] or the German 'Supply Chain Act'[3]—have attracted great public attention as potential mechanisms for improving working conditions in the garment industry (see, e.g., Beckers et al. 2021; Clerc 2021; Maihold et al. 2021). However, far less attention has been paid to the role of workers in global garment producing countries as agents capable of improving their own working and living conditions (Kumar 2019a: 351; Wells 2009).

---

[1] The terms 'garment' and 'apparel' are used synonymously in this study.

[2] The French 'Loi de Vigilance' (engl. Vigilance Law) was passed in 2017. The law requires French companies with more than 5000 employees to implement and publish a so-called Vigilance Plan setting out proactive measures to prevent health, safety and environmental risks in their subsidies as well as across their subcontractors and suppliers (Clerc 2021).

[3] The German 'Act on Corporate Due Diligence Obligations in Supply Chains' was passed in 2021 and will come into force starting 2023. It establishes due diligence obligations for companies with a main seat or a branch office in Germany. These due diligence obligations apply to a company's entire supply chains, and compliance failures may be penalised with a fine of up to 2% of a company's global annual turnover (German Federal Ministry of Labour and Social Affairs [2021]).

A notable exception is provided by a growing corpus of studies combining Global Commodity Chain (GCC)/ Global Value Chain (GVC)/ Global Production Network (GPN) analysis with perspectives from Labour Process Theory and labour geography, which have been at the forefront of exploring the conditions and strategies for the agency of workers and unions in garment producing countries (see, e.g., Anner 2015b; Doutch 2021; Kumar 2014, 2019a; Ruwanpura 2015; Zajak 2017). These studies form part of two broader debates in economic and labour geography concerned with labour control and labour agency in GVCs/GPNs (Coe et al. 2008; Coe and Jordhus-Lier 2011; Cumbers et al. 2008; Newsome et al. 2015; Rainnie et al. 2011; Taylor et al. 2015). Drawing on Marxist theories, studies on labour control and labour agency in GVCs/GPNs highlight the exploitative nature of capitalist production as the root cause for 'indecent work' and stress that lasting improvements can only be achieved through collective worker organisation in garment producing countries (Selwyn 2013; Kumar 2019b: 351f.). A central concern of these studies has hence been to examine mechanisms of exploitation as well as conditions and strategies for the collective resistance of workers in garment producing countries.

Studies on labour control in the garment GVC/GPN have contributed in particular to our understanding of constraints for collective worker and union agency in garment producing countries. Inter alia, these studies have highlighted the presence of insti-tutionalised labour control regimes that ensure the process of capital accumulation at specific nodes of the garment GPN through what Baglioni (2018: 111) refers to as "the interplay of labour exploitation and disciplining". Whereas exploitation refers to the extraction of surplus value from 'living labour' in the labour process, disci-plining refers to preventing, mitigating and repressing labour resistance (Baglioni 2018: 114). Even though studies on labour control in the garment GPN have focussed primarily on mechanisms of exploitation, these studies have also highlighted capital and state actors' various disciplining mechanisms and practices undermining collec-tive worker organisation in the garment industry (see, e.g., Anner 2015a; Ruwanpura 2015; Smith et al. 2018).

Against this backdrop, *studies on labour agency in the garment GVC/GPN* have tended to highlight the 'constrained' nature of the agency of workers and unions in garment producing countries (Coe and Jordhus-Lier 2011). Several studies have high-lighted that workers and unions 'at the bottom' of the garment GVC lack domestic bargaining power "since they can easily be replaced and their wages are not expected to provide effective demand for the goods they produce" (Hauf 2017: 1001; see also Kumar 2014; Tsing 2009; Zajak et al. 2017). In face of these barriers for building domestic bargaining power, studies on labour agency in the garment GVC/GPN have repeatedly emphasised strategies of 'up-scaling' or 'scale-jumping' as central for workers in garment export countries to improve their own working and living conditions (Anner 2015b; Merk 2009; Wells 2009). In particular, these studies have stressed opportunities for local unions to 'up-scale' workplace conflicts through engaging with consumer campaigning networks or multi-stakeholder initiatives in the Global North. These actors can pressure brands and retailers to harness their lead firm power and to request improvements in working conditions from their suppliers (see, e.g., Anner 2011, 2015a; Armbruster-Sandoval 2005; Kumar 2014; Merk 2009).

However, accounts of successful 'up-scaling' of labour struggles by unions in garment producing countries have been accompanied by more critical writing, pointing out two important limits of cross-border campaigning strategies for achieving sustainable improvements of workers' rights and conditions (Anner 2015b; Fink 2014; Fütterer and López Ayala 2018; López and Fütterer 2019; Zajak 2017): First, campaigning strategies are always reactive and tend to be effective only in severe cases of labour rights violations, in which the cost of reputational damage is higher for lead firms and suppliers than the cost of corrective action (Fütterer and López Ayala 2018: 21; Kumar 2019a: 347). Second, particularly when the leverage of geographically distant consumers and lead firms is not underpinned by strong local worker organisation, the success achieved through lead firms' and consumers' 'top-down' pressure tends to be rather short-lived—or in the words of Ross (2006: 78) "a temporary rescue, fragile and vulnerable to employers' attacks".

Existing research at the intersection of GPN analysis and labour geography clearly states that far-reaching and sustained improvements for workers in garment producing countries can only be achieved where strong workplace and industry-level unions exist. At the same time, we still lack a systematic understanding of the various factors and conditions that enable and constrain the building of strong unions and worker bargaining power in garment producing countries. First insights into the conditions that curb union building and collective worker organisation have been provided by studies of labour control regimes in garment producing countries (see e.g. Anner 2015a; Ruwanpura 2016; Wickramasingha and Coe 2021). However, the conditions, relationships and practices that enable garment workers and unions to shift the capital-labour power balance in favour of labour remain underexplored. This gap can be attributed inter alia to a general tendency of labour geography to produce rather descriptive accounts of workers' and unions' strategic (up-scaling) actions in the context of specific labour struggles. These accounts have however not systematically embedded workers' and unions' actions within broader structural conditions that shape, enable and constrain the agency of labour (c.f. Coe and Jordhus-Lier 2011: 213). In this light, recent studies have pointed out the need for a deeper exploration not only of how labour control regimes in garment producing countries shape the terrain for labour agency, but also of the structural effects that 'Networks of Labour Activism'—such as networks with consumer groups and NGOs—have on unions' practices and internal relations, thereby shaping unions' capacities to build bargaining power vis-à-vis employers (see, e.g., Hauf 2017; Zajak et al. 2017).

In this light, this study aims to contribute to a better understanding of the conditions that constrain and enable the building of strong local unions in garment producing countries. In view of this, the central research question guiding this study can be formulated as follows: *Which conditions enable and constrain the capacities of local unions in garment producing countries to build bargaining power vis-à-vis employers and the state and thereby to bring about sustained improvements for workers in the garment industry?*

To answer this question, I develop *a practice-oriented, relational research approach to labour control and labour agency in GPNs* that allows us to look beyond 'isolated' labour struggles and instead to analyse the agency of unions as embedded within broader networks of social, cultural and economic relations (c.f. Berndt and Fuchs, 2002; Coe and Jordhus-Lier 2010; Coe 2015). To this end, I link academic debates on labour control and labour agency in GPN and reconceptualise central conceptual frameworks from a relational, practice-oriented meta-theoretical perspective (c.f. Amin 2004; Jones and Murphy 2010, Martin 2010; Massey 1994). On the one hand, I build on work at the intersection of Labour Process Theory and GVC/GPN analysis (Newsome et al. 2015) as well as on studies of (local) labour control regimes in GPNs (Jonas 1996; Smith et al. 2018) to develop a practice-oriented, relational approach for analysing labour control regimes at specific nodes of a GPN as place-specific articulations of multiple *processual relations of labour control* stretching across various distances with localised labour processes. These networked processual relations are, in turn, constructed through intertwined exploiting and disciplining practices performed by a variety of capital and state actors.

On the other hand, I build on the analytical frameworks of 'union power resources' (Schmalz et al. 2018) and 'Networks of Labour Activism' (Zajak et al. 2017) to conceptualise the agency strategies of local unions at specific nodes of the GPN as emerging from the intersection of three relational *'spaces of labour agency'* constructed by workers and unions themselves: (1) *spaces of organising* linking union organisers, workers and union members; (2) *spaces of collaboration* linking local unions to other external labour and non-labour actors in solidary ways; and (3) *spaces of contestation* constructed by unions around specific labour struggles through practices of targeting capital and state actors, on the one hand, and through engaging with allies—such as consumer networks or other labour actors—to plan and execute solidary action.

Figure 1.1 summarises the theoretical underpinnings and contributions of this book. It visualises the meta-theoretical perspective that guides this study, the two academic debates in which this study situates itself, and the specific theoretical frameworks that this study builds on. Marked in italics are the conceptual contributions of this book to debates on labour control and labour agency in GPN.

Following the relational, practice-oriented analytical approach of this book, two subordinate research questions can be derived from the central research question:

1. *How do labour control regimes at specific nodes of the garment GPN—constituted through place-specific articulations of processual labour control with localised labour processes—shape and constrain the terrain for the agency of workers and unions in garment producing countries?*
2. *Which relationships and routine interactions enable unionists and workers in garment producing countries to develop strategic capacities and power resources that allow them to shift the capital-labour power balance in favour of workers?*

I approach the formulated research questions through the lens of a qualitative, single embedded case study (Yin 2014) of the labour control regime and union agency in the Bangalore export-garment cluster. The justification for selecting Bangalore's export-garment cluster as the case for this study is provided in the following section.

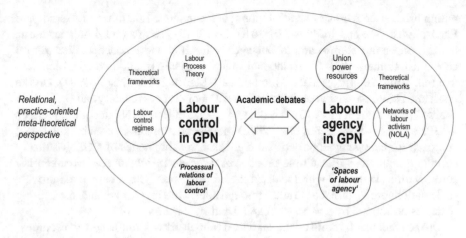

Fig. 1.1 Theoretical underpinnings and contributions of this book. *Source* Author

1.2   Empirical Case Study of This Book: The Export-garment Cluster in Bangalore, India

I have chosen the Bangalore export-garment cluster as an empirical case for this study for the following two reasons. First, the Bangalore export-garment cluster represents an important node in the garment GPN: Garments manufactured in Bangalore account for roughly 60% of India's garment exports, with India itself being the fifth largest export country of garments on the global market (SLD and AFWA 2013: 18; WTO 2020: 119). Compared to other export-garment clusters in India, Bangalore has an exceptionally high presence of large tier one factories acting as strategic suppliers for US and EU fashion retail companies. For example, out of H&M's 253 tier one supplier factories in India, 53 factories—or one-fifth of all factories—are located in and around Bangalore (H&M 2021). The strategic importance of the Bangalore cluster as a node in EU and US retailers' production networks is also exemplified by the fact that many of these retailers maintain local production offices in Bangalore.

Second, compared to other garment production hubs in India, the Bangalore garment industry is characterised by a high level of union activity. Three politically independent, local grassroots unions have been organising workers in the Bangalore export-garment sector since 2006. Over the past 15 years, these unions have achieved important benefits for workers, such as significant minimum wage increases and—most recently—the first factory-level collective bargaining agreement in the Indian garment industry. Nevertheless, at the same time, unionisation rates remain low at around five per cent, reflecting the existence of a tight labour control regime posing severe challenges for collective worker organisation. Hence, the Bangalore export-garment cluster provides a rich empirical case to study the constraining and enabling conditions for the agency and bargaining power of local unions in garment producing countries.

1.3 Structure of This Book

Following this introduction, in Chapter 2, I position this study within geographical debates on labour in GPNs and provide an overview of existing analytical approaches and empirical studies. To this end, in Sect. 2.1, I first introduce the GCC, GVC and GPN frameworks and illustrate the evolution of labour as a study object within GVC/GPN analysis. After that, in Sect. 2.2, I outline the most important characteristics of two contrasting analytical approaches to labour in GVCs/GPNs—the 'Decent Work'-approach and the 'Marxist Political Economy' approach—and place this study within the latter approach. Sections 2.3 and 2.4 then introduce the major research strands on labour control and labour agency in GVCs/GPNs and highlight the central contributions and shortcomings of existing studies. After the literature review, Chapter 3 introduces the core tenets of a practice-oriented, relational analytical perspective and develops heuristic frameworks for studying labour control regimes and union agency at specific nodes of a GPN from a practice-oriented, relational approach. Chapter 4 introduces the single embedded case study research design underpinning this study and discusses the different methods used for data collection and data analysis. Thereafter, Chapter 5 situates the empirical case of this study—the Bangalore export-garment cluster—within the garment GPN. In Sect. 5.1, I first lay out central characteristics of the garment GPN before describing the historical and geographical evolution of the Bangalore export-garment cluster in more detail in Sect. 5.2. After this introduction of the study area and case study, Chapters 6 and 7 finally present the empirical analysis of the labour control regime and unions' agency strategies in the Bangalore export-garment cluster. Chapter 6 analyses how the labour control regime in the Bangalore export-garment industry emerges from the place-specific articulations of different processual labour control relations stretching over various distances with the localised labour process. Chapter 7, in turn, scrutinises the agency strategies of three local garment unions active in the Bangalore export-garment cluster regarding their potential for building sustained bargaining power vis-à-vis employers and the state. After the empirical analysis, Chapter 8 concludes by answering the posed research questions and discussing the theoretical contributions of this study to current debates on labour control and agency in GPNs as well as for GPN analysis and practice-oriented research in economic geography more generally.

References

Amin A (2004) Regions unbound: towards a new politics of place. Geogr Ann Ser B, Hum Geogr 86:33–44. https://doi.org/10.1111/j.0435-3684.2004.00152.x
Anner M (2015a) Labor control regimes and worker resistance in global supply chains. Labor Hist 56:292–307. https://doi.org/10.1080/0023656X.2015.1042771

Anner M (2015b) Social downgrading and worker resistance in apparel global value chains. In: Newsome K, Taylor P, Bair J, Rainnie A (eds) Putting labour in its place: labour process analysis and global value chains. Palgrave Macmillan, London/New York, pp 152–170

Anner M (2019) Predatory purchasing practices in global apparel supply chains and the employment relations squeeze in the Indian garment export industry. Int Labour Rev 158:705–727. https://doi. org/10.1111/ilr.12149

Anner M (2020) Squeezing workers' rights in global supply chains: purchasing practices in the Bangladesh garment export sector in comparative perspective. Rev Int Polit Econ 27:320–347. https://doi.org/10.1080/09692290.2019.1625426

Anner MS (2011) Solidarity transformed: labor responses to globalization and crisis in Latin America. ILR Press, Ithaca

Armbruster-Sandoval R (2005) Globalization and cross-border labor solidarity in the Americas: the anti-sweatshop movement and the struggle for social justice. Routledge, New York

Baglioni E (2018) Labour control and the labour question in global production networks: exploitation and disciplining in Senegalese export horticulture. J Econ Geogr 18:111–137. https://doi. org/10.1093/jeg/lbx013

Bartley T, Egels-Zandén N (2015) Responsibility and neglect in global production networks: the uneven significance of codes of conduct in Indonesian factories. Glob Netw 15:S21–S44. https:// doi.org/10.1111/glob.12086

Beckers A, Kühlert M, Liedtke C, Micklitz H-W (2021) A legal framework for global value chains: impulse 2021. https://wupperinst.org/fa/redaktion/downloads/misc/Legal_Framework_Global_ Value_Chains.pdf. Accessed 30 December 2021

Berndt C, Fuchs M (2002) Geographie der Arbeit. Plädoyer für ein disziplinübergreifendes Forschungsprogramm. Geogr Z 90:157–166

CareRatings (2019) Indian readymade garments (apparel) industry overview: April 17, 2019, Industry Research. http://www.careratings.com/upload/NewsFiles/Studies/Indian%20Ready% 20Made%20Garments%20(Apparel)%20Industry.pdf

Clerc C (2021) The French 'duty of vigilance' law: lessons for an EU directive on due diligence in multinational supply chains. SSRN J. https://doi.org/10.2139/ssrn.3765288

Coe NM, Dicken P, Hess M (2008) Global production networks: realizing the potential. J Econ Geogr 8:271–295. https://doi.org/10.1093/jeg/lbn002

Coe NM (2015) Labour and global production networks: mapping variegated landscapes of agency. In: Newsome K, Taylor P, Bair J, Rainnie A (eds) Putting labour in its place: labour process analysis and global value chains. Palgrave Macmillan, London/New York, pp 171–192

Coe NM, Jordhus-Lier DC (2010) Re-embedding the agency of labour. In: Bergene AC, Endresen SB, Knutsen HM (eds) Missing Links in Labour Geography. Ashgate Pub, Farnham, pp 29–42

Coe NM, Jordhus-Lier DC (2011) Constrained agency?: re-evaluating the geographies of labour. Prog Hum Geogr 35:211–233. https://doi.org/10.1177/0309132510366746

Cumbers A, Nativel C, Routledge P (2008) Labour agency and union positionalities in global production networks. J Econ Geogr 8:369–387. https://doi.org/10.1093/jeg/lbn008

Dicken P (2015) Global shift: mapping the changing contours of the world economy. Guilford Pr, New York, NY

Doutch M (2021) A gendered labour geography perspective on the Cambodian garment workers' general strike of 2013/2014. Globalizations 18:1406–1419. https://doi.org/10.1080/14747731. 2021.1877007

Esbenshade JL (2004) Monitoring sweatshops: workers, consumers, and the global apparel industry. Temple University Press, Philadelphia

Fink E (2014) Trade unions, NGOs and transnationalization: experiences from the ready-made garment sector in Bangladesh. Asien: 42–59

Fütterer M, López Ayala T (2018) Challenges for organizing along the garment value chain. Experiences from the union network TIE ExChains. https://www.rosalux.de/en/publication/id/39369/ challenges-for-organizing-along-the-garment-value-chain/. Accessed 5 April 2022

Gereffi G (1994) The organization of buyer-driven global commodity chains: how U.S. retailers shape overseas production network. In: Gereffi G, Korseniewicz M (eds) Commodity chains and global capitalism. Praeger Publishers, Westport, pp 95–122

Gereffi G (1999) International trade and industrial upgrading in the apparel commodity chain. J Int Econ 48:37–70

German Federal Ministry of Labour and Social Affairs (2021) Act on corporate due diligence obligations in supply chains. https://www.csr-in-deutschland.de/EN/Business-Human-Rights/Supply-Chain-Act/supply-chain-act.html. Accessed 30 December 2021

H&M (2021) H&M Group supplier list. https://hmgroup.com/sustainability/leading-the-change/supplier-list.html. Accessed 23 May 2021

Hale A, Wills J (eds) (2005) Threads of labour: garment industry supply chains from the workers' perspective. Antipode book series. Blackwell, Malden, Mass, Oxford

Hauf F (2017) Paradoxes of transnational labour rights campaigns: the case of play fair in Indonesia. Dev Chang 48:987–1006. https://doi.org/10.1111/dech.12321

Hess M (2013) Global production networks and variegated capitalism: (self-)regulating labour in Cambodian garment factories: Better work discussion paper series: no. 9. http://betterwork.org/global/wp-content/uploads/DP-9.pdf

Hughes A, Wrigley N, Buttle M (2008) Global production networks, ethical campaigning, and the embeddedness of responsible governance. J Econ Geogr 8:345–367. https://doi.org/10.1093/jeg/lbn004

ILO (2015) Strong export and job growth in Asia's garment and footwear sector: Asia-Pacific Garment and Footwear Sector Research Note, Issue 1, November 2015. ILO, Bangkok

ILO (2018) From obligation to opportunity: scrutinizing the garment and textiles value chains to foster better working conditions in Vietnam and Indonesia. ILO Publications, Geneva

Indian Ministry of Textiles (2018) Annual report 2017–2018. http://texmin.nic.in/sites/default/files/AnnualReport2017-18%28English%29.pdf. Accessed 2 May 2020

Jenkins J, Blyton P (2017) In debt to the time-bank: the manipulation of working time in Indian garment factories and 'working dead horse.' Work Employ Soc 31:90–105. https://doi.org/10.1177/0950017016664679

Jonas AEG (1996) Local labour control regimes: uneven development and the social regulation of production. Reg Stud 30:323–338. https://doi.org/10.1080/00343409612331349688

Jones A, Murphy JT (2010) Practice and economic geography. Geogr Compass 4:303–319. https://doi.org/10.1111/j.1749-8198.2009.00315.x

Kühl B (2006) Protecting apparel workers through transnational networks: the case of Indonesia. Zugl.: Kassel, Univ., Diss., 2005. Ibidem-Verl., Stuttgart

Kumar A (2014) Interwoven threads: building a labour countermovement in Bangalore's export-oriented garment industry. City 18:789–807. https://doi.org/10.1080/13604813.2014.962894

Kumar A (2019a) A race from the bottom? Lessons from a workers' struggle at a Bangalore warehouse. Compet Chang 23:346–377. https://doi.org/10.1177/1024529418815640

Kumar A (2019b) Oligopolistic suppliers, symbiotic value chains and workers' bargaining power: labour contestation in South China at an ascendant global footwear firm. Glob Netw 19:394–422. https://doi.org/10.1111/glob.12236

Lindholm H, Egels-Zandén N, Rudén C (2016) Do code of conduct audits improve chemical safety in garment factories? Lessons on corporate social responsibility in the supply chain from Fair Wear Foundation. Int J Occup Environ Health 22:283–291. https://doi.org/10.1080/10773525.2016.1227036

López T, Fütterer M (2019) Herausforderungen und Strategien für den Aufbau gewerkschaftlicher Verhandlungsmacht in der Bekleidungswertschöpfungskette: Erfahrungen aus dem TIE-ExChains-Netzwerk. In: Ludwig C, Simon H, Wagner A (eds) Bedingungen und Strategien gewerkschaftlichen Handelns im flexiblen Kapitalismus. Westfälisches Dampfboot, Münster, pp 175–191

Lund-Thomsen P, Lindgreen A (2018) Is there a sweet spot in ethical trade? A critical appraisal of the potential for aligning buyer, supplier and worker interests in global production networks. Geoforum 90:84–90. https://doi.org/10.1016/j.geoforum.2018.01.020

Maihold G, Müller M, Saulich C, Schöneich S, Stiftung Wissenschaft und Politik (2021) Responsibility in supply chains: Germany's due diligence act is a good start: SWP Comment No. 21 March 2021. https://www.swp-berlin.org/publications/products/comments/2021C21_Responsibility_Supply_Chains.pdf. Accessed 15 January 2022

Make in India (n.d.) Textiles and garments. https://www.makeinindia.com/sector/textiles-and-garments. Accessed 20 May 2020

Martin J (2010) Limits to 'thinking space relationally.' Int J Law Context 6:243–255. https://doi.org/10.1017/S1744552310000145

Massey DB (1994) Space, place and gender. Polity Press, Cambridge

Maurer M (2011) Skill formation regimes in South Asia, a comparative study on the path-dependent development of technical and vocational education and training for the garment industry. Komparatistische Bibliothek, vol 21. Peter Lang, Frankfurt am Main

Merk J (2009) Jumping scale and bridging space in the era of corporate social responsibility: cross-border labour struggles in the global garment industry. Third World Q 30:599–615. https://doi.org/10.1080/01436590902742354

Mezzadri A (2017) Sweatshop regimes in the Indian garment industry. Cambridge University Press

Newsome K, Taylor P, Bair J, Rainnie A (eds) (2015) Putting labour in its place: labour process analysis and global value chains. Palgrave Macmillan, London/ New York

Rainnie A, Herod A, McGrath-Champ S (2011) Review and positions: global production networks and labour. Compet Chang 15:155–169. https://doi.org/10.1179/102452911X13025292603714

Ross RJS (2006) A tale of two factories: successful resistance to sweatshops and the limits of firefighting. Labor Stud J 30:65–85. https://doi.org/10.1177/0160449X0603000404

Ruwanpura KN (2015) The weakest link?: Unions, freedom of association and ethical codes: a case study from a factory setting in Sri Lanka. Ethnography 16:118–141. https://doi.org/10.1177/1466138113520373

Ruwanpura KN (2016) Garments without guilt? Uneven labour geographies and ethical trading—Sri Lankan labour perspectives. J Econ Geogr 16:423–446. https://doi.org/10.1093/jeg/lbu059

Schmalz S, Ludwig C, Webster E (2018) The power resources approach: developments and challenges. Glob Labour J 9:113–134. https://doi.org/10.15173/glj.v9i2.3569

Selwyn B (2013) Social upgrading and labour in global production networks: a critique and an alternative conception. Compet Chang 17:75–90. https://doi.org/10.1179/1024529412Z.00000000026

SLD, Society for Labour and Development, AFWA, Asia Floor Wage Alliance (2013) Wage structures in the Indian garment industry. https://asia.floorwage.org/resources/wage-reports/wage-structures-in-the-indian-garment-industry. Accessed 16 March 2018

Smith A, Barbu M, Campling L, Harrison J, Richardson B (2018) Labor regimes, global production networks, and European union trade policy: labor standards and export production in the Moldovan clothing industry. Econ Geogr 94:550–574. https://doi.org/10.1080/00130095.2018.1434410

Taylor P, Newsome K, Bair J, Rainnie A (2015) Putting labour in its place: labour process analysis and global value chains. In: Newsome K, Taylor P, Bair J, Rainnie A (eds) Putting labour in its place: labour process analysis and global value chains. Palgrave Macmillan, London/New York, pp 1–26

Tsing A (2009) Supply chains and the human condition. Rethink Marx 21:148–176. https://doi.org/10.1080/08935690902743088

Wells D (2009) Local worker struggles in the global south: Reconsidering northern impacts on international labour standards. Third World Q 30:567–579. https://doi.org/10.1080/01436590902742339

Wickramasingha S, Coe N (2021) Conceptualizing labor regimes in global production networks: uneven outcomes across the Bangladeshi and Sri Lankan apparel industries. Econ Geog:1–23. https://doi.org/10.1080/00130095.2021.1987879

WTO (2020) World Trade Statistical Review 2020. https://www.wto.org/english/res_e/statis_e/wts 2020_e/wts2020_e.pdf. Accessed 25 March 2021

Yin RK (2014) Case study research: design and methods. Sage, Los Angeles

Zajak S (2017) International allies, institutional layering and power in the making of labour in Bangladesh. Dev Chang 48:1007–1030. https://doi.org/10.1111/dech.12327

Zajak S, Egels-Zandén N, Piper N (2017) Networks of labour activism: collective action across Asia and beyond. An introduction to the debate. Dev Chang 48:899–921. https://doi.org/10.1111/dech. 12336

Part II
Theoretical Framework

Chapter 2
From a 'Decent Work' Approach to a Marxist Analysis of Labour Control and Labour Agency in Global Production: Reviewing Research on Labour in GPNs

Abstract This chapter reviews literature on labour in GVCs and GPNs. It argues that within the interdisciplinary literature on labour issues in GVCs/GPNs, two parallel research strands have emerged that are characterised by very different conceptual approaches: (1) a 'Decent Work' approach underpinned by the institutionalist perspective of the ILO Decent Work Agenda and (2) a 'Marxist Political Economy' approach, which is based on the assumption that the exploitation of labour is an inherent structural feature of capitalist production systems. Situating this study within the second research strand, this chapter then reviews the contributions and shortcomings of existing literature on labour control and labour agency in GVCs/GPNs. In doing so, the chapter highlights the limitations of existing scalar approaches for studying labour control and labour agency in GVCs/GPNs, which have not paid enough attention to how dynamics of labour control and labour agency at different levels influence each other. Against this background, this chapter argues that to gain a more nuanced understanding of the 'architectures of labour control' underpinning specific GPNs as well as of workers' and unions' networked agency strategies, a relational analytical approach can be beneficial.

Keywords Global production networks · Global value chains · Workplace regimes · Labour control regimes · Labour agency · Union power resources

This study aims to contribute to and establish itself within the broader debate of economic geography on labour in global production networks (GPNs). In particular, it aims to contribute to two strands of research on labour in GPNs that are concerned, firstly, with structures and mechanisms of labour control in GPNs; second, with the conditions and strategies for the agency of workers and unions in GPNs. In this regard, it is important to note that rather than exclusively working with the GPN framework, contributions to these debates have also worked with the Global Commodity Chain (GCC) and the Global Value Chain (GVC) framework. Therefore, in the following literature review, I include studies focussing on labour control and labour agency working with either of these approaches. When developing my own relational approach to labour control and labour agency in Chapter 3, I will, however, draw on the GPN framework since its network perspective fits best with the relational perspective adopted in this book.

T. López, *Labour Control and Union Agency in Global Production Networks*, Economic Geography, https://doi.org/10.1007/978-3-031-27387-2_2

In the remainder of this chapter, I will first demarcate the conceptual features of the GCC, GVC and GPN approaches and introduce central analytical concepts of the GPN approach (Sect. 2.1). After that, I outline two basic conceptual approaches to labour in GPNs—the 'Decent Work' approach and the Marxist Political Economy approach—and position this study within the latter approach (Sect. 2.2). Sections 2.3 and 2.4 then give an overview of the current state of research on labour control and labour agency in GPNs. Each chapter sketches three main strands of research on labour control and labour agency, respectively, summarising their conceptual and empirical contributions and highlighting their shortcomings.

2.1 From Linear Commodity Chains to Relational Production Networks: Opening Up Analytical Space for the Role of Labour

In this section, I introduce the three main analytical frameworks that have under-pinned past studies of labour in global production systems: the Global Commodity Chain (GCC), the Global Value Chain (GVC) and the Global Production Network (GPN) framework. Whereas the GPN framework was the first framework to explicitly open up analytical space for labour, an increasing number of GCC/GVC studies has also tackled the enabling and constraining conditions for union and worker agency in global production (see e.g. Anner 2015b; Riisgaard and Hammer 2008, 2011; Selwyn 2012, 2013, 2015). This development is in line with the general trend towards closer integration of GCC, GVC and GPN studies over the last two decades, with the three approaches coming to form one wider interdisciplinary research community (Coe and Yeung 2019: 775). Therefore, when displaying the state of the broader research debate on 'labour in GPN' in the remaining sections of this literature review chapter, I will include conceptual approaches concerned with the control and agency of labour in global production systems working with either of the three approaches. Never-theless, important conceptual differences remain between the GCC, the GVC and the GPN framework, particularly with regard to their potential for producing rela-tional accounts of the constraining and enabling factors for the agency of workers and unions within global production systems. I argue that when it comes to under-standing the agency of workers as embedded in and constituted through manifold networked relationships, the GPN framework provides the best conceptual tools (c.f. Cumbers 2015: 136). To illustrate this argument, in the following, I briefly introduce the central ontological assumptions and conceptual tools of the GCC, the GVC and the GPN approach.

The *Global Commodity Chains (GCC) framework* was introduced by the Amer-ican sociologist Gary Gereffi (1994, 1999) in the 1990s and subsequently found wide application within economic geography. The declared aim of the GCC framework was to provide a framework for analysing different modes of organising interna-tional "production systems that give rise to particular patterns of coordinated trade"

(Gereffi 1994: 96). As the name suggests, the GCC framework is underpinned by a linear chain ontology that conceptualises international production systems through the lens of their sequential 'input-output structure', defined as "the set of products and services linked together in a sequence of value-adding activities" (Gereffi 1994: 97). As two other central dimensions for analysis, Gereffi (1994) has introduced *'territoriality'*, referring to the specific geographical distribution of the activities involved in the production, distribution and sales of a specific commodity, and *'governance structure'*, referring to the power relations between firms in a specific production chain that determine "how financial, material and human resources are allocated and flow within a chain" (Gereffi 1994). The most important conceptual contribution of the GCC framework has been in the dimension of governance with the distinction between buyer-driven and producer-driven commodity chains (Gereffi 1999: 41f.). Whereas *producer-driven commodity chains* are controlled by large industrial multinational enterprises and found typically in capital and technology-intensive sectors, *buyer-driven commodity chains* are typically controlled by large retailers and branded merchandisers selling labour-intensive mass consumption commodities. Retailers and branded merchandisers—typically originating from the Global North—act as lead firms. They capture the biggest share of added value in the commodity chain by controlling the higher value-adding steps of design, marketing and retail, while outsourcing the typically labour-intensive production to independent suppliers in low-wage countries (Gereffi 1994: 97). At the same time, retailers and branded merchandisers exercise significant control over the parameters of the production process by providing detailed specifications regarding product design and quality as well as by setting prices and lead times (Gereffi 1999: 55). Considering that the global garment chain is a prototypical buyer-driven chain, the distinction between buyer-driven and producer-driven value chains has been a focal point for studies concerned with inter-firm power relations and its impact on labour in the global garment sector.

Notwithstanding the important to date influence of the distinction between producer-driven and buyer-driven chains provided by the GCC framework, it was also subjected to various critiques, particularly by economic geographers. Firstly, critics pointed out the GCC framework's almost exclusive emphasis on 'governance' as an analytical dimension, which was perceived to make an overly crude distinction with its exclusive focus on buyer-driven and producer-driven commodity chains as two sole modes of governance (Dicken et al. 2001). Second, critics argued that due to its linear chain ontology and the resulting focus on inter-firm relations, the GCC framework possessed limited capacity for analysing the role of state, labour and civil society actors in shaping global production systems (Dicken et al. 2001: 100). Lastly, economic geographers criticised the GCC framework for paying insufficient attention to the 'specific social and institutional contexts' at the national, regional and local level into which firms are embedded and hence to account for how commodity chains shape local dynamics and vice versa (Henderson et al. 2002: 441).

Reacting to the first critique, Gereffi et al. (2005) presented the *Global Value Chains (GVC) framework* as a further development of the GCC framework, which provided a more nuanced typology of inter-firm power relationships in international production systems. The GVC framework hence distinguished between five types of value chains, each characterised by a specific type of inter-firm relationships: (1) markets, i.e. value chains that are characterised by arm's length relationships between lead firms and suppliers; (2) modular value chains, in which 'turn-key' suppliers carry out technology-intensive production of specialised inputs or 'modules' for lead firms; (3) relational value chains characterised by complex interactions, high levels of trust and co-dependence between lead firms and suppliers; (4) captive value chains, in which a large number of rather small suppliers are 'transactionally dependent' on large buyers; and (5) hierarchy, a type of value chain that is characterised by vertical integration of all steps of production (Gereffi et al. 2005: 83f.). While already demonstrating a more fine-grained understanding of power relations within GVCs, the GVC framework, however, continued to focus on inter-firm relations around linear input-output structures, while largely ignoring other sets of relationships in international commodity production systems (Bair 2008; Coe et al. 2008: 275).

It is against this backdrop that a group of economic geographers of the so-called Manchester school set out in the early 2000s to develop the GPN framework as an alternative heuristic approach for analysing relationships in global production systems (Dicken et al. 2001; Coe et al. 2004, 2008; Henderson et al. 2002). As opposed to the linear ontology of the GCC and GVC framework, the GPN framework is underpinned by a relational ontology. This relational ontology conceptualises global production systems as constituted through "highly complex *network* structures in which there are intricate links – horizontal, diagonal, as well as vertical – forming multi-dimensional, multi-layered lattices of economic activity" (Henderson et al. 2002: 442, emphasis in original). Hence, the GPN framework recognises that each production network inevitably contains a *vertical value chain dimension,* i.e. a set of relationships linking actors throughout the linear sequence of stages manufacturing stages to distribution and consumption. At the same time, the GPN framework highlights that actors at each stage of the vertical dimension are also embedded into various sets of relationships at the *horizontal dimension that* constitute place-specific local, regional and national political economies (Coe et al. 2008: 274ff.) (Fig. 2.1).

In this light, Coe (2015: 185) has argued that the ontology of the GPN framework can best be characterised as *'territorial cum relational',* since it integrates network relationships that link actors at different 'nodes' of the GPN on the one hand, and territorially embedded, institutionalised multi-scalar regulatory dynamics on the other. As a result of this particular ontology, the GPN framework can furthermore include a wide range of non-firm actors, such as labour, governments, civil society organisations and consumers, as "constituent parts of the overall production system" (Coe et al. 2008: 275). Through the relational lens of the GPN framework, it is from the interactions of these societal and state actors with firm actors at multiple levels that global production systems emerge in the form of networks. These networks are at the same time relational and structural: "Networks are structural, in that the composition and interrelation of various networks constitute structural power relations, and they are relational because they are constituted by the interactions of variously powerful social actors" (Dicken et al. 2001: 94). In this line, GPNs are understood as "contested organisational fields", in which various actors with their own interests "struggle over

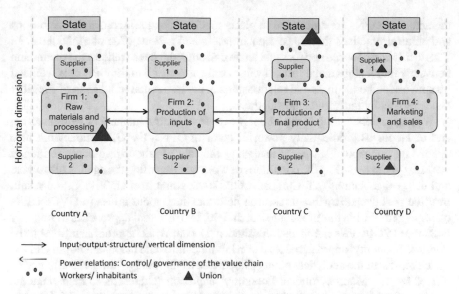

Fig. 2.1 Horizontal and vertical dimension in the GPN framework. *Source* Author's own elaboration based on Coe et al. (2008)

the construction of economic relationships, governance structures, institutional rules and norms, and discursive frames" (Levy 2008: 944).

The GPN framework has introduced three central analytical dimensions: value, power and embeddedness (Henderson et al. 2002: 448ff.). The analytical dimensions of '*value*' examines how different actors within the production network create, enhance and capture value. The GPN framework understands value in this context as encompassing "both Marxian notions of surplus value and more orthodox ones associated with economic rent" (Henderson et al. 2002: 448). '*Power*' as an analytical dimension in turn raises questions about which actors exercise power in which ways to secure or increase their share of value within the production system (Henderson et al. 2002: 450). As opposed to the GCC/GVC framework, which conceptualises power very narrowly as lead firm power over suppliers, the GPN framework recognises various actors as potentially capable of exercising power within GPNs, including, for example, multinational and domestic firms, local and national state agencies, international organisations, trade unions and consumer organisations (Henderson et al. 2002: 450f.). Lastly, the analytical dimension '*embeddedness*' introduces two different types of embeddedness that characterise GPNs: network embeddedness and territorial embeddedness (Henderson et al. 2002: 453f.). '*Network embeddedness*' refers to the fact that GPNs link actors across territorial boundaries "regardless of their country of origin or local anchoring in particular places" (Henderson et al. 2002: 453). '*Territorial embeddedness*' in turn refers to the fact that the actors and activities that GPNs' links are at the same time 'grounded' in specific places for two reasons. First, most economic activities are spatially fixed in particular locations due to the immobility of the needed production infrastructure or labour force. Second, the actors in GPNs are embedded in place-specific social relationships, institutions and cultural practices that shape their interests and actions (Coe et al. 2008: 279). The distinction between network and territorial embeddedness reflects the distinction between relationships at the vertical value chain dimension of the GPN characterised by network embeddedness and relationships at the horizontal dimension of the GPN characterised by territorial embeddedness.

With its understanding of the labour process as a central moment of value creation and of labour as a potentially powerful actor in GPNs, the GPN framework also opened up conceptual space for analysing the role of labour in global production systems—both regarding the conditions of work and regarding the agency of workers and their organisations (c.f. Cumbers et al. 2008; Rainnie et al. 2011). As a result, over the past decade a vibrant research field tackling labour issues in GVCs/GPNs has emerged. Within this broader research field, two different approaches to studying labour in GVCs/GPNs can be distinguished: a 'Decent Work' approach and a 'Marxist Political Economy' approach. Debates on labour control and labour agency in GPN—which represent the analytical point of departure of this study—are generally underpinned by the 'Marxist Political Economy' approach. Therefore, to demarcate the research field that this study aims to contribute to, in the next Sect. 2.2 I briefly sketch the main theoretical-philosophical assumptions and concepts informing both approaches.

2.2   Contrasting Approaches to Analysing Labour in GVCs/GPNs: The 'Decent Work' Approach and the 'Marxist Political Economy' Approach

With labour issues coming into the focus of both GVC and GPN analysis over the past decade, two parallel research strands emerged that are characterised by very different conceptual approaches to analysing labour in global production systems. In this book, I refer to these two approaches as the 'Decent Work' approach and the 'Marxist Political Economy' approach. As the name suggests, the *'Marxist Political Economy' approach* is underpinned by a Marxist political economy perspective. This perspective focuses on the relations between labour, capital, the state and consumers and is based on the assumption that the exploitation of labour is an inherent structural feature of capitalist production systems (c.f. Swyngedouw 2003; Rainnie et al. 2011). The *'Decent Work' approach* to analysing labour in GVCs/GPNs in turn is underpinned by the institutionalist perspective of the ILO Decent Work agenda and focuses on the links between value chain governance and working conditions as a means for promoting social development (Barrientos et al. 2011b; Mayer and Pickles 2010). As a result of these different perspectives, the 'Marxist Political Economy' research approach and the 'Decent Work' research approach are characterised by rather different agendas and concepts of labour (Fig. 2.2).

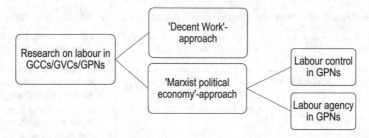

Fig. 2.2 Research approaches to labour in GVCs/GPNs—'Marxist political economy' approach and 'Decent Work' approach. *Source* Author

In the following two sections, I first outline the philosophical foundations, the conception of labour, central theoretical concepts, main research interests and criticisms of the 'Decent Work' approach (Sect. 2.2.1) and then of the 'Marxist Political Economy' approach (Sect. 2.2.2). It is important to note that the representation of these approaches in this chapter is a stylised one that aims to highlight their different conceptual and philosophical underpinnings. In this light, Table 2.1 gives an overview of the characteristics of each research approach to labour in GPN.

2.2.1 The 'Decent Work' Approach to Labour in GVCs/GPNs

The 'Decent Work' approach to labour in GVCs/GPNs was originally conceived by scholars from the GVC-school to provide policy-oriented conceptual tools for promoting the ILO Decent Work Agenda (Barrientos 2007; Mayer and Pickles 2010; Barrientos et al. 2011a). The ILO Decent Work Agenda was introduced in 1998 to tackle the increasing informalisation, deregulation and flexibilisation of labour markets under globalisation, especially in the growing export industries of developing economies in the Global South (Lerche 2012: 18). Whereas the ILO and GVC scholars saw the integration of developing economies into global value chains as an important motor for employment and economic development, they also recognised the need to improve the conditions of work and employment in newly industrialising countries (Barrientos 2007: 1; ILO 1999).

In this context, the ILO Decent Work Agenda was introduced to achieve a fair globalisation and poverty reduction through promoting rights at work, employment and income opportunities, social protection and social security as well as social dialogue (ILO 1999). 'Decent Work' was conceived in this context as a term that should converge the ILO's four strategic objectives and thus designate work that takes place "under conditions of freedom, equity, security and dignity, in which rights are protected and adequate remuneration and social coverage is provided" (Barrientos 2007: 1). In the seminal report from 1999, the ILO Director-General pleaded that the concept of Decent Work "must guide [the ILO's] policies and define its international role in the near future". In this light, the ILO Decent Work Agenda

Table 2.1 Characteristics of the 'Decent Work' approach and the 'Marxist political economy' approach to labour in GPN

	'Decent Work' approach	'Marxist political economy' approach
Philosophical foundations	• ILO Decent Work Agenda • Development Theories • Institutional Economics	• Marxian political economy, sociology and history • Marxist economic geography • Labour process theory • Labour geography
Conception of labour	• Labour as a productive factor • Workers as socially embedded agents with capabilities and entitlements	• Labour as social process of surplus extraction • Labour as 'living' commodity • Labour as active agent shaping geographies of capitalism
Central theories and concepts	• Economic and social upgrading • Regulatory governance in GVCs/GPNs • Concept of Decent Work	• Factory regimes • Labour control regimes • Labour agency in GPNs
Central research interests	• Quantity and quality of employment generated by GVCs/GPNs • Promotion of decent work in developing and emerging economies • Relationship between economic and social upgrading • Effectiveness mechanisms of social regulation, e.g. MSIs, CSR, ILO standards, consumer campaigns	• Nature and forms of capitalist exploitation • Capital's strategies and mechanisms of labour control • Potentials, barriers and strategies for labour agency • Multi-scalar and spatial forms of worker and union agency
Criticisms/blind spots	• Neglects the exploitation process at the heart of capitalist production • Neglects conflict of interest between capital and labour • Discounts agency of workers and trade unions in improving workers' conditions	• Theoretical concept of labour agency remains under-theorised • Neglects internal and external factors shaping collective labour agency/union agency

Source Author

must be interpreted as the main pillar of the broader ILO development agenda, which sought to promote the ILO's principles and rights at work in international development policies and initiatives (ILO 1999; Lerche 2012: 18ff.). As the main strategy to achieve its goals, the ILO's Decent Work Agenda proposed institution-building, particularly in the fields of worker participation and representation, social dialogue and social protection. Intellectually, the ILO Decent Work Agenda thus drew from debates in development theory and policies about promoting globalisation or development 'with a human face' as well as on institutional economics (ILO 1999).

GVC scholars started to engage with the ILO Decent Work Agenda in the 2000s to complement the perceived one-sided focus of the GVC framework on economic development with a social dimension (c.f. Gereffi and Korseniewicz 1994; Gereffi 1994, 1999, 2005; Gereffi et al. 2005). This engagement gave rise to a research strand within GVC and GPN studies concerned with identifying potentials, barriers and strategies for promoting decent work in GVCs/GPNs to promote economic and social development. Barrientos et al. (2011b: 320) formulate as the central question of the 'Decent Work' approach "how to improve the position of both firms and workers within GPNs". On the one hand, the 'Decent Work' approach is hence interested in the impact of individual firms' strategic choices on working conditions and labour rights. On the other hand, the 'Decent Work' approach to labour in GVC/GPN aims to conceptualise workers "beyond their role as factors of production, highlighting them as human beings with capabilities and entitlements" (Barrientos et al. 2011b: 322). In this view, the well-being of workers mainly depends on access to rights and resources that enhance their well-being. Access to these rights and resources is, however, mediated through institutional arrangements encompassing employers as well as government institutions and communities (ibid.).

In this line, the 'Decent Work' approach to labour in GVCs/GPNs proposed has put forth two main analytical concepts: the concept of regulatory governance and the concept of social upgrading. Mayer and Pickles (2010: 2) introduced the concept of *regulatory governance* as a type of governance that "constrains the behaviour of profit-seeking firms that might otherwise tend to exploit workers, leading to poor working conditions, lack of job security, constraints on worker organisation, and general downgrading of industrial relations systems and practices". Regulatory governance can take on the form of *public governance* comprising governmental rules, regulations and policies and *private governance* comprising inter alia social norms, corporate codes of conduct, CSR initiatives, consumer campaigns, social movements and other non-governmental institutions (ibid.)

To promote Decent Work in GVCs/GPNs, Mayer and Pickles (2010) argued that it is central to overcome the present 'governance deficit' in the globalised economy resulting from "limited [governance] capacities in the emerging economies, weak international institutions, increasingly challenged institutions in advanced industrial countries and everywhere greater emphasis on facilitation than on regulation" (Mayer and Pickles 2010: 3). According to Mayer and Pickles, this '*governance deficit*' has led to the deterioration of working conditions and to a shift away from formal, regular and secure employment towards more flexible, informal and insecure employment in global production. Pressure by consumer movements has then led to the rise of new private forms of regulatory governance, such as corporate Codes of Conduct (CoC), Multi-Stakeholder Initiatives (MSI) or Global Framework Agreements. There are, however, significant limits to the potential of private regulatory governance mechanisms alone improving workers' rights and working conditions. Against this background, Mayer and Pickles argue that private governance mechanisms need to be complemented by initiatives that aim to strengthen public governance at the national and international level (Mayer and Pickles 2010: 14f.).

The concept of social upgrading, in turn, was introduced by Barrientos et al. (2011a) as a counterpart to the concept of economic upgrading, informing GVC studies until then. Whereas *economic upgrading* refers to the process by which economic actors move from lower value-added activities to higher value-added activities (Gereffi 2005: 171ff.), *social upgrading* designates the "process of improvement in the rights and entitlements of workers as social actors, which enhances the quality of their employment" (Barrientos et al. 2011b: 324). The central assumption of the concept of 'social upgrading' is that economic upgrading does not automatically lead to social upgrading (Barrientos et al. 2011b; Milberg and Winkler 2011; Rossi 2013). Whether social upgrading occurs in GVCs mainly depends on which market pressures prevail and on which competitive strategy firms adopt: Firms may compete via quality—a strategy that depends on a skilled workforce and thus is likely to promote social upgrading processes. Or firms may compete via price, which would most probably hinder processes of social upgrading or even effect social downgrading, e.g. deterioration of working conditions and workers' rights (Barrientos et al. 2011b: 333).

Central research topics of studies adopting a 'Decent Work' approach to labour in GPN have been the quantity and quality of employment generated in GVCs/GPNs, with a particular focus on: export sectors in developing and emerging economies (Barrientos et al. 2011a), the relationship between economic and social upgrading (Milberg and Winkler 2011; Rossi 2013; Pyke and Lund-Thomsen 2016), as well as the effectiveness of private governance mechanisms such as codes of conduct (Locke et al. 2007; Egels-Zandén and Merk 2014).

Studies engaging with labour in GVCs/GPNs with a 'Decent Work' approach need to be given credit for problematising the initial implicit assumption of the GVC framework that processes of economic upgrading automatically bring about processes of social development. Moreover, they have produced relevant insights into the limitations of private governance mechanisms for improving workers' conditions. However, research adopting a 'Decent Work' approach to labour in GVCs/GPNs has also encountered criticism, especially by scholars advocating a Marxist political economy approach for labour in GVCs/GPNs (Arnold and Hess 2017; Rainnie et al. 2011; Selwyn 2013; Werner 2012).

Marxist scholars have argued that the 'Decent Work' approach to labour in GVCs/GPNs and especially its conceptualisation of social upgrading present two crucial analytical and political weaknesses. According to Marxist scholars, the first and major analytical weakness of the 'Decent Work' approach is its inability to understand the systemic processes of exploitation characterising capitalist social relations

as root cause for indecent work (Selwyn 2013: 75f.). Given its intellectual foundations in institutional theory, the 'Decent Work' approach in GVC/GPN is based on
the assumption that "given the right institutional context, capital does not exploit
labour" (Selwyn 2013: 82). According to Marxist scholars, this assumption is prob-
lematic because "the upgrading analytic centres industrial change on the relations of
power and dynamics of competition among firms, rendering the social relations that
mediate the production of exploitable workers and the conditions of their exploitation
marginal to the analysis" (Werner 2012: 407). Politically, these scholars criticise that
the emphasis on promoting social dialogue in the 'Decent Work' approach serves
the interests of capital by de-legitimising adversarial bargaining and more militant
forms of labour agency (Arnold and Hess 2017: 2191; Standing 2008).

As a second weakness, Marxist scholars hold that the 'Decent Work' approach
diminishes the role of the agency of labour (Selwyn 2013: 83). In the institutional
notion of governance promoted by the 'Decent Work' approach, labour is conceived
as just one among many institutional actors that could possibly work together
to strengthen regulatory governance and, thus, constrain "the behaviour of profit-
seeking firms" (Mayer and Pickles 2010: 2). At the same time, the 'Decent Work'
approach is based on the assumption that firms cooperate with other stakeholders and
engage in regulatory governance, not necessarily because they are forced to do so by
labour and/or other actors, but because "capital can be persuaded that workers are
vital to its reproduction […] and by extension deserving of socio-economic rights"
(Arnold and Hess 2017: 2191). Due to this assumption of possible shared interests between capital and labour, according to Marxist scholars, the 'Decent Work'
approach ignores that in reality the institutional arrangements regulating the labour
process are often the result of potential or real struggles between capital and labour
(Selwyn 2013: 83). As a result, studies approaching labour through a 'Decent Work'
approach have tended to produce top-down strategies for improving workers' conditions, which focus on the collaboration between elite bodies (such as lead firms,
governments and international organisations), while discounting the role of workers
and trade unions (Selwyn 2013: 76).

In summary, in the perspective of the 'Decent Work' approach, improvements
in workers' conditions can be brought about through market pressures on firms to
develop workers' capacities on the one hand, and through strengthening regulatory
governance institutions, particularly at the national and international level, on the
other. Therefore, even though labour is conceived as a 'social agent' entitled to
rights, workers and their organisations only play a minor role as active agents in the
framework of the 'Decent Work' approach (Selwyn 2016: 792ff.).

Building on the criticisms of the 'Decent Work' approach, I argue in this book
that to understand the nature of 'indecent work' and to conceive effective strategies
for improving workers' conditions, we need to give analytical priority to labour in
two regards. First, when analysing the roots of 'indecent work', we need to prioritise
to the conditions of the labour process "as a fundamental process of creating surplus
value under capitalism that is at the heart of all systems of commodity production"
(Cumbers et al. 2008: 371). Secondly, to develop effective strategies for improving

workers' rights, priority needs to be given to the agency of workers and their organisations, i.e. trade unions. To summarise, "the social relations of production, class conflict and resistance" (Cumbers et al. 2008: 372) and the "politically contested state-capital-labour relations" (Arnold and Hess 2017: 2184) should be at the core of analysing the conditions and role of labour in GPNs.

At the same time, these two arguments are the main ideological tenets of the "Marxist Political Economy" approach introduced in the next section.

2.2.2   The 'Marxist Political Economy' Approach to Labour in GVCs/GPNs

Studies on labour in GVCs/GPNs adopting a 'Marxist Political Economy' approach are underpinned by the assumption that social and economic phenomena are shaped significantly (yet not exclusively) by the nature of capital-labour relations (Swyngedouw 2003: 44). As a result, studies on labour in GVCs/GPNs taking on a 'Marxist Political Economy' approach are generally concerned with delivering a critical analysis of social relations of production, class conflict and resistance, and of the resulting material conditions under capitalism (Cumbers et al. 2008: 372). Production is defined in this context in its most general sense as "any human activity of formation and transformation of nature and includes physical, material, and social processes as well as the human ideas, views and desires through which this transformation takes place" (Swyngedouw 2003: 44). The term production, thus, represents all forms of economic activity, not only limited to the production of physical goods. The production process is furthermore conceptualised as an integral part of a set of wider social, political and environmental processes and relations, of which capital-labour relations are the most decisive ones in capitalist societies (ibid.). The 'Marxist political economy' approach, thus, adopts a fundamentally relationist view of the economy, which is thought of as an interlinked network of processes and relations of production, exchange and consumption (MacKinnon and Cumbers 2011: 29).

The intellectual foundations of the 'Marxist Political Economy' approach to labour in GVCs/GPNs lie—as the name suggests—in Marxian political economy, history and theory, as well as in several academic strands that build on the work of Marx, such as Marxist economic geography, labour geography and Labour Process Theory (LPT). While recognising that capitalist relations are historically and spatially contingent and intersect with culturally specific relations of gender, ethnicity, etc. (MacKinnon and Cumbers 2011), geographical studies taking on a 'Marxist Political Economy' approach still adopt six rather universalist theoretical assumptions—originating from Marxian theory—about the nature of social and economic relations under capitalism.

The first and most important assumption is that under capitalism, the individual and collective form of production is characterised by "a fundamental social division between those owning the means of production (capitalists), and those only owning their labour, which they need to sell to as labour force to capitalists in order to secure their own short- and medium-term survival" (Swyngedouw 2003: 44).

Second, the socially accepted goal and driving force of processes of production, exchange and consumption under capitalism is profit-making. Hence, the capitalist market economy is necessarily expansionary and growth oriented (Swyngedouw 2003: 47).

Third, profit in the form of surplus value is generated in the labour process through the transformation of labour power (or abstract labour) into actual work (or concrete labour) (Thompson 2010). Although surplus is generated by living labour, it is appropriated in the form of profit by the owners of capital. Therefore, the labour process under capitalism is inherently exploitative in nature since workers only receive a part of the value that they generate in the form of wage or salary (Swyngedouw 2003: 47).

From this follows the fourth assumption, that capital-labour relations under capitalism are inherently antagonistic and conflictive due to opposed interests: Whereas capitalists seek to maximise the generation of surplus to ensure profits and investments and thus the process of capital accumulation, workers seek to ensure their own reproduction, i.e. their short-term and mid-term survival for which they need means in the form of salaries or wages. This conflict of interests leads to continuous tension and potential labour unrest (Swyngedouw 2003: 47f.).

Fifth, in addition to the *inter*-class struggle between labour and capital, economic relations under capitalism are characterised by the *intra*-class struggle between individual capitalists competing over the conditions of surplus production, appropriation and transfer (Swyngedouw 2003: 48).

From this double nature of inter- and intra-class struggle then results the sixth assumption that to ensure the process of capital accumulation, capital needs to exercise control over labour (Cumbers et al. 2008: 370). To ensure continued generation of profits, capital needs to overcome the 'indeterminacy' of labour, which results from two characteristics of labour as a special production factor. On the one hand, as Thompson and Smith (2009: 924) point out, "hiring labour power does not guarantee an automatic outcome or product for the buyer, as the capacity to work remains within the person of the worker". On the other hand, workers under capitalism have the burden and freedom to decide to which capitalist they want to sell their labour power. As a result, capital needs to strive to control labour time in the production process, on the one hand, and over the deployment of labour in labour markets on the other hand (ibid.).

Starting from these six basic assumptions, Marxist economic geographers such as David Harvey (1982), Doreen Massey (1984), Jamie Peck (1989, 1992) and Andrew Jonas (1996), and labour geographers such as Andrew Herod (1997, 2001b) and Noel Castree (2007) have further developed Marxian theory by theorising the role of space, place and scale in constituting and shaping capital-labour relations. These scholars have pointed at the variety of geographically and historically specific forms of capitalist production systems in which class relations intersect with locally specific 'cultural' relations, such as relations of ethnicity or gender (see e.g. Hudson 2004; Jonas 1996; Massey 1984). Moreover, Marxist economic geographers and labour geographers have advanced our understanding of how spatial asymmetries are crucial in constructing and reproducing asymmetrical capital-labour power relationships (see e.g. Castree et al. 2004; Herod 2001a): Under globalised capitalism, the spatial asymmetry between capital and labour, i.e. the relative mobility of capital in comparison to the relative immobility of labour, has been aggravated due to new developments in logistics and information and communication technologies, allowing capital to set up, manage and control geographically dispersed global production networks. However, as Harvey (1982) points out, even in the era of globalisation, capital is never completely mobile since it depends on locally fixed material infrastructure and institutional settings for production and the labour process to take place. Capital is thus caught in a permanent tension between the need for being fixed in one place for a sustained period, and the need for mobility to seek locations offering more cost-efficient conditions for production.

Drawing on these central assumptions about capital-labour relations under capitalism, the 'Marxist Political Economy' approach to labour in GPNs is based on a two-fold notion of labour. From a 'Marxist Political Economy' perspective, the notion of labour encompasses, on the one hand, *the labour process* as "fundamental process of creating surplus value under capitalism". On the other hand, the notion of labour refers to *workers and their organisations as sentient socio-economic actors*, who actively shape the geographies of capitalism (Cumbers et al. 2008: 371f.). This two-fold notion of labour has in turn given rise to two distinct strands of work within the 'Marxist Political Economy' approach to labour in GVCs/GPNs: (1) a strand of work concerned primarily with the dynamics of the labour process and mechanisms of labour control in GVCs/GPNs; (2) a strand of work concerned with conditions and strategies for labour agency in GVCs/GPNs. Studies within the first strand of work concerned with *labour control* in GVCs/GPNs have focussed on institutionalised labour control dynamics at the local, regional and national level within GVCs/GPNs that ensure the reproduction of the labour process at specific nodes of a GPN (see e.g. Baglioni 2018; Pattenden 2016; Smith et al. 2018; Wickramasingha and Coe 2021). Labour control dynamics can be defined most broadly in this context as encompassing, on the one hand, dynamics of exploitation, i.e. dynamics that ensure the production of surplus value, and, on the other hand, dynamics of disciplining, i.e. dynamics that mitigate or prevent workers' resistance (c.f. Baglioni 2018). Studies within the second strand concerned with the conditions and strategies for *labour agency* in GVCs/GPNs have in turn focussed on workers' and unions' strategies for improving their material conditions within GVCs/GPNs (see e.g. Alford et al. 2017; Cumbers et al. 2008; Hastings 2019; Pye 2017).

Whereas both strands are underpinned by the assumptions of the 'Marxist Political Economy' introduced above, the research strands on labour control in GVCs/GPNs and on labour agency in GVCs/GPNs, however, share intellectual properties with rather different schools within (neo-)Marxist research. *Studies concerned with labour control in GVCs/GPNs* draw predominantly on work from Marxist economic geography and Labour Process Theory (Jonas 1996; Kelly 2001, 2002; Thompson and Smith 2009; Thompson 2010. *Studies concerned with labour agency in GVCs/GPNs* in turn closely engage with work from Labour Geographies, a sub-discipline within economic geography that is concerned with highlighting the active role of workers in shaping economic landscapes (Herod 1997, 2001a; Castree et al. 2004; Castree 2007; Coe and Jordhus-Lier 2011).

It can probably be attributed to the different intellectual traditions of these strands that hitherto interaction between studies on labour control and on labour agency in GVCs/GPNs has been rather limited.[1] I argue here that a closer engagement between both schools can be fruitful for advancing our understanding of the constraining and enabling conditions for local union agency in garment-producing countries. Analysing institutionalised dynamics of labour control can help us to better understand the conditions that constrain collective worker organisation and processes of building union bargaining power at a particular node within the garment GPN. Moreover, an enhanced understanding of labour control dynamics can help us to unveil the specific practices of capital and state actors that (re-)produce labour exploitation in GPNs—and therefore allow us to identify potential target points for labour action. In turn, a focus on labour agency—i.e. on the practices and actions of workers and unions—is useful to identify strategic approaches that enable workers and unions to build sustained bargaining power, allowing them to successfully contest practices of labour control.

With this taken into consideration, this book contributes to the 'Marxist Political Economy'—strand of research on labour in GVCs/GPNs by linking dynamics of labour control and strategies for labour agency. To this end, this study draws on central ideas and concepts from both research on labour control and labour agency in GVCs/GPNs and develops a relational approach to labour control regimes and labour agency in GVCs/GPNs. However, before doing so in Chapter 3, the next section first provides an overview of the conceptual and empirical contributions, and of the shortcomings of existing studies on labour control (Sect. 2.3) and labour agency (Sect. 2.4).

2.3 Research on Labour Control in GVCs/GPNs

Labour control emerged as a popular research subject within social sciences in the 1970s and 1980s, with the inception of Labour Process Theory (LPT) by critical industrial sociologists. The research agenda of LPT can be summarised in most general terms as explaining the "nature and transformation of labour power under capitalism" (Thompson 2010). LPT highlights labour control as a central condition for the transformation of labour power in the labour process, following the assumption that market mechanisms alone cannot hedge the 'indeterminacy of labour' (see Sect. 2.2.2). As a result, there is a 'control imperative' which compels capital to implement management systems to reduce the 'indeterminacy gap' (Thompson 2010: 10).

---

[1] Notable exceptions are provided by Anner (2015a) and Wickramasingha and Coe (2021) who highlight interrelations between labour agency and labour control regimes in various export-garment countries.

One of the most influential and founding contributions to LPT, which has also been influential in debates on labour control in GVCs/GPN, has been made by Michael Burawoy's (1979, 1985) typology of different 'factory regimes'. In this typology, Burawoy (1985) links control dynamics of the labour process with external factors such as dynamics of inter-firm competition, mode of reproduction of labour power and forms of state intervention. He distinguishes between two basic types of factory regimes based on two distinct modes of including workers into the labour process: despotic factory regimes based on coercive work and hegemonic factory regimes based on consenting work. *Despotic factory regimes* tend to exist within political economic systems characterised by 'market despotism', in which capital-labour relations are predominantly mediated through labour markets, with the state being absent as a regulating instance. *Hegemonic factory regimes*, in contrast, emerge under political economic systems characterised by 'hegemonic regimes', in which the state plays an active role in mediating capital-labour relations through the provision of welfare and social security, labour rights and legislation for collective bargaining. As a result, in hegemonic regimes, capital is compelled to coordinate its interests with those of labour and to take measures to persuade workers to take part in the labour process and consent to their own exploitation. Whereas historically under capitalism, despotic factory regimes prevailed during the period of industrialisation, hegemonic factory regimes prevailed under Fordism. For the current period of globalised capitalism, Burawoy argues that a third, new type of factory regime characterised by '*hegemonic despotism*' is likely to emerge, which is characterised by hybrid elements of coercion and consent under new, harsh market conditions (Burawoy 1985: 122–129). This new factory regime under globalised capitalism is hegemonic in so far as that consent is more dominant than coercion in the labour process. However, it is at the same time despotic since capital uses its relatively higher mobility to extract concessions from relatively immobile labour. The capitalist period of 'hegemonic despotism' is thus characterised by an increasing shift of production to developing and newly industrialising countries as well as by an accompanying process of undermining and undercutting of labour standards (Kelly 2001: 3).

Contrary to Burawoy's early attempts to connect the dynamics of the labour process to the wider political economy, subsequent studies in LPT have, however, adopted a rather narrow focus on dynamics of control, consent and resistance at the point of production, while neglecting dynamics outside of the workplace. This narrow focus on the workplace in mainstream contemporary LPT can be attributed to the paradigm of the 'relative autonomy of the labour process' underpinning contemporary LPT studies. This paradigm is based on the assumption that "similar external situations can produce different internal labour process outcomes because of the distinctiveness and peculiarities of particular workplaces" (Taylor et al. 2015: 4; see also Edwards 1990). As a result, contemporary LPT has for a long time been perceived as "less equipped to address, […] the varieties of (often informal or unwaged) types of work, [and] temporal and spatial dimensions" of labour control (Thompson and Smith 2009: 917).

Over the past decade, however, a dialogue between LPT and GVC/GPN literature that is swiftly gaining more traction has addressed these shortcomings, while also sharpening the attention of GVC/GPN studies for the social relations of production (Cumbers et al. 2008; Newsome et al. 2015; Rainnie et al. 2011; Selwyn 2013). Marxist economic geographers and LPT theorists have repeatedly called for a greater focus on labour process dynamics as crucial for the understanding of the structure and functioning of GPNs (Cumbers et al. 2008: 371f.; Hammer and Riisgaard 2015: 89; Rainnie et al. 2011: 160; Selwyn 2013: 87). In this line, Cumbers et al. (2008), for example, point out that the entire rationale for capital restructuring in the form of outsourcing and setting up GPNs is the need for capital to overcome the indeterminacy of labour:

> In the abstract, capital restructuring is always [...] about being 'in flight from labour' or rather is a response to the problems capital comes up against in extracting surplus value through exploitation of labour in production. Whether through the imposition of new technical or spatial fix (Harvey, 1982), capital is viewed as responding to the problem of labour control. (Cumbers et al. 2008: 372)

This revived interest in labour control from GVC/GPN scholars has given rise to a significant body of work analysing mechanisms, dynamics and frameworks of labour control in the context of GVCs/GPNs over the past ten years. In particular, work by economic geographers drawing on the concept of (local) labour control regimes (Jonas 1996) has contributed to broadening the notion of labour control: Whereas LPT-informed studies tended to focus on intersections between value chain dynamics and workplace labour control dynamics, studies of labour control regimes shifted the analytical focus to capital and state strategies directed at securing the broader conditions for capital accumulation at the local and national level (see e.g. Baglioni 2018; Neethi 2012; Smith et al. 2018). In this vein, Baglioni (2018) has stressed that all GPNs are underpinned by complex 'architectures of labour control' that emerge from the interplay between dynamics of exploitation and disciplining at various levels.

In the light of the diversification of analytical perspectives and concepts within the broader research field on labour control in GVCs/GPNs, I propose that we can distinguish between three sub-strands: (1) a research strand at the intersection of GVC analysis and LPT that is concerned with how value chain dynamics shape labour control in the workplace; (2) a research strand drawing on the concept of labour control regimes that is concerned with how the relation between 'global capital' and 'local labour' is mediated by local and national actors; and (3) a research strand that combines the concept of labour control regimes with the multi-scalar perspective of the GPN framework. Figure 2.3 provides a graphic representation of these three research strands on labour control in GVCs/GPNs.

The following sections introduce the theoretical frameworks and assumptions of each research strand and point out their main empirical insights and limitations regarding their potential for identifying constraints and potential target areas for labour agency in the context of GVCs/GPNs.

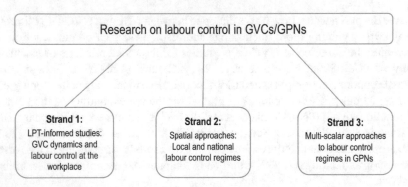

Fig. 2.3   Research strands on labour control in GVCs/GPNs. *Source* Author

2.3.1   Studies from Labour Process Theory: Approaching Labour Control in GVCs/GPNs with a Focus on the Workplace

The first research strand, which is concerned with labour process and control dynamics in the workplace and how these are shaped by broader dynamics of GVC governance, is primarily informed by the assumptions and research agenda of Labour Process Theory (LPT). As aforementioned, LPT has traditionally focussed on the dynamics of the labour process and of control, consent and resistance at the workplace. In this vein, the rather young strand of LPT studies focussing on labour control in GVCs/GPNs aims to reveal how managerial control practices and labour processes in specific sectors are shaped by dynamics of value chain restructuring in the face of new competitive pressures resulting from globalisation.

In principle, it can be said that LPT-informed studies share two basic assumptions about the role of the labour process under globalised capitalism: First, to remain competitive under globalised capitalism, firms must organise and coordinate labour processes at the different stages of the value chain in a way that ensures maximum surplus value (Hammer and Riisgaard 2015: 97). Second, the restructuring of GVC governance and inter-firm relations has led to greater competition among suppliers and generated increased cost pressure in many sectors. As a result, GVC restructuring has affected how relations of production and control over the labour process are coordinated around the globe (Hammer and Riisgaard 2015: 84). In particular, LPT-informed studies have highlighted three interrelated processes that shape the relations of production and the organisation of the labour process under globalised capitalism: First, the increased fragmentation of production on a global scale has led to a reordering of how different labour processes are linked and compete with each other. Second, new competitive challenges for firms resulting from outsourcing and upgrading processes lead to a re-segmentation of the workforce along lines of employment status, type of contract, etc. As a result, we can, third, observe an

increased 'tiering' of the workforce whereby workers performing essentially equiv-
alent tasks are divided by a range of different employment statuses (Hammer and
Riisgaard 2015: 90). From the perspective of LPT, this fragmentation and segmentation of the labour process and of the relations of production represent capital strategies
that ensure continued value extraction from 'living labour' under changing structural
conditions in the era of global capitalism (Bair and Werner 2015: 131).

With this in mind, LPT-informed studies of labour control in GVCs/GPNs have
made three important contributions to advancing our understanding of labour control
dynamics in GVCs/GPNs at the workplace scale. First, LPT-informed studies have
produced valuable insights into the effects of lead firms' practices and strategies
of outsourcing, off-shoring and subcontracting on employment relations and labour
processes at supplier firms. Particularly in captive value chains, lead firms frequently
exercise significant price pressure over suppliers or subcontractors, which in turn
leads to the rationalisation and flexibilisation of labour processes (Flecker and Meil
2011). Haidinger and Flecker (2015) point out in this regard that increased workforce segmentation resulting from the combination of various types of outsourcing—
including subcontracting firms, temporary employment agencies or self-employed
work—also serves as a disciplining mechanism since it hampers collective worker
organisation.

With regard to empirical insights into *the influence of lead firms in the garment
GVC/GPN on labour processes and employment relations at suppliers*, a particularly
comprehensive study has been provided by Anner (2019). He illustrates how fashion
retailers' 'predatory purchasing practices'—including 'price squeeze', demands for
shorter lead times, fluctuations in order volumes and changes to product specifications
at short notice—lead to an 'employment relations squeeze' in the Indian export-
garment industry. To respond to cost pressures and fluctuations in demand, Indian
garment manufacturers rely on several practices of 'squeezing' workers, including
informal employment and piece-rate work, gender-based forms of exploitation and
'wage theft' practices (Anner 2019: 707f.).

The second contribution of LPT-informed studies of labour control in GVCs/GPNs
lies in their description of the 'new factory regimes' and practices of managerial
control in the labour-intensive export-manufacturing sectors of the Global South.
This contribution has been centrally informed by Burawoy's (1985: 263ff.) argument that many countries of the periphery are characterised by "political orders
which would nurture repressive factory regimes". Hence, according to Burawoy,
labour control in peripheral countries tends to be exercised through "brutal coercion at the point of production"—as opposed to the mix of consent and coercion
prevalent under 'hegemonic despotism' in industrialised countries (Burawoy 1985:
265). Drawing from this argument, various LPT-informed studies of labour control
in GVCs/GPNs have drawn attention to 'new factory regimes' in labour-intensive
export-manufacturing of the Global South. These new factory regimes are characterised by "strategies of control [..] that go far beyond the fairly regulated terrain of
the workplace and the employment relationships in the Global North" (Hammer and
Riisgaard 2015: 91; Anner 2015b; Jenkins and Blyton 2017).

With regard to *new factory regimes' in garment-producing countries*, Anner (2015a) develops a three-fold typology of factory regimes in the garment industry (which he calls 'labour control regimes'[2]) that distinguishes between state, market and employer regimes. In state labour control regimes, labour is controlled by a system of legal and extra-legal mechanisms that prevent or curtail worker organisation and collective action. State labour control regimes can, thus, be found particularly in countries with authoritarian governments, such as China or Vietnam. In market labour control regimes, in turn, labour is disciplined by unfavourable market conditions installing fear of job loss and resulting un- or underemployment in workers, as is the case in countries with a large 'reserve army' of labour, such as India, Bangladesh or Indonesia. Lastly, employer labour control regimes are characterised by highly repressive employer actions against workers, including the use or threat of violence. It is important to note, however, that although Anner proposes this typology to distinguish between different labour control regimes, he also stresses that the three forms of control (i.e. state, market and employer control) are not mutually exclusive or static. Rather all countries have elements of each system.

As a third contribution, LPT-informed studies concerned with labour control in GVCs/GPNs have drawn attention to the subjectivity of labour exploitation and disciplining (Burawoy 1985). The subjectivity of labour exploitation and disciplining results from their intersection with other elements of workers' identity, such as gender, age, religion or ethnicity. Employers can strategically deploy these identity features to duplicate disciplinary structures rooted in wider social relationships within the workplace. In this light, several LPT studies have sought to expose how "the creation of value from heterogeneous living labour depends upon […] [the] ideological power that is effected through constructions […] of gender, […] racialisation, heteronormativity, and other forms of social difference" (Werner 2012: 408; see also McGrath 2013).

Concerning *the subjectivity of labour exploitation and disciplining in the export-garment industry*, LPT-informed studies have pointed in particular at the intersections between gender and exploitation in the labour process. In this vein, Werner (2012), for example, illustrates how the upgrading from assembly to full-package production in a Dominican garment factory is linked to a restructuring of the labour process, which gives rise to a new segmentation of the workforce along intersecting lines of gender, skill- and pay levels. Jenkins (2015) adds to Werner's observations by highlighting how employers in the South Indian export-garment industry base their competitive strategies on the use of 'dis-empowerment' as a mechanism of labour control. Employers hire predominantly women because they are associated with greater distance from access to employment and representation rights and are thus perceived to be less likely to organise or to cause unrest at the workplace.

---

[2] Anner (2015b) employed the term 'labour control regimes'. However, Anner's approach is in line with Burawoy's concept of 'factory regimes' than with the spatial approach to labour control regimes from Marxist geography, which is introduced in the next chapter. Whereas Anner's typology seeks to point out generalizable patterns of labour control, geographical studies of labour control regimes highlight the place-specific nature of labour control regimes.

In summary, LPT-informed studies have brought the workplace back into GVC/GPN analysis and thereby advanced our understanding of labour control dynamics in GVCs/GPNs in three ways. Firstly, LPT-informed studies have opened up analytical space for exploring the interrelations between lead firm strategies under globalised capitalism and new work organisation and managerial control forms. Second, LPT-informed studies have sharpened our understanding of the coercive managerial strategies that characterise 'new factory regimes' in the export industries of the Global South 'at the bottom' of buyer-driven GVCs. Lastly, LPT-informed studies drawing from newer feminist and anthropological perspectives have highlighted the intersections of exploitation processes at the point of production with wider social relations of gender or race.

Regarding our understanding of the *constraining factors and potential target areas for labour agency in garment-producing countries*, LPT-informed studies hence have two important implications. First, findings from LPT-informed studies highlighting the influence of lead firms' practices on suppliers' labour processes imply that workers and unions at specific nodes of a GPN need to simultaneously target factory managers' and lead firms' practices to achieve long-lasting changes in the labour process. Second, by highlighting the diverse disciplining mechanisms that hinder collective labour organisation in garment-producing countries—including labour market pressures, employer and state repression and managers' gendered 'disempowerment' strategies—LPT-informed studies have provided important insights into the factors that constrain labour power in garment-producing countries.

Notwithstanding these critical contributions, LPT-informed studies show two crucial limitations when it comes to understanding the nature of the 'labour control architectures' underpinning GPNs. First, due to their rather exclusive focus on value chain and workplace dynamics, LPT-informed studies have neglected an important dimension of labour control in GPNs. Labour control not only encompasses employers' and lead firms' exercise of direct control over labour, it also encompasses capital and state strategies directed at securing the broader conditions that allow lead firms and domestic firms to reproduce exploitative labour processes (c.f. Neethi 2012: 1241). Considering this, it seems necessary to broaden the analysis of labour control in GVCs/GPNs to account for the role of capital and state actors at the horizontal dimension, who secure the broader conditions and social relations for capital accumulation at specific nodes of a GPN. Second, LPT-informed studies of labour control in GVCs/GPNs have paid little attention to the spatial characteristics of labour control. For example, LPT studies have provided little insight into how labour control dynamics at the workplace may vary across various nodes of GPNs due to the territorial embeddedness of labour processes into place-specific social relations and regulatory frameworks.

To address these shortcomings, I argue that a more geographical and therefore spatially sensitive approach to labour control in GVCs/GPNs can be helpful. Such an approach has been developed by the second strand of work concerned with labour control in GVCs/GPNs. This strand builds on the concept of labour control regimes as a heuristic for analysing the architectures of labour control underpinning GPNs. It will be introduced in the next section.

2.3.2   Spatial Approaches to Labour Control: National and Local Labour Control Regimes

The second research strand on labour control in GVCs/GPNs draws on the conceptual framework of *labour control regimes* to explore territorially embedded, institutionalised frameworks for capital accumulation at specific nodes of a GPN (Azmeh 2014; Neethi 2012; Padmanabhan 2012; Smith and Pun 2006). The theoretical concept of labour control regimes has been originally developed by Jonas (1996, 2009) and Peck (1992, 1996). In the most general manner, labour control regimes can be defined as stable institutional frameworks for accumulation and labour regulation constructed around national and local labour market reciprocities (Jonas 1996: 323).

Studies concerned with labour control regimes in the context of GPNs share two central assumptions. The first assumption is that—despite the ability of global capital to move to (and between) countries in the periphery—global capital still depends on 'spatial fixes' to realise value extraction from labour (Harvey 1982, 2001). These spatial fixes can only be realised through the engagement of global capital with actors in production countries, who ensure labour supply, regulation and disciplining (Kelly 2001: 3). The relation between 'global capital' and 'local labour' hence needs to be understood as enabled and mediated through various actors, relationships and institutions at different levels (Kelly 2001: 2). Against this backdrop, the concept of labour control regimes was also introduced explicitly in response to Burawoy's (1985) thesis that coercive labour control in the periphery is enabled by repressive state environments. Kelly (2001: 3) points out in this regard that "rather than simply being sites of oppression and coercion, [...] new destinations for global capital require a new and more or less stable regime of social regulation for labour control to be put in place". These regimes are not solely based on providing "the cheapest and most unregulated economic environment" for lead firms and suppliers, but also encompass strategies of active regulation of the labour market, for example through measures directed at promoting "productivity enhancement, skill development and innovation" (ibid.).

The second shared assumption of studies of labour control regimes in GPNs is that labour control is an "irretrievably [...] spatial process" (Jonas 1996: 328) for two reasons. On the one hand, labour control is spatial because it is territorially embedded in labour control regimes that emerge in place-specific form from reciprocities constructed between the spheres of production, regulation and reproduction in a specific locale (Jonas 1996: 325). As a result, labour control regime literature rejects Burawoy's (1985) thesis of a universal regime of 'despotic hegemonism' under globalised capitalism. Instead, studies of labour control regimes emphasise that "there is no one grand institutional fix to the problem of labour control but rather multiple fixes constructed in different ways in different places (and at different scales) by different agencies" (Jonas 1996: 331). On the other hand, labour control is also an irretrievably spatial process because the control of space is a central tool to control labour under capitalism (Kelly 2001: 1). Since workers are 'free' to sell their labour power to the employer of their choice, regulating and restricting the mobility of labour in space becomes a central instrument of labour control. In this context, broad attention has been given by studies of labour control regimes in GPNs to the role of capital and state-controlled migration flows in producing spatial and institutional fixes for ensuring labour supply (see e.g. Azmeh 2014; Mezzadri 2008, 2017; Padmanabhan 2012).

Hence, whereas LPT-informed studies of labour control focus on the intersection of lead firm and workplace control dynamics, studies of labour control regimes shift the analytical focus from workplaces to 'work-places'. They focus on local and national actor networks and institutional frameworks securing the broader conditions for capital accumulation at specific nodes of a GPN (Jonas 2009: 64). Figure 2.4 illustrates the two analytical emphases underpinning by LPT-informed studies of labour control in GVCs on the one hand, and studies of labour control regimes in GPNs on the other.

Studies of labour control regimes can be further divided into two sub-strands according to their main scale of analysis and the features of labour control that they emphasise. The first sub-strand of work on labour control regimes is inspired by Peck's (1989, 1992, 1996) work on national labour market regulation and takes on a *national regulatory approach to labour control regimes*. This strand emphasises the role of national regulatory frameworks and institutions in ensuring the broader conditions for capital accumulation. In this light, studies adopting a regulatory approach to labour control regimes tend to pay particular attention to the state's role in balancing labour supply and demand, e.g. through training programmes, employment measures and welfare systems. Whereas these regulatory mechanisms usually function at a national level, Peck, however, stresses that concrete local labour market structures

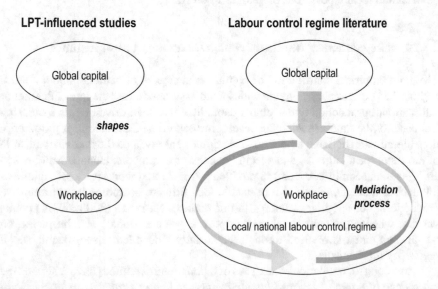

Fig. 2.4 Analytical emphases of LPT-influenced literature and labour control regimes literature. *Source* Author

may show variations in different places. Nevertheless, the heuristic entry point for regulationist studies of labour control regimes typically consists in scrutinising the wider institutionalised regulatory frameworks at the *national* (or international) level and how these materialise in specific (work)places.

The second sub-strand of work on labour control regimes in GPNs, in turn, draws on Jonas' (1996) concept of the *'local labour control regime'*, which—as the name suggests—takes informal practices and networks at the *local* level as heuristic entry point for analysing institutionalised labour control frameworks. As opposed to Peck's focus on national formal regulatory institutions, Jonas emphasises the concrete *localised, informal practices and relationships* constructed around specific workplaces or local industries as constitutive of local labour control regimes. According to Jonas (2009: 61), "there is a tendency for labour control to stabilise around place-specific social practices, which affect the social integration of labour inside the workplace but influence conditions outside it as well. Whether firm-specific or industry-wide, these practices are locally constructed and become routinised and institutionalised in time and space". As a result of this tendency for labour control to stabilise around place-specific social practices, local labour control regimes emerge in form of "historically contingent and territorially embedded set[s] of mechanisms which coordinate the time-space reciprocities between production, work, consumption and labour reproduction within a local market" (Jonas 1996: 325). The analytical movement of studies concerned with local labour control regimes thus starts at the workplace or local level. It extends the analysis from there to the wider sphere of the labour market.

The remainder of this section outlines the contributions of the two sub-strands on national and local labour control regimes in GPNs.

2.3.2.1 'Regulationist' Studies of National Labour Control Regimes

The main empirical contribution of regulationist studies of national labour control regimes has been to challenge the widespread assumption that the intensification of transnational production under global capitalism has been driven by an increasing 'rollback' of the state, especially in developing countries. Contradicting this assumption, literature on national labour control regimes has revealed the fundamental "role of the state, local cultures, and specific classes of employers, managers and workers" (Ngai and Smith 2007: 29) in producing fragmented working classes and flexible labour markets, which in turn enable lead firms to outsource labour-intensive production to developing countries. In this context, Mezzadri (2008: 603ff.) points out that whereas global capital may impose a 'general diktat' of 'cheapness' on labour, this diktat is realised through territorially embedded, place-specific social regulatory mechanisms.

In this context, regulationist analyses of labour control regimes have reiterated the emergence of *national migrant labour regimes in major garment-producing countries*. Ngai and Smith (2007), for example, illustrate how the competitiveness of the export-manufacturing sector in China is rooted in a specific type of national migrant labour regime, which they designate the 'dormitory labour regime'. This labour regime captures the labour force of rural migrants for short-term use by Chinese and foreign export factories located in Special Economic Zones as well as in urban industrial areas. Ngai and Smith highlight the role of the Chinese state in regulating the mobility of rural migrants by providing housing and accommodation in the form of state-owned dormitories, which factory-owners can rent for their migrant workers. Under the agency of the state and local capitalists who actively promote, regulate and channel labour mobility from rural to industrial zones, "a hybrid, transient workforce is created, circulating between factory and countryside, dominated by employers' control over housing needs and state controls over residency permits" (Ngai and Smith 2007: 31).

In her extensive work on the *Indian export-garment industry*, Mezzadri (2010, 2016, 2017), in turn, highlights the *role of the state and local capital in ensuring international competitiveness by promoting the fragmentation and informalisation of labour relations*. According to Mezzadri (2017), the increasing informalisation of production has been actively facilitated and bolstered by the Indian state through two (historical) regulation mechanisms. Firstly, by allocating quotas for garment production to (initially only small and medium) individual enterprises, the state has actively promoted the regionalisation and fragmentation of production. As a result, the Indian export-garment production is now scattered across multiple localised networks of small firms and subcontracting units. Second, while formally maintaining relatively strong labour laws restricting, for example, the use of contract work or retrenchments, the state has since the 1970s increasingly allowed capital to circumvent those laws in practice, thereby actively promoting the informalisation of labour relations.

Altogether, regulationist studies of labour control regimes have made an important contribution to our understanding of labour control in garment-producing countries (and other export industries 'at the bottom' of GVCs/GPNs) by highlighting the role of the state and of local capitalists in mediating the relationship between 'global capital' and 'local labour'. They have emphasised how labour exploitation and control at the workplace is shaped by broader regimes of labour regulation at the national level. In particular, they have highlighted the important role of migrant labour regimes and the informalisation of employment relationships as broader conditions for reproducing exploitative labour processes in garment-producing countries. By exposing the active role of the state in producing the conditions for labour exploitation, regulationist studies of labour control regimes have highlighted state policies as a field of contestation for workers and unions in garment-producing countries. Moreover, by showcasing the spatial fragmentation of the workforce caused by circulatory migration and informal, home-based work, regulationist studies of labour control regimes have exposed barriers for collective labour organisation in garment-producing countries beyond direct repression by employers and state actors.

However, critics have argued that due to their focus on formal labour regulation and their analytical priority for dynamics at the national level, regulationist studies of labour control regimes have paid too little attention to the highly localised informal relationships and practices that are "often contrary to the formal provisions of national-level labour regulations" (Kelly 2001: 21). This criticism might not be entirely justified when reviewing recent literature on national labour control regimes, which has studied informal exploitation practices of local employers (see e.g. Mezzadri 2016, 2017). Nevertheless, it remains true that the second sub-strand of work on labour control regimes in GPNs drawing on Jonas' concept of local labour control regimes has shown a much more explicit focus on informal networks and practices. The following section provides an overview of the empirical insights produced with this analytical focus.

2.3.2.2   Studies of Local Labour Control Regimes

As opposed to regulationist studies of labour control regimes, which predominantly choose the *national* level as analytical entry point, literature concerned with *local* labour control regimes focuses on institutionalised frameworks of labour control at the sub-national level. In doing so, studies of local labour control regimes have made two key contributions to advancing our understanding of the 'architectures of labour control' underpinning GPNs. First, studies of local labour control regimes have highlighted the importance of everyday informal social and cultural practices (as opposed to formal regulatory mechanisms) for the constitution of institutionalised frameworks of labour control at specific nodes of a GPN (Jonas 1996: 327). Thereby, studies of local labour control regimes have opened analytical space to account for the role of discursive practices in shaping and justifying labour exploitation and disciplining (Coe and Kelly 2002).

Besides highlighting the role of informal cultural practices as constituent parts of labour control regimes, studies of local labour control regimes have, second, underlined the relational nature of local labour control regimes which are conceptualised as emerging from the 'collective interaction' of a wide variety of actors (Kelly 2001: 7). As a result, local labour control regimes are at the same time relatively stable, yet open to contestation and change over time (Jonas 1996: 328f.). The concept of local labour control regimes is hence underpinned by a deeply relational understanding of 'the local', which considers the "material, social and political form" of the local as "determined by its interactions with local and wider power geometries" (Jonas 1996: 328; see also Massey 1992, 1993).

Several studies have provided empirical insights into these characteristics of local labour control regimes, with particular attention on export industrial clusters in newly industrialising regions in Asian countries (Coe and Kelly 2002; Kelly 2001, 2002; Neethi 2012; Padmanabhan 2012).

With regard to *local labour control regimes in the export-garment industry*, Padmanabhan (2012) provides insights into the *direct practices and indirect mechanisms constituting the local labour control regime* in an export-garment park in Kerala, South India. The 'direct practices' of labour control in the workplace include inter alia restrictions for leave-taking as well as a strict monitoring of production rates. The 'indirect methods' of labour control consist of tailoring recruitment practices to the specific characteristics of the local labour market. Employers exploit the labour market dependence of young, less educated women from economically vulnerable families in rural areas by specifically targeting recruitment strategies at this segment of labour. Workers stay in park-run hostels, allowing employers to dispose flexibly over labour's time even beyond the official working hours. Padmanabhan also highlights informal networks extending into workers' sites of reproduction as indirect mechanism of labour control. For garment firms in Kerala's export promoting parks, fostering relationships with local leaderships of surrounding villages is crucial for ensuring labour supply since in most cases communities collectively take the decision to send young women to garment export parks.

In a nutshell, whether adopting a focus on the formal regulation of labour markets or on informal relationships and practices as means of labour control, studies of labour control regimes have advanced our understanding of labour control in GPNs by highlighting the role of regulatory frameworks, actors, processes and relationships at the national and local level as mediators between 'global capital' and 'local labour'. Moreover, literature on labour control regimes has refined our understanding of labour control dynamics in GPNs by highlighting the geographical variations between labour control regimes constructed around different national and local labour markets, and by exposing the control of space as an essential tool of labour control under capitalism. By highlighting the place-specific dynamics and networks at the horizontal dimension of the GPN, studies of labour control regimes complement studies at the intersection of GVC analysis and LPT, which have tended to neglect the broader political economies into which workplaces are embedded. Regarding the constraining factors and the territory for the agency of local unions in garment-producing countries, the insights from studies on (local) labour control regimes suggest that we should take a more critical stance towards formal labour law as potential power source for local unions. In reality, a variety of informal interactions and practices of state actors and employers tend to make these frameworks ineffective in practice (see e.g. Kelly 2001; Mezzadri 2010). Taking this into consideration, the state appears—besides employers—as a relevant target for local union action.

Despite these important contributions, there are also some shortcomings with studies of national and local labour control regimes in GPNs. Particularly with regard to the originally relational concept of local labour control regimes as embedded within and shaped by 'wider power geometries' (Jonas 1996: 329), empirical studies of labour control regimes have not yet lived up to this conception. Despite the explicit framing of (local) labour control regimes as mediating institutions between 'global capital' and 'local labour', labour control studies have paid little to no attention to how value chain dynamics or lead firm practices shape national or local labour control regimes. Instead, global capitalist dynamics remain in the background of these studies as abstract structural forces that impose the 'diktat' to produce cheap and flexible labour on national, regional and local actors (Mezzadri 2008: 614).

Addressing this gap, over the past five years, a still small number of studies has sought to integrate the labour control regimes framework more explicitly with the multi-level heuristic of the GPN framework (Baglioni 2018; López Ayala 2018; Pattenden 2016; Smith et al. 2018; Wickramasingha and Coe 2021). The following section gives an overview of this relatively new branch of research promoting an explicitly 'multi-scalar' approach to labour control regimes.

2.3.3 Multi-Scalar Approaches to Labour Control Regimes

Multi-scalar approaches to labour control regimes in GPNs have emerged since 2016 in an attempt by economic geographers to explore the role of dynamics, actors and practices at *various* levels in constructing labour control regimes at specific nodes of a GPN in a more in-depth manner (Baglioni 2018; López Ayala 2018; Pattenden 2016; Smith et al. 2018; Wickramasingha and Coe 2021). The first scholar to propose an explicitly multi-scalar framework for analysing labour control regimes was Pattenden (2016: 1813), who suggested a 'three way approach to labour control regimes'. Pattenden's approach distinguishes between the scales of the macro-labour control regime, the local labour control regime and the labour process. The macro-labour regime—comprising social relations, the state agenda and formal regulations at the national level—shapes the material conditions of the local labour control regime as well as of the labour process through 'pro-capital state practices that create and maintain informal flexible working places, dangerous labour processes and poor living conditions' (Pattenden 2016: 1823).

While Pattenden's 'three way approach' stops at the national level, Baglioni (2018), Smith et al. (2018) and Wickramasingha and Coe (2021) go one step further by integrating the 'global level' into their approaches to LCRs in GPNs. In Baglioni's (2018) concept of the LCR, global dynamics of capital restructuring provide the broader context within which states and transnational firms construct labour control regimes at the national and local level. Driven by global competitive pressures, national states construct regulatory frameworks and development agendas, which in turn pave the way for the agency of transnational firms. Transnational firms, in return, construct locally specific labour control regimes by linking labour exploitation at the firm level with strategies of spatial and social disciplining extending to the reproductive sphere (Baglioni 2018: 113ff.).

In contrast, Smith et al.'s (2018) 'nested scalar approach' to labour control regimes in GPNs takes labour control dynamics at the workplace as an analytical starting point. Smith et al. then suggest analysing how the workplace labour control regime—including the labour process, wage relations and forms of worker representation—is shaped on the one hand by dynamics of the 'regional and national political economy of labour control' and by global lead firm dynamics, on the other. Similar to Jonas' (1996) conception of the local labour control regime, Smith et al. (2018: 558) conceptualise the 'regional political economy of labour control' as comprising dynamics of social reproduction and local labour markets as well as local political contexts. The 'national political economy of labour control' is constituted through formal regulatory mechanisms such as national labour laws, labour inspectorates and state policies relating to labour standards as well as through 'the balance of class forces' (ibid.). The workplace labour regime and the regional and national political economy of labour control are in turn shaped by 'lead firm dynamics' comprising supply chain pressures and governance dynamics as well as lead firms' codes of conduct and lobbying practices (ibid.) Fig. 2.5 visualises the different levels of labour control and their articulations in Smith et al.'s 'nested scalar approach' to labour control regimes.

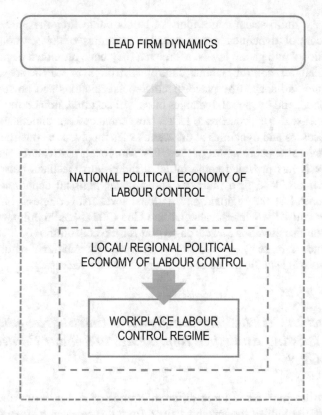

Fig. 2.5 'Nested scalar approach' to labour control regimes by Smith et al. (2018). *Source* Adapted from Smith et al. (2018: 558)

From an empirical point of view, Smith et al.'s (2018) work has contributed to our understanding of the *complex, multi-scalar constitution of labour control regimes in garment-producing countries*. Applying their conceptual framework to the Moldovan export-garment industry, Smith et al. illustrate how workplace regimes in Moldovan garment factories are characterised by 'poverty wages', extensive and regular use of overtime, intensification of the labour process and employer resistance to union formation. Smith et al. then identify how these workplace regimes are shaped by lead firms' demands at the global level and by regional and national dynamics and institutions of labour regulation. On the one hand, lead firms' demands for lower prices, higher product flexibility and shorter production times require manufacturers to maximise productivity while keeping labour costs down. On the other hand, poor workplace conditions are shaped by the weakening of labour inspectorates and low national minimum wages due to the asymmetric power balance between employer associations, the state and labour. With their analysis, Smith et al. provide a good understanding of how more general 'network dynamics' intersect with place-specific labour control dynamics at the national and workplace level.

In summary, multi-scalar conceptions of labour control regimes have embedded workplace control dynamics within broader dynamics of labour control at the regional, national and global levels. Moreover, they have promoted a spatial under-standing of labour control: labour control structures in GPNs are territorially embedded into and shaped by place-specific social relations and power structures on the one hand, and by global dynamics of capital accumulation on the other hand. Thereby, multi-scalar approaches to LCRs have enhanced our understanding of the multiple processes and dynamics at different scales involved in constructing labour control regimes at specific nodes of the GPN. Moreover, particularly Smith et al.'s (2018) analysis has provided some initial insights into the links between labour process dynamics, lead firm practices and broader political economic dynamics of labour control at the regional and national level. Nevertheless, only recently a pioneering study by Wickramasingha and Coe (2021) has highlighted the links between dynamics at various levels as central for the constitution of labour control regimes. Hence, a more systematic engagement with the relational nature of labour control regimes at specific nodes of the garment GPN is necessary.

2.3.4 Interim Conclusion: Contributions and Shortcomings of Existing Analytical Approaches to Labour Control in GPNs

This section has introduced three different strands of research on labour control in GPNs and highlighted the insights gained from these research strands as well as their shortcomings: (1) an LPT-informed research strand focussing on the intersection between value chain and workplace control dynamics; (2) a research strand focussing on national and local labour control regimes; and (3) multi-scalar approaches to labour control regimes at specific nodes of a GPN.

In a nutshell, these research strands have enhanced our understanding of the constraining conditions and the terrain for worker and union agency in global produc-tion countries in two regards. First, existing research on labour control in GVCs/GPNs has highlighted the interrelations between global value chain dynamics and work-place exploiting and disciplining dynamics. In particular, past studies have illustrated how lead firm pressures for lower prices and shorter production times give rise to workplace regimes characterised by informalised employment relations, low wages, gender-based exploitation and violations of workers' collective rights.

Second, in particular studies concerned with local, national or multi-scalar labour control regimes at specific nodes of a GPN have drawn attention to the fact that structures of exploitation and disciplining are not only produced by lead firms at the global scale and by employers in the workplace. They are also produced by various other actors at multiple scales who fulfil central functions in reproducing the broader social relations that secure capital accumulation at specific nodes of a GPN. Hence, to understand the terrain and constraining factors for the agency of local unions in garment-producing countries, we need to look at the practices of these actors and scrutinise their implications for worker and union agency—either concerning the specific constraints they pose for building local unions´ bargaining power or as potential target areas for union interventions.

Despite these critical overall contributions to the literature on labour control in GVCs/GPNs, I argue that three aspects of the 'labour architectures underpinning GPNs' remain underdeveloped in existing studies of labour control (regimes) in GPNs. First, while existing research on labour control in GPNs has revealed the diversity of actors and dynamics at multiple scales involved in the production of labour control regimes, the interrelations and interdependences between labour control dynamics at different levels have so far remained under-researched. I hold that this analytical gap results at least in part from the scalar heuristic underpinning past studies of labour control regimes: The focus on a priori-defined scalar categories creates an artificial analytical separation between processes and dynamics conceptualised as located at different levels (c.f. Marston et al. 2005: 442). However, as Wickramasingha and Coe (2021: 6) have recently stressed, place-specific labour (control) regimes are in reality constituted through "a mix of geographically distant and proximate relations across different scales". In this light, I argue that to gain a more nuanced understanding of the 'architectures of labour control' underpinning specific GPNs and of the implications for labour agency, a relational analytical perspective that emphasises connectivity rather than a priori defined scalar scaffolds can be beneficial.

Second, scalar studies of labour control regimes in GPNs tend to assume a universal socio-spatial order as characterising labour control regimes in GPNs (c.f. Latham 2002: 138). This assumption is also closely intertwined with the pre-supposed 'nested multi-scalar' structure of labour control regimes (c.f. Smith et al. 2018). I argue that such a pre-given assumption of a universal socio-spatial order of labour control regimes constrains the analytical space for empirically carving out the place-specific articulations of geographically more proximate and more distant relationships that constitute labour control regimes at specific nodes of a GPN (c.f. López 2021; Wickramasingha and Coe 2021).

Lastly, the nested multi-scalar perspective informing existing studies of labour control regimes in GPNs has tended to reproduce a hierarchical 'top-down' conceptualisation of 'global/local' dynamics, in which labour control dynamics at higher levels are unilaterally imposed on local labour (Hastings and MacKinnon 2017: 105). Such a 'top-down' understanding, however, completely ignores the role of workers' as sentient social actors who actively shape economic landscapes, including labour control regimes (Herod 1997; Wickramasingha and Coe 2021). Under repressive state and employer regimes, workers and unions may not have the power to challenge institutionalised frameworks for labour control in their totality; however, workers and unions may still succeed in shaping and transforming specific elements of the labour control regime by stopping selected practices of exploitation. Thus, I argue that to gain a better understanding of how workers and unions in garment-producing countries can achieve lasting improvements for workers, we need an enhanced understanding of the conditions that enable workers to achieve even 'small transformations' of labour control regimes at specific nodes of the GPN (c.f. Latham 2002).

Table 2.2 on the next page provides an overview of the central research questions, contributions and shortcomings of the three research strands/approaches introduced in this section.

To tackle these shortcomings in existing research on labour control in GPNs, and to produce a more nuanced understanding of labour control regimes as relational-structural contexts for worker and union agency in garment-producing countries, I develop a relational analytical approach to labour control regimes in GPNs (see Chapter 3). This approach conceptualises labour control regimes at specific nodes of a GPN as emerging from the place-specific articulations of multiple processual relations of labour control that stretch across various distances and link actors through practice relations. I argue that such a practice-oriented, relational approach to labour control regimes in GPNs bears two central benefits for generating a better understanding of the conditions that constrain and enable the agency of workers and unions in garment-producing countries: First, a relational approach conceptualises labour control regimes by means of networked practices and relationships and therefore allows for a closer examination of the complex interdependencies and interrelations between labour control dynamics at various scales. Thereby, a relational approach is well equipped to shed light on the challenges for union agency resulting from the complex networks of practices and relations that constitute the labour control regime. Second, the relational approach developed in this book highlights that the labour control regime is primarily constituted through practices (as opposed to abstract capitalist forces or mechanisms). As a result, it is able to shed light on the 'small transformations' of labour control structures achieved by workers and unions.

Before proceeding to develop my relational approach to labour control regimes in Chapter 3, in the next section, I first turn to review a second body of literature concerned with the agency of workers and unions in GPNs. While research on labour control in GPNs has provided important insights into the constraining conditions and the broader terrain for the agency of labour in garment-producing countries, it has told us little about the conditions that *enable* workers and unions to build the capacities and power to actively challenge labour control structures. Therefore, in the next section, I introduce a second research thread that explores the strategies and enabling conditions for the agency of workers and unions in GPNs.

2.4   Research on Labour Agency in GVCs/GPNs

Labour agency first became popular as a research subject among Marxist geographers at the end of the 1990s with the emergence of labour geography as a geographical sub-discipline (Lier 2007; Coe and Jordhus-Lier 2011: 213). Labour geography emerged as a critique of the until then predominant conceptualisations of labour in neoclassical and Marxist economic geography, which tended to conceptualise labour either as a mere production factor or as a passive victim of the strategies of capital (1997: 1ff.).

Table 2.2 Overview of different research strands on labour control in GVCs/GPNs introduced in this section

Research strand	Central research question(s)	Contributions for our understanding of labour control 'at the bottom' of GVCs/GPNs	Shortcomings
LPT-informed studies of labour control in GVCs/GPNs	• How are workplace labour control dynamics influenced by broader GVC dynamics? • What are central mechanisms of managerial labour control in the new export sectors of the Global South?	• Influence of value chain restructuring and lead firm practices on labour processes and employment relations • Characteristics of 'new factory regimes' in export industries of Global South • Interrelations of workplace labour control dynamics with broader relations of gender and race	• Neglects the role and influence of actors, practices and institutions at local, regional and national level • Narrow conceptualisation of labour control as direct control by capital and state actors over workers • No attention paid to spatial characteristics of labour control dynamics (e.g. place-specific variations, control of space as disciplining mechanism)
Spatial approaches to labour control in GVCs/GPNs: National & local labour control regimes	• Which actors, institutions and regulatory mechanisms at the local, regional and national mediate the relationship between 'global capital' and 'local labour'? • Through which mechanisms are the broader conditions for capital accumulation at specific nodes of a GPN secured?	• Geographical variations in labour control regimes at different national or local nodes of a GPN • Control of space as an essential tool of labour control, e.g. controlling migration flows and workers' reproductive spaces • Importance of national regulatory mechanisms and everyday informal practices as constituents of institutionalised national and local frameworks of labour control	• No attention paid to the role of value chain dynamics or lead firm practices in shaping labour control regimes at the national, regional, local and workplace scale

(continued)

Table 2.2 (continued)

Research strand	Central research question(s)	Contributions for our understanding of labour control 'at the bottom' of GVCs/GPNs	Shortcomings
Multi-scalar approaches to labour (control) regimes	• Which dynamics, actors and relations at *multiple* scales play a role in the production of labour control regimes at specific nodes of a GPN?	• Workplace control dynamics embedded within broader multi-scalar dynamics of labour control at the regional, national and global level	• Focus on nested scales as pre-given, discrete containers for social action obstructs analysis of interrelations and interdependencies between dynamics at the different scales

Source Author

In response, early labour geographers such as Andrew Herod (1997, 2001a) and Jane Wills (2005) sought to shift the attention to "working class people as sentient social beings who both intentionally and unintentionally produce economic geographies through their actions" (Herod 1997: 3).

The research objective of labour geography in its initial phase was to highlight how workers and their organisations shape the geographies of capitalism. Early studies in labour geography tended to conceptualise 'labour agency' in narrow terms as the organised, collective agency exercised by trade unions in manufacturing sectors in developed countries (Coe and Jordhus-Lier 2011: 213). Consequently, early studies in labour geography were predominantly characterised by narrative accounts of "isolated success stories of workers with strong capacities to act and enhance their position vis-à-vis capital" (ibid.). As a result, the theoretical concept of agency initially remained underdeveloped. As Taylor et al. (2015: 10) put it: labour agency "had simply come to mean any meaningful manifestation of collective worker *activity*" (ibid., emphasis in original).

Responding to these shortcomings, labour geographers have, over the last decade, sought to 're-embed' worker agency within the broader structures and social relations that shape the terrain for the agency of unions and workers, particularly in GPNs (Coe and Jordhus-Lier 2010, 2011; Cumbers et al. 2008; Riisgaard and Hammer 2011). These studies have stressed workers' and unions' embeddedness into a set of intersecting 'structural forces' at the vertical and horizontal dimensions of the GPN that enable and constrain collective labour strategies (Arnold 2013). At the *vertical dimension*, these 'structural forces' comprise inter-firm power relations and dynamics of value chain governance, including the CSR and sourcing practices of lead firms (Lund-Thomsen and Coe 2013: 277). At the *horizontal dimension*, these structural forces entail "the formations of capital, the state, the community and the labour market in which workers are incontrovertibly yet variably embedded" (Coe and Jordhus-Lier 2011: 214). The terrain for the agency of workers and unions at a

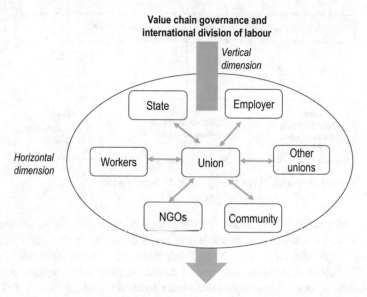

Fig. 2.6 Intersecting structural forces and relations at the vertical and the horizontal level that shape the terrain for worker and union agency in GPNs

specific node of the GPN is hence shaped by value chain governance dynamics on the one hand and by territorially embedded social relations (including those constituting local labour control regimes) on the other hand. Figure 2.6 illustrates the intersecting structural forces and relations at the vertical and the horizontal level that shape the terrain for worker and union agency in GPNs.

In this context, Coe and Jordhus-Lier (2011) have coined the notion of 'constrained agency' to highlight the paradigmatic shift in labour geography from affirmative research telling 'isolated success stories' of worker struggles towards a more relational and holistic study of the complex socioeconomic practices and relationships shaping the terrain for labour agency in different places and industries.

As mentioned earlier, the central objective of this book is to develop a more nuanced understanding of the conditions that enable and constrain union agency in garment-producing countries, following the assumption that only by shifting the local capital-labour balance sustained improvements for workers can be achieved. With this in mind, the remainder of this section gives an overview of the various concepts and analytical approaches adopted so far by studies at the intersection of labour geography and GVC/GPN analysis to study the collective agency strategies of workers and unions in GPNs. It is important to note that, in this book, collective worker agency or union agency are by no means limited to the agency of formal, multi-level trade union organisations that prevail in the Global North. The focus of this study is on worker organisations in garment-producing countries—and specifically in India. As a result, this book adopts a broader notion of collective worker agency as has been forwarded in debates on 'social movement unionism'. These debates have

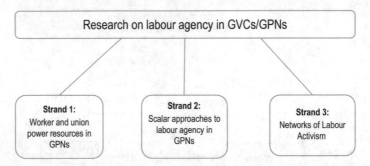

Fig. 2.7 Research strands on labour agency in GPNs. *Source* Author

highlighted the variety of collective worker organisations in the Global South, which also encompass politically independent grassroots unions or worker-led community organisations (see e.g. Fairbrother 2008; Fairbrother and Webster 2008; Moody 1997; Nowak 2017). Within these organisations, boundaries between workplace organising and community organising and between class and other social categories are increasingly blurred (Castree et al. 2004: 225). The notion of labour agency adopted in this study hence includes various forms of collective worker organising and is sensitive to the intersection of class with other lines of social differentiation, such as gender or race (c.f. Valentine 2007).

In the remainder of this section, I review existing studies on collective worker and union agency in GPNs. I classify existing studies on labour agency in GPNs into three main research strands working with different conceptual and analytical approaches: (1) a research strand drawing centrally on the *concept of worker or union power resources*; (2) a research strand adopting a *scalar approach to labour agency;* and (3) a research strand concerned with the constitution and structural effects of *'Networks of Labour Activism'* (Fig. 2.7).

In the following, I will introduce the major conceptual and empirical contributions made by each strand of work and point out research gaps, before explaining how this book contributes to closing these gaps.

2.4.1   Approaching Union Agency Through the Lens of Worker and Union Power Resources

The 'power resources approach' has been widely used by labour geographers as a conceptual tool to assess the conditions that constrain and enable labour and union agency in GPNs (Schmalz et al. 2018; Webster et al. 2008; Webster 2015). The power resources approach is based on the assumption that labour can successfully advance its own interests vis-à-vis capital by collectively and strategically mobilising different types of power resources (Schmalz et al. 2018: 114). 'Power' is generally defined in the context of the power resource approach as "the ability of actor A to make actor B do something that B would not otherwise do" (Zajak 2017: 1012; Knight 1992). The groundwork for this conceptual approach was laid by Wright's (2000) distinction between *structural power* deriving from workers' position within the economic system and *associational power* deriving from workers' capacities to form collective organisations. This primary distinction has subsequently been refined and expanded by other labour scholars. Silver (2003), for example, has proposed to distinguish between *marketplace bargaining power* and *workplace bargaining power* as two subtypes of structural power. Brookes (2013) and Schmalz and Dörre (2014) have introduced the notion of *institutional power* to refer to worker and union power resources that are derived from formal and informal rules, regulations and mechanisms, such as labour laws or collective bargaining mechanisms.

Debates on strategies for trade union renewal in the Global North (Dörre et al. 2009) as well as debates on 'social movement unionism' in newly industrialising countries of the Global South (Moody 1997) have further added to the concept of worker and union power resources. In this context, scholars have pointed at several 'new' trade union power resources (Webster 2015), which might be able to compensate for workers' decreasing structural and associational power under globalised capitalism. For example, authors like Brookes (2013) or Schmalz et al. (2018) have pointed at *coalitional power* as a broader form of associational power that workers and unions may activate through building relationships of external solidarity with community organisations, social movements or consumer organisations. Lastly, several authors have pointed to new resources of *moral power,* which unions may activate by creating awareness for labour rights violations or struggles in public and media discourses (Webster et al. 2008; Schmalz et al. 2018). Webster (2015) has proposed to subsume coalitional and moral power under the category of 'societal power', since both derive from workers' or unions' collaboration with other societal groups. Figure 2.8 gives an overview of the different types of power resources that workers and unions may potentially activate in GPNs.

In the remainder of this section, I will introduce each power resource in more detail and summarise findings from existing studies regarding the potential of different power resources to help local unions build sustained bargaining power vis-à-vis employers.

2.4.1.1 Structural Power: Marketplace and Workplace Bargaining Power

Structural power derives from the specific position of workers within the economic system (Wright 2000: 962). When analysing the structural power of workers in GPNs, we can distinguish between marketplace and workplace bargaining power as two sub-forms of structural power (c.f. Silver 2003). *Marketplace bargaining power* accrues to workers positioned in labour markets characterised by low unemployment, who possess scarce skills, or who have access to savings, social security or other income enabling them to withdraw from the labour market (Silver 2003: 3). The exercise of marketplace bargaining power is thus not necessarily tied to collective

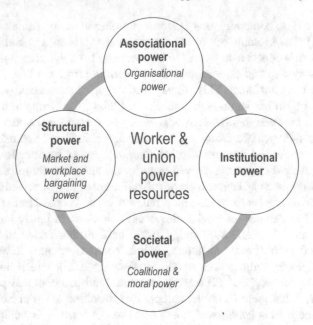

Fig. 2.8 Worker and union power resources. *Source* Author

forms of agency, but instead it could also be exercised by individual workers. In contrast, *workplace bargaining power* accrues to workers "who are enmeshed in tightly integrated production processes, where a localised work stoppage in a key node can cause disruptions on a much wider scale than the stoppage itself" (Silver 2003: 13). Whether or not workers and unions at a specific node of a GPN are able to exercise structural power hence depends on their position within intersecting value chain dynamics at the vertical dimension and labour market dynamics at the horizontal dimension of the GPN.

Past studies have stressed, first and foremost, the *constraints for workers in garment-producing countries to leverage structural power resources.* On the one hand, workers' workplace bargaining power tends to be severely curbed by the geographical fragmentation of production networks and retailers' ability to relocate their sourcing activities in reaction to wage increases or strikes (López 2021). On the other hand, workers in the export-garment industries of the Global South generally possess low marketplace bargaining power due to the little skill and knowledge intensity of the larger share of tasks in garment production (c.f. López et al. 2021). However, Kumar (2019a, 2019b) argues that, in more recent years, the emergence of 'oligopolistic suppliers' in garment GPNs has endowed garment workers with new structural power resources. These new structural power resources result from lead firms' increasing dependence on a smaller number of large 'oligopolistic' tier one full-package suppliers. These full-package suppliers increasingly take on strategic logistics and inventory management functions for lead firms (see also Azmeh and Nadvi 2014; Merk 2014). Consequently, Kumar (2019a: 355) argues that for workers and unions at these tier one suppliers' "structural power (…) increases alongside the degree of market spatial inflexibility". Since these large suppliers can capture greater value shares, unions at these suppliers are more likely to succeed in pressuring employers into collective bargaining through work stoppages and other forms of collective action.

The emphasis on disruptions to the production process through work stoppages inherent to the notion of 'structural power', makes it clear that to mobilise structural power resources, workers need the strategic capacities to plan and perform collective action. Therefore, 'structural power' cannot be thought independent from the second basic form of workers' power, which is associational power.

2.4.1.2 Associational and Organisational Power

Associational power accrues to workers from their capacities to form collective organisations and to act collectively (Wright 2000: 962). The mobilisation of associational power resources thus requires organising and training processes directed to build workers' capacities to develop and execute collective action strategies (Silver 2003: 13ff.). Since associational power is tied to workers' abilities to act collectively, trade unions and other forms of collective worker organisations are a central driving force for the exercise of associational power. However, it is important to note that the formal existence of a trade union is not enough for workers to exercise associational power. Unions as organisations need to foster specific types of relationships and capacities for agency among their members (c.f. Hauf 2017: 1003; Lévesque and Murray 2002). In particular, unions need to cultivate and synthesise the 'social capital' of their members "so that they identify themselves as part of a collectivity and support its purpose and its policies" (Gumbrell-McCormick and Hyman 2013: 30).

This qualitative understanding of associational power has led Gumbrell-McCormick and Hyman (2013) to introduce the notion of *organisational power* as a sub-form of associational power. Organisational power refers to the capacities of a union to build the social capital and strategic capacities of its members and functionaries as well as to foster a 'cohesive collective identity' among their members. Hence, to build organisational power, unions need to foster a solidary mindset among union members that goes beyond individual motives of obtaining personal benefits or protection through the membership. This solidary mindset is achieved through stimulating lively communication and relationships among members and through actively involving members in union life (Lévesque and Murray 2010: 336f.). Moreover, unions need to foster internal democratic structures and practices to develop their members' strategic capacities. To this end, Gumbrell-McCormick and Hyman (2013: 30) stress that unions need to nurture "a culture favouring discussion between rank and file and officials and [through] educational work to ensure that policies are well understood and reflect the conditions experienced on the ground".

Regarding the *ability and opportunities of workers and unions in garment-producing countries for building associational and organisational power,* past research has highlighted the various constraints resulting from repressive employer and state regimes (Anner 2015a, b; López 2021; Ruwanpura 2015). Given the price pressure by lead firms in global garment value chains, employers in export-garment sectors of the Global South frequently rely on disciplining practices to suppress collective worker organising. These include threatening, dismissing or co-opting union activists. Likewise, public rallies or demonstrations for higher wages are not seldom shattered by the police (e.g. Anner 2015a, b; Padmanabhan 2012). In addition, various studies have highlighted constraints for building workers' associational power—particularly in Asian countries—resulting from the internally fragmented nature of labour movements (see e.g. Arnold 2013; Hauf 2017). Established trade union federations in Asian countries are often affiliated to political parties and dominated by rent-seeking union leaders (ibid.). As a result, over the last decades, in many Asian countries, an increasing number of 'independent' trade unions and worker organisations have emerged (Kumar 2014; Jenkins 2015). This proliferation of worker organisations often hampers the ability of workers in garment-producing countries to exercise associational power through coordinated joint action in the political sphere, for example in the context of national minimum wage negotiations (Arnold 2013).

2.4.1.3   Institutional Power

Next to associational and structural power, institutional power represents the third 'traditional' power resource that workers in GPNs may be able to draw on. Institutional power accrues to workers from their ability to invoke formal and informal rules and mechanisms that structure capital-labour relationships. Sources of institutional worker power are represented, for example, in national labour legislation and in institutionalised dispute settling, wage setting or collective bargaining mechanisms (Brookes 2013: 188). However, in the context of global capitalism, 'traditional' sources of workers' institutional power are dwindling due to neoliberal state policies of labour market flexibilisation (see e.g. Cumbers et al. 2016; Fairbrother and Webster 2008; Webster et al. 2008). This is particularly true for many garment-producing countries, where states have pushed global market integration through regulatory reforms that have severely curtailed institutionalised labour representation. Governments of garment-producing countries have, for example, frequently set up so-called Special Economic Zones for export industries, which are exempt from national minimum wage regulations or from the right to collective bargaining and freedom of association (FoA) (Ruwanpura 2015).

Against this backdrop, studies concerned with (potential) power resources for unions in the garment GPN have pointed at transnational private or public-private social regulation mechanisms, such as *codes of conduct (CoC), multi-stakeholder initiatives (MSI) or Global Framework Agreements as potential 'new' institutional power sources for workers and unions in garment-producing countries* (see e.g. Anner 2012; Lund-Thomsen and Coe 2013; Zajak 2017). In response to sustained pressure from anti-sweatshop movements, almost all international garment brands and retailers have introduced corporate CoC that set supplier labour standards. In addition, many retailers and brands are members in MSI, which include NGOs as independent monitoring organisations to increase firms' accountability (Fütterer and López Ayala 2018: 15ff.). It is important to note, however, that—as opposed to national labour laws and institutionalised bargaining mechanisms—(semi-)private transnational regulatory mechanisms are not the outcome of prior labour struggles, in which workers successfully shifted the capital-labour power balance through exercising associational and structural power. (Semi-)private transnational regulatory mechanisms, such as CoC or MSI, are voluntary initiatives by retailers, which are usually introduced as a response to consumer criticisms from the Global North (Hauf 2017: 1003). As a result, rather than institutionalising labour power, CoC and MSI seek to institutionalise 'business models of ethics' (Scheper 2017: 1086) based on discourses of corporate responsibility and conflict management.

In view of this, several empirical studies pointed at the limits of CoC, and related auditing and complaint mechanisms to be used as sources of institutional power by workers and unions in garment-producing countries. For example, Anner (2012) and Egels-Zandén and Merk (2014) studied two prominent transnational MSIs in the garment GPN: the Fair Labour Association and the Fair Wear Foundation. They found that local unions' complaints about garment manufacturers' union-busting practices in most cases had no effect. Lund-Thomsen and Coe (2013) in turn highlight that CoC may represent an effective institutional power resource for unions in garment-producing countries when used as an organising tool or as a reference in public campaigns. They illustrate how local unions in Pakistan achieved improvements in working conditions by using Nike's CoC as a reference in worker organising campaigns and in public media campaigns conducted in collaboration with several national NGOs. Hence, as opposed to 'traditional' institutional power sources such as labour laws, which workers and unions can deploy autonomously, to deploy private regulatory mechanisms as a power source, workers and unions depend in part on other societal actors to advocate for their rights.

In the light of the limitations for workers to leverage traditional institutional power resources, several scholar have argued that workers' abilities to build coalitions with other societal actors and to influence public discourses are central for leveraging new 'societal power resources' (see e.g. Webster et al. 2008; Webster 2015).

2.4.1.4   Societal Power: Coalitional and Moral Power

Webster (2015: 1) employs the term 'societal power' to refer to workers' and unions' capacities to build coalitions with social movements and to influence public and media discourses. Societal power can be further divided into coalitional power and moral power as two sub-forms. *Coalitional power* is defined by Schmalz et al. (2018: 122) as the ability of unions to build relationships of external solidarity with other social actors. These social actors may be mobilised for public campaigns or they may provide other types of support (e.g. financial support, capacity-building). *Moral power*, in turn, refers to workers' and unions' ability to discursively frame labour issues and solutions "in line with prevailing views of morality" and therefore able to attract public support (Schmalz et al. 2018: 123).[3] This framing may be achieved, for example, by invoking notions "of the struggle of 'right' against 'wrong', providing a basis for an appeal to both, the public and politicians, as well as to allies in civil society" (Webster et al. 2008: 12).

Research on labour agency in garment-producing countries has debated in particular the *possibilities for unions in garment-producing countries to deploy coalitional and moral power resources to effectively compensate for the lack of structural and institutional power resources* (Zajak et al. 2017: 907). This debate has been motivated by the widespread argument in labour studies and social movement literature that workers and unions 'at the bottom' of GPNs can shift the local power balance by harnessing the so-called boomerang effect (Wells 2009; Merk 2009). The term 'boomerang effect' refers to generating "Northern pressure to support workers' rights in the South", usually through transnational consumer campaigns (Wells 2009: 571). However, empirical studies of union agency in the garment GPN have found that moral power through consumer campaigns remains limited when unions lack a strong local associational and structural power base (Anner 2015b; Kumar 2014, 2019b; Zajak 2017). Anner (2015b), for example, found that the transnational campaigns carried out by local unions at college apparel suppliers in Honduras and El Salvador in collaboration with student anti-sweatshop movements in the US were only effective in cases of particularly crude and violent labour rights violations. In these cases, "the very extreme nature of the threat of bodily harm" allowed anti-sweatshop activists to frame labour rights violations "in a way that resonates with larger values of human decency" and hence to mobilise broad public support (Anner 2015b: 165). In his research on union agency in the Indian garment industry, Kumar (2014, 2019a) similarly finds that unions without a strong associational power base have only occasionally achieved some 'isolated victories' by activating moral power resources through transnational consumer campaigns. According to Kumar, unions were more successful in achieving lasting improvements for workers when they used moral power from transnational consumer campaigns to strategically support workplace organising and collective bargaining processes (Kumar 2019a: 360ff.).

In summary, existing empirical studies on how unions in garment-producing countries use 'new' societal power resources have shown that "only if strength is established domestically does a positive reinforcement effect across different power sources become possible" (Zajak 2017: 1008f.). However, it remains a central issue for empirical studies to develop a more nuanced understanding of how specifically unions can use moral and coalitional power resources to support the building of associational power in the workplace. I argue that to generate such an understanding, we still need more insights into the interplay between different power resources as well as into the structural effects that transnational consumer campaigns have on

---

[3] Literature from labour studies uses various terms to refer to this type of power resource. The term 'moral power' that I adopt in this book has been coined by Chun (2009) and Schmalz et al. (2018) originally refer to this type of power resource as 'discursive power'. A third common term for this type of power resource is 'symbolic power' (see, e.g. Webster 2015).

local garment unions' organising practices and internal relations. First insights by Hauf (2017) and Zajak (2017), for example, illustrate how close cooperation with transnational consumer campaigning networks can also constrain unions' capacities for building associational and organisational power: Relying on the 'borrowed' moral power of consumer campaigning networks can, for example, decrease unions' investments into members' strategic capacities and into fostering internal union democracy (Zajak 2017: 1019f.; Hauf 2017: 997).

Motivated by these observations, this study aims to better understand how different forms of mobilising coalitional and moral power resources may enhance or constrain associational power building. I argue that we can benefit from adopting a relational perspective on union agency to gain such an understanding. Taking on such a perspective, we can conceptualise the agency strategies of local unions in garment-producing countries as emerging at the intersection of various relational spaces of labour agency that workers and unionists construct themselves. These spaces for labour agency comprise the union itself as a relational space as well as networks with other unions and NGOs at various levels. In this context, it is essential to pay attention to the spatial and scalar dynamics shaping unions´ interactions with other actors (Coe 2015). In the next section, I therefore introduce a second branch of work that approaches agency strategies of unions in GPNs through the lens of scale as a central spatial analytical category.

Before doing so, to conclude this section, Table 2.3 summarises the definitions of the various power resources and their sub-forms (marked with →) that have been introduced in this section.

2.4.2 'Scaling' Worker and Union Agency in GPNs: 'Scale-Jumping' and 'Up-Scaling'

While the concept of worker and union power resources has been progressed primarily by labour sociologists and industrial relations scholars, labour geographers have highlighted workers' and unions' scalar strategies. According to Coe and Jordhus-Lier (2011: 219), "scale is particularly useful [as an analytical concept] as that it captures the double nature of the spatiality of labour agency: not only do the actions of labour play out in complex social geographies, but they can be understood as spatial phenomena in themselves. In other words, both the conditions and the strategies of labour agency are spatial". However, scale as a theoretical concept has been scarcely theorised in research on labour agency in GPNs. Generally, Coe and Jordhus-Lier (2011) argue that we can distinguish between three analytical moments in which the concept of scale has been employed in research on labour agency in GPNs: to define the level at which bargaining takes place, to refer to the group of workers for whom decisions are made, and to demarcate the territories across which solidarity is being sought.

Table 2.3 Union and worker power resources and their definitions

Power resource	Definition
Structural power	Derives from workers' strategic position within the economic system
→ Marketplace bargaining power	Accrues to workers who are positioned in labour markets that are characterised by low unemployment, possess scarce skills or have access to alternative types of income
→ Workplace bargaining power	Derives from workers' ability to cause disruptions to the broader production network through localised work stoppages
Associational power	Accrues to workers from their capacity to form collective organisations and to act collectively
→ Organisational power	Accrues to unions from their capacity to build the social capital, strategic capacities and 'cohesive collective identity' of their members
Institutional power	Accrues to workers and unions from their ability to invoke formal and informal rules and mechanisms that structure capital-labour relationships
Societal power	Derives from workers' and unions' capacity to mobilise the forces of other societal actors (e.g. NGOs, consumer organisations, community organisations) for their cause
→ Coalitional power	Derives from workers' and unions' ability to build solidary relationships with other social actors to access resources (e.g. financial, training etc.) or to mobilise their support
→ Moral power	Derives from workers' and unions' ability to mobilise support from political and societal actors for their cause by appealing to broader moral concepts of 'right' or 'wrong', e.g. through public campaigns

Source Author's elaboration drawing on Brookes (2013), Schmalz et al. (2018), Silver (2003), Webster et al. (2008) and Wright (2000)

The engagement of research on labour agency in GPNs with the concepts of space and scale hence reflects the approach to these concepts adopted in labour geography more generally. As mentioned before, research on labour agency in GPNs has developed at the intersection of labour geography and GPN analysis. It has been driven to a large extent by labour geographers' concern for understanding the spatial conditions and strategies of labour agency in the context of globalised production. Originally, labour geographers' concern with scale has been informed by the empirical observation that capital—while seeking to 'up-scale' its own operations—is simultaneously interested in 'downscaling' labour relations: capital aims to contain the bargaining of wages and working conditions at the local or workplace level as a means to play workers in different locations off against each other (Merk 2009: 603). Hence, labour geographers have frequently pointed out that 'jumping scale' and 'bridging space' are important strategies for labour agency under global capitalism since they allow workers to match capital's organisational scales (Castree et al. 2004; Herod 2001a; Merk 2009).

Labour geographers' concern for workers' scalar and spatial strategies has been informed by the political assertion that "workers actively produce economic spaces and scales in particular ways" and hence "shape the location of economic activity and the economic geography of capitalism" (Herod 1997: 24f.). In this vein, labour geographers have explicitly distinguished their theoretical perspective from the one forwarded by regulatory international political economy, which has traditionally been dominant in GPN analysis. Studies from the field of regulatory political economy conceptualise the "spatial scales at which social life is organised" as constructed and shaped predominantly by capital and state actors (Herod 2007: 29). Herod (2007) argues that such a perspective leads to a view of local, national and international scales for labour action as "areal containers of discrete absolute spaces" that are premade by capital and state actors and which contain workers' activities. According to Herod (2007), this perspective "denies [labour] actors the social agency to construct the geography of global capitalism in different and varied ways, for it suggests that the global scale of capital organisation is something that simply exists, waiting to be discovered and used, rather than something that had to be made and is constantly remade through the actions of diverse social actors".

Against this background, it has been a central concern of research at the intersection of GPN analysis and labour geography to enhance our understanding of the different scalar strategies that workers and unions employ to "reconfigure political landscapes and [to] renegotiate social hierarchies in ways which are more beneficial to the interests of workers" (Coe and Jordhus-Lier 2011: 219). In this light, various studies from labour geography have analysed the 'multi-scalar strategies' deployed by unions in globalised industries or service sectors (see e.g. Alford et al. 2017; Anderson 2009; Tufts 2007; Wills 2002). These studies have enhanced our understanding of workers' scalar agency in GPNs in two important ways. First, these studies have highlighted that while transnational alliances can help to increase workers' leverage vis-à-vis globally organised capital, there is no substitute for local organising (Herod 2001a; Wills 2002). Second, these studies have highlighted the diverse scales of labour organising and alliance building within GPNs, which range between workplaces, communities, cities, regions or the globe (Alford et al. 2017).

Which scales of organising and alliance building are most apt for workers at a specific node of a GPN depends on the spatial configurations of capital organisation and on workers' and unions' positionality within the broader production system. Particularly for workers in the industrial export sectors of the Global South, where transnational lead firms shape working conditions, literature has stressed the importance for workers to engage in transnational organising and campaigning to exercise leverage on geographically distant lead firms (Anner 2015b; Hale and Wills 2005; Kumar 2014; Merk 2009). With this in mind, Merk (2009), for example, illustrates how local unions in garment-producing countries can use the 'urgent appeal system' provided by the Clean Clothes Campaign (CCC)—a campaigning network led by European consumer organisations—as tool for 'up-scaling' workplace struggles.

The urgent appeal system allows workers or unions in garment-producing countries to submit a complaint about labour rights violations in garment factories to the CCC together with a request for action. The CCC member organisations then engage in various actions to harness brands' leverage over local garment manufacturers. Actions carried out by the CCC range from sending letters or emails to brands' management teams, to planning and carrying out full public campaigns. This includes publishing fact-finding reports and the naming and shaming of brands embroiled with the factory where the labour rights violations occurred (Merk 2009: 607). According to Merk, the CCC's urgent appeal mechanism "provides a grassroots based system to build labour solidarity across space, which may help [workers in garment-producing countries] to regain leverage over capital" (Merk 2009: 599).

However, studies focussing on the 'up-scaling' of workplace conflicts through transnational consumer campaigning have, at the same time, received criticism from various scholars. These scholars argue that the narrow focus on 'up-scaling' through transnational campaigning has shifted the analytical focus from the agency of workers to the agency of consumer groups or NGOs in the Global North. As a result, the latter are perceived and presented as the main agents of change (see e.g. Hauf 2017: 989; Wells 2009). Contradicting this notion, Wells (2009: 568) stresses that the mobilisation of extra-local leverage through transnational campaigns merely represents *one* of *many* strategic actions that workers and unions in the Global South usually employ when leading a labour struggle. To achieve sustained improvements in working conditions, he argues, workers' and unions' strategic actions at the workplace or at the community level are equally important (Wells 2009: 577). Similarly, Wills (2002) and Tufts (2007) emphasise that 'up-scaling' specific labour struggles should not be seen as an isolated strategy but as part of a broader ensemble of networks and institutional arrangements with which unions engage. They, therefore, stress that analyses of 'multi-scalar' agency strategies should not be limited to identifying the most appropriate scale for action. Instead, they argue, studies of 'multi-scalar' labour agency strategies in GPNs should explore the different types of resources that unions can leverage through actions at *multiple* scales (Tufts 2007: 2387).

In summary, scalar approaches to labour agency in GPNs have contributed to enhancing our understanding of the diverse spatial strategies at various scales that workers and unions in GPNs may employ. However, we still need a better understanding of how workers and unions can *strategically link* actions at various levels in the context of specific struggles. I argue that we need a better understanding in particular of the following three aspects of workers' and unions' multi-scalar strategies. First, we need a better grasp of the different power resources that unions can activate through building relationships at *different* scales—from intra-union relationships over building alliances with local community organisations to engaging with transnational consumer campaigning networks (c.f. Nicholls 2009; Tufts 2007). Second, I contend that we need a better understanding of the different resources and capacities that unions may access and develop in different relationships and networks at the *transnational* scale. Studies of the 'multi-scalar' agency of workers and unions in GPNs have tended to conflate all transnational networks and relationships constructed by local unions under the notion of 'up-scaling', be it relationships with consumer

campaigning networks or with international worker networks. However, this conflation has obscured the fact that different kinds of transnational networks may be
constituted through very different practices, therefore enabling unions and workers
to access very different types of resources (c.f. Lohmeyer et al. 2018; Fütterer and
López Ayala 2018; López and Fütterer 2019). Third, we need a better understanding
of how unions' strategies and networks at different scales not only complement but
also shape and influence each other. Understanding the interplay of unions' actions at
various scales is important particularly in the light of the findings of studies working
with the power resources approach. These studies found that unions' practices of
engaging in transnational alliances can directly impact unions' internal organisational
practices and workplace organising strategies (see Sect. 2.4.2).

   To tackle these gaps, in this book, I develop a practice-oriented, relational approach
to union agency in GPNs that emphasises practices and relationships instead of a
priori-defined scalar categories (see Sect. 3.3). To this end, I build on a third strand
of work within research on labour agency in GPNs focussing on 'Networks of Labour
Activism', which is introduced in the next section.

2.4.3 Network Approaches to Labour Agency in GPN: 'Networks of Labour Activism'

The 'Networks of Labour Activism' (NOLA) approach was first introduced by a
group of interdisciplinary scholars in 2017. The approach centres explicitly on the
practices, actions and relationships of workers and labour organisations in the Global
South positioned "on the bottom rungs of the globally networked economy" (Zajak
et al. 2017: 916; see also contributions to Forum Debate 2017 of Development and
Change Vol. 48, Nr. 5). The NOLA approach is motivated by the observation that
labour-intensive export sectors in Asian countries "have also become the testing
ground for new forms of networked worker agency and activism" (Zajak et al.
2017: 900). These new forms of networked worker agency are based on workers'
construction of solidarity networks and relationships with a broad range of actors
in varying geographical distances. Actors with whom workers construct solidarity
networks range from local community organisations over national union coalitions
to transnational consumer activist networks (Zajak et al. 2017: 901).

   Contrary to scalar approaches to labour agency in GPNs, which have focussed
rather one-sidedly on workers' transnational 'up-scaling'-strategies, the NOLA
approach conceptualises workers' agency in GPNs as multi-directional and multi-
layered. The agency of workers and unions is *multi-directional* since it is performed
in and through relationships with a variety of actors, including potential targets as
well as potential allies. In addition, workers' agency is *multi-layered* since it is
performed through relationships that stretch across various distances—i.e. relationships at multiple scales (Zajak et al. 2017: 904). Nevertheless, the NOLA approach
still emphasises *'cross-border strategising'* as a central feature of networked labour

agency in GPNs (Zajak et al. 2017: 905). Cross-border strategising is essential for workers 'at the bottom' of GPNs due to the structural context of supply chain capitalism within which they are embedded. Supply chain capitalism refers to global value chains as a system of capital accumulation that is based on Northern lead firms' practices of dis-embedding the labour-intensive production steps from regulated, unionised environments through outsourcing (Tsing 2009). Therefore, to tackle geographically distant lead firms, which ultimately determine labour conditions along the supply chain, workers need to engage in cross-border strategising. Cross-border strategising refers here to strategic practices of building solidary relationships with potential allies in different countries on the one hand, and to strategic practices of targeting actors or institutions located in a foreign country on the other hand (Zajak et al. 2017: 905).

It needs to be noted here that a central difference to 'up-scaling' literature—which has also been concerned with workers' cross-border strategies—consists in the explicitly constructivist perspective of the NOLA approach. Whereas 'up-scaling literature' has put emphasis on how workers and unions *use* the leverage of transnational alliances to enhance their position vis-à-vis local capital (or the state), the NOLA approach shifts the attention to the practices and relationships through which workers and unions *construct* these alliances in the first place (Zajak et al. 2017: 903). It is within these alliances that workers and unions develop specific strategic capacities and power resources which may in turn contribute to strengthening workers' domestic associational power base (Zajak et al. 2017: 1009). At the same time, the NOLA approach recognises that 'networks of labour activism' become "their own structural forces" (Zajak et al. 2017: 1020). In other words, the interactions and learning processes within these networks shape the behaviour of the actors involved in it.

Empirical studies using the NOLA approach have provided a range of important insights on the structural effects of various types of solidarity networks on the practices of unions in garment-producing countries. Hauf (2017), for example, has studied the effect of local unions' engagement with the Play Fair Campaign in Indonesia, an MSI led by Northern NGOs and global trade union federations. Hauf found that participation of Indonesian unions in the transnational Play Fair campaign helped unions to forge new networks with other local garment unions, since they had an incentive to develop a common position. However, he also finds that the potential of these new solidary networks at the local and national level for building sustained associational power was limited. Practices of collaboration between unions were strictly confined to interactions within the framework of the Play Fair campaign. Consequently, these interactions did little to bridge the political and ideological differences between local unions in the Indonesian garment sector and thus to establish the foundation for collaborations on other issues (Hauf 2017: 988).

Hauf's findings are further supported by Zajak (2017) in her study of Bangladeshi garment unions' engagement with the Bangladesh Accord for Fire and Building

Safety.[4] She finds that local unions' interactions with the transnational stakeholder network of the Bangladesh Accord have, on the one hand, strengthened local unions' organisational power. For example, trainings conducted by these stakeholders have enabled local unionists to develop important strategic capacities for reframing workplace issues so that they fall under the mandate of the Accord (Zajak 2017: 1017). But, on the other hand, she finds that local unions' increased engagement with international organisations and Northern NGOs has also led to new divisions and relations of competition between local unions and hence to a weakening of associational power at the national level (Zajak 2017: 1019).

Whereas Hauf (2017) and Zajak (2017) have focussed on the practices through which local garment unions engage with transnational multi-stakeholder networks, Lohmeyer et al. (2018) have analysed how Asian garment unions engage with transnational labour networks. In their study of the TIE ExChains network, which links Asian garment workers with German fashion retail workers, they illustrate how solidary relationships within the network are constructed through practices of facilitating shared experiences and of developing a common interpretation of labour rights violations along the chain. At the same time, Lohmeyer et al. (2018) illustrate potential tensions between practices of performing cross-border solidarity campaigns within the network and practices or local workplace organising. These tensions arise because cross-border solidarity campaigns are usually centred on specific issues and therefore have a rather short-term outlook. Workplace-organising practices are, in turn, usually embedded within a unions' long-term strategy. Against this background, Lohmeyer et al. (2018) stress that a careful strategic alignment of transnational campaigning practices with unions' long-term strategic goals and organising strategies is necessary to ensure that transnational campaigning strengthens unions' local associational power base instead of undermining it (Lohmeyer et al. 2018: 417).

In summary, network-centred approaches to worker and union agency have enhanced our understanding of union agency in GPNs in three ways. First, networked approaches to labour agency have contributed to breaking up the 'black box' of unions as collective actors. This has been achieved by making visible how "union strategies evolve through contested socio-spatial relations both within unions themselves and with other social actors" (Cumbers et al. 2008: 369). Second, through engaging with the power resources approach, studies of workers' networked agency have opened up analytical space for assessing the potential of different 'networks of labour activism' for strengthening local unions' associational power base. Third, network-centred studies of labour agency in GPNs have enhanced our understanding of how workers and unions forge their own relational geographies and add new layers to existing GPNs.

---

[4] The Bangladesh Accord Bangladesh Accord for Fire and Building Safety is an independent, legally binding agreement between international fashion retailers and global and Bangladeshi trade unions to improve building safety in the garment industry in Bangladesh. The agreement was negotiated and signed in response to international pressures by consumer and labour organisations in the aftermath of the Rana Plaza building collapse on 24 April 2013, which killed more than 1300 garment workers (Zajak 2017: 1008f.).

Notwithstanding these critical insights into the agency of local workers and unions in garment-producing countries, a central dimension of labour agency in GPNs has hitherto remained underexplored in NOLA studies: the spatial dimension of labour's agency. The neglect of space as an analytical category is reflected, first, in Zajak et al.'s (2017) proposal of 'cross-border strategising' a central characteristic of networks of labour action. 'Cross-border' refers, on the one hand, to networks that connect actors across borders. On the other hand, it also refers to a situation in which workers build alliances with other social groups at the local or national level to target actors located in a different country. However, this subsumption of local, national and transnational networks under the same category of 'cross-border action' ignores that very different practices and mechanisms are involved in constructing networks at different scales. Second, subsuming networks that span across different territorial extensions under the same category of 'cross-border strategising' neglects the fact that networks forged *within* particular places and networks forged *across* great distance may play distinct but complementary roles in building local union power (c.f. Nicholls 2009). Last but not least, the neglect of the spatial dimension of labour agency is reflected in the little attention paid by studies working with the NOLA approach to the place-specific structures of labour control that constrain labour's spaces and options for agency.

Therefore, in this book, I build on the network perspective of the NOLA approach and integrate relational perspectives from economic geography on the one hand, and with literature on labour control regimes on the other. Drawing on these different bodies of literature, I develop a relational approach to labour control and labour agency in GPNs. Before doing so, in the next section I first summarise the contributions and limitations of existing research on labour agency in GPNs.

2.4.4   Interim Conclusion: Contributions and Shortcomings of Existing Analytical Approaches to Labour Agency in GPNs

This section has introduced three different research strands on labour agency in GPNs: (1) a research strand drawing on the concept of worker and unions power resources; (2) a research strand adopting a scalar approach with a focus on workers' 'up-scaling' strategies; and (3) a research strand that analyses labour agency in GPNs through the lens of 'Networks of Labour Activism'. Table 2.4 on the next page provides a synthesised overview of the central research questions, main insights and shortcomings of each research strand.

In a nutshell, I argue that we can learn three critical lessons from existing research on labour agency in GPNs regarding the conditions that constrain and enable the agency of workers and unions in garment-producing countries:

First, by re-embedding the agency of workers and unions into structural sets of social and economic relations—encompassing value chain dynamics at the vertical

Table 2.4 Overview of different research strands on labour agency in GVCs/GPNs

Research strand	Central research question(s)	Central insights on constraining and enabling conditions for labour agency 'at the bottom' of GVCs/GPNs	Shortcomings
Worker and union power resources	• Which strategies enable workers and unions 'at the bottom of GPNs' to build or activate different types of power resources? • To what extent can workers and unions 'at the bottom of GPNs' compensate for a lack of structural, associational and institutional power through the leverage of 'new' coalitional or moral power resources?	• For workers in industrial export sectors 'at the bottom' of buyer-driven GVCs significant barriers for building and leveraging 'traditional' (structural, associational and institutional) power resources exist • Moral and coalitional power resources through alliances with NGOs and consumer campaigns from the Global North can be deployed to reinforce associational power resources?	• So far only rudimentary understanding of how unions can deploy moral and coalitional power resources to build associational and organisational power resources
Scalar approaches to labour agency	• At which scales do workers and unions in globalised industries exercise strategic actions vis-à-vis capital?	• Transnational alliance building and campaigning can help to build leverage vis-à-vis global capital but it cannot substitute local 'on the ground' organising • Workers and unions in GPNs deploy actions at a diversity of scales (e.g. workplace, communities, regions, the globe)	• Narrow focus on transnational up-scaling strategies and neglect of workers' actions at other scales • Need for a better understanding of which (power) resources unions can leverage through actions at different levels

(continued)

Table 2.4 (continued)

Research strand	Central research question(s)	Central insights on constraining and enabling conditions for labour agency 'at the bottom' of GVCs/GPNs	Shortcomings
Networks of labour activism	• Through which practices and relations do workers and unions construct solidarity networks across varying distances? • Which types of power resources can workers and unions develop within different networks of labour activism?	• Different types of solidarity networks have varying (and potentially mixed) structural effects for building local unions' associational power base • Unions' engagement with NGOs and consumer campaigning networks from the Global North can also pose constraints for building associational power 'on the ground'	• Neglects spatial dimension of labours' agency, e.g. different roles of local and transnational networks • Neglects the influence of place-specific labour control regimes in shaping workers' practices of constructing 'networks of labour activism'

dimension and state, labour market and cultural relations at the horizontal dimension—existing research on labour agency in GPNs has highlighted *potential structural constraints for building local union power* 'at the bottom of GPNs'. The analysis of different worker and union power resources in combination with a focus on 'constrained' labour agency has contributed to a more fine-grained understanding of the conditions that curb local unions' opportunities to build sustained bargaining power vis-à-vis employers. In this regard, past studies have highlighted how value chain governance dynamics at the vertical dimension of the GPN curtail the structural power of workers and unions in garment-producing countries. Moreover, repressive state and employer labour control regimes at the horizontal level significantly constrain workers' opportunities for collective organising and building associational power resources, as well as workers' opportunities for leveraging institutional power sources, such as legal frameworks.

Second, existing research on labour agency in GPNs has also provided insights into *potential enabling conditions for building local union power* in garment-producing countries. Since geographically distant lead firms represent the locus of power in these GPNs, existing research has stressed opportunities for local unions in garment-producing countries to 'up-scale' local labour struggles to the transnational level. 'Up-scaling' labour struggles can enable unions to leverage coalitional and moral power resources through consumer campaigns in the Global North and thereby to exert pressure over lead firms and manufacturers. Nevertheless, existing research has also pointed out that activating coalitional and moral power resources through 'up-scaling' strategies cannot substitute associational power resources on the ground. In this line, past studies of labour agency in GPNs have highlighted that transnational consumer campaigns only contribute to building local union power, when local unions strategically deploy them to create spaces for collective organising and bargaining on the ground.

Third, in particular the NOLA approach has contributed to our understanding of the constraining and enabling conditions for building local union power in garment-producing countries. It has highlighted that not only capital, state and labour market relations but also the networks constructed by workers and unions themselves represent enabling and/or constraining contexts for labour agency. On the one hand, workers and unions can develop and build capacities and power resources through and within these networks. On the other hand, unions' and workers' embeddedness within 'networks of labour activism' can also pose constraints for unions' capacities of building associational power. In this line, studies using the NOLA approach have highlighted how the participation of local unions in transnational NGO-led multi-stakeholder networks can hamper internal union democracy and create inter-union divisions within garment-producing countries.

Notwithstanding these important contributions of past research on labour agency in GPNs, three important aspects of union agency remain underdeveloped and require further research. First, for a more nuanced understanding of how local unions in garment-producing countries can build sustained bargaining power, we still need a better understanding of how unions can link actions at *various* scales to exercise leverage over employers. In this context, labour geographers' argument that workers and unions can activate and develop different power resources through actions at different scales requires further exploration. Labour geographers have stressed that scales of labour action traditionally encompassed the workplace and the international level and also multiple scales in between, such as the community, the city or the national level (c.f. Anderson 2009; Tufts 2007; Wills 2002). In particular when studying unions in the Global South, where non-traditional forms of 'social movement unionism' have emerged, a stronger analytical focus on the implications of unions' engagement with community actors for building bargaining power vis-à-vis employers is necessary.

Second, past research on labour agency in GPNs has tended to conflate the variety of practices through which local unions engage with transnational networks under the same notions of 'up-scaling' or leveraging 'coalitional power'. However, this conflation has concealed essential differences in the everyday practices and power relations that constitute local unions' relationships with different types of transnational networks. There are important differences, for example, between the practices through which consumer campaigning networks exercise solidarity compared to grassroots worker networks (see Lohmeyer et al. 2018). Consequently, if we take the lesson from NOLA studies seriously that also the networks constructed by unions themselves have structural effects on unions' associational power resources, we need to be more sensitive to the different practices and relationships that constitute different 'networks of labour activism'. Only an analytical perspective that is sensitive to such differences will allow us to evaluate to which extent the engagement in *different* transnational networks allows local unions to build strategic capacities and power resources.

Third and last, studies concerned with the agency of labour in GPNs have only superficially engaged with studies of labour control (regimes) in GPNs. Studies on labour agency in GPNs have generally highlighted that repressive labour control regimes in the export industries of the Global South constrain workers' and unions' opportunities for building and leveraging associational and institutional power resources. However, these studies have not yet provided a more fine-grained analysis of the specific capital and state practices that constrain the agency of workers and unions. Moreover, studies on labour agency in GPNs have so far neglected the role of workers and local unions 'at the bottom' of GPNs in co-shaping and potentially transforming labour regimes at specific nodes of a GPN. Therefore, to fully understand the dialectical relationship between labour's agency and labour control regimes as structural contexts within GPNs, we need to look closely at how unions tackle and transform specific state and capital practices that constitute labour control regimes.

To tackle these shortcomings and to further develop our understanding of the constraining and enabling factors for building sustained union power in garment-producing countries, in the next section, I develop a relational approach for studying labour control regimes and labour agency in GPNs.

References

Alford M, Barrientos S, Visser M (2017) Multi-scalar labour agency in global production networks: contestation and crisis in the South African fruit sector. Dev Chang 48:721–745. https://doi.org/10.1111/dech.12317

Anderson J (2009) Labour's lines of flight: rethinking the vulnerabilities of transnational capital. Geoforum 40:959–968. https://doi.org/10.1016/j.geoforum.2009.08.002

Anner M (2012) Corporate social responsibility and freedom of association rights. Polit Soc 40:609–644. https://doi.org/10.1177/0032329212460983

Anner M (2015a) Labor control regimes and worker resistance in global supply chains. Labor Hist 56:292–307. https://doi.org/10.1080/0023656X.2015

Anner M (2015b) Social downgrading and worker resistance in apparel global value chains. In: Newsome K, Taylor P, Bair J, Rainnie A (eds) Putting labour in its place: labour process analysis and global value chains. Palgrave Macmillan, London and New York, pp 152–170

Anner M (2019) Predatory purchasing practices in global apparel supply chains and the employment relations squeeze in the Indian garment export industry. Int Labour Rev 158:705–727. https://doi.org/10.1111/ilr.12149

Arnold D (2013) Workers' agency and re-working power relations in Cambodia's garment industry. http://www.capturingthegains.org/publications/workingpapers/wp_201324.htm. Accessed 24 Apr 2020

Arnold D, Hess M (2017) Governmentalizing Gramsci: topologies of power and passive revolution in Cambodia's garment production network. Environ Plann A 49:2183–2202. https://doi.org/10.1177/0308518X17725074

Azmeh S (2014) Labour in global production networks: workers in the qualifying industrial zones (QIZs) of Egypt and Jordan. Global Netw 14:495–513. https://doi.org/10.1111/glob.12047

Azmeh S, Nadvi K (2014) Asian firms and the restructuring of global value chains. Int Bus Rev 23:708–717. https://doi.org/10.1016/j.ibusrev.2014.03.007

Baglioni E (2018) Labour control and the labour question in global production networks: exploitation and disciplining in Senegalese export horticulture. J Econ Geogr 18:111–137. https://doi.org/10.1093/jeg/lbx013

Bair J (2008) Analysing global economic organization: embedded networks and global chains compared. Econ Soc 37:339–364. https://doi.org/10.1080/03085140802172664

Bair J, Werner M (2015) Global production and uneven development: when bringing labour in isn't enough. In: Newsome K, Taylor P, Bair J, Rainnie A (eds) Putting labour in its place: labour process analysis and global value chains. Palgrave Macmillan, London and New York, pp 119–134

Barrientos S (2007) Global production systems and decent work. Working Paper No. 77. http://ilo.org/wcmsp5/groups/public/---dgreports/---integration/documents/publication/wcms_085041.pdf

Barrientos S, Mayer F, Pickles J, Posthuma A (2011a) Decent work in global production networks: framing the policy debate. Int Labour Rev 150:297–317. https://doi.org/10.1111/j.1564-913X.2011.00118.x

Barrientos S, Gereffi G, Rossi A (2011) Economic and social upgrading in global production networks: a new paradigm for a changing world. Int Labour Rev 150:319–340. https://doi.org/10.1111/j.1564-913X.2011.00119.x

Brookes M (2013) Varieties of power in transnational labor alliances. Labor Stud J 38:181–200. https://doi.org/10.1177/0160449X13500147

Burawoy M (1979) Manufacturing consent: changes in the labor process under monopoly capitalism. University of Chicago Press, Chicago

Burawoy M (1985) The politics of production: factory regimes under capitalism and socialism. Verso, London

Castree N (2007) Labour geography: a work in progress. Int J Urban and Reg Res 31. https://doi.org/10.1111/j.1468-2427.2007.00761.x

Castree N, Coe NM, Ward K, Samers M (2004) Spaces of work: global capitalism and the geographies of labour. Sage, London

Chun JJ (2009) Organizing at the margins: the symbolic politics of labor in South Korea and the United States. Cornell University Press

Coe NM (2015) Labour and global production networks: mapping variegated landscapes of agency. In: Newsome K, Taylor P, Bair J, Rainnie A (eds) Putting labour in its place: labour process analysis and global value chains. Palgrave Macmillan, London and New York, pp 171–192

Coe NM, Jordhus-Lier DC (2010) Re-embedding the agency of labour. In: Bergene AC, Endresen SB, Knutsen HM (eds) Missing links in labour geography. Ashgate Publishing, Farnham, pp 29–42

Coe NM, Jordhus-Lier DC (2011) Constrained agency? Re-evaluating the geographies of labour. Prog Hum Geogr 35:211–233. https://doi.org/10.1177/0309132510366746

Coe NM, Kelly PF (2002) Languages of labour: representational strategies in Singapore's labour control regime. Polit Geogr 21:341–371. https://doi.org/10.1016/S0962-6298(01)00049-X

Coe NM, Yeung HW-C (2019) Global production networks: mapping recent conceptual developments. J Econ Geogr 19:775–801. https://doi.org/10.1093/jeg/lbz018

Coe NM, Hess M, Yeung HW-C, Dicken P, Henderson J (2004) "Globalizing" regional development: a global production networks perspective. Trans Inst Br Geogr 29:468–484. https://doi.org/10.1111/j.0020-2754.2004.00142.x

Coe NM, Dicken P, Hess M (2008) Global production networks: realizing the potential. J Econ Geogr 8:271–295. https://doi.org/10.1093/jeg/lbn002

Cumbers A, Nativel C, Routledge P (2008) Labour agency and union positionalities in global production networks. J Econ Geogr 8:369–387. https://doi.org/10.1093/jeg/lbn008

Cumbers A (2015) Understanding labour's agency under globalization; embedding GPNs within an open political economy. In: Newsome K, Taylor P, Bair J, Rainnie A (eds) Putting labour in its place: labour process analysis and global value chains. Palgrave Macmillan, London and New York, pp 135–151

Cumbers A, Featherstone D, MacKinnon D, Ince A, Strauss K (2016) Intervening in globalization: the spatial possibilities and institutional barriers to labour's collective agency. J Econ Geogr 16:93–108. https://doi.org/10.1093/jeg/lbu039

Dicken P, Kelly PF, Olds K, Yeung HW-C (2001) Chains and networks, territories and scales: towards a relational framework for analysing the global economy. Global Netw 1:89–112

Dörre K, Holst H, Nachtwey O (2009) Organising—a strategic option for trade union renewal? Int J Action Res 5:33–67

Edwards PK (1990) Understanding conflict in the labour process: the logic and autonomy of struggle. In: Knights D, Willmott H (eds) Labour process theory. Palgrave Macmillan Limited, Basingstoke, pp 125–152

Egels-Zandén N, Merk J (2014) Private regulation and trade union rights: why codes of conduct have limited impact on trade union rights. J Bus Ethics 123:461–473

Fairbrother P (2008) Social movement unionism or trade unions as social movements. Employ Responsib Rights J 20:213–220. https://doi.org/10.1007/s10672-008-9080-4

Fairbrother P, Webster E (2008) Social movement unionism: questions and possibilities. Employ Responsib Rights J 20:309–313. https://doi.org/10.1007/s10672-008-9091-1

Flecker J, Meil P (2011) Organisational restructuring and emerging service value chains—implications for work and employment. Work Employ Soc 24(4):1–19

Fütterer M, López Ayala T (2018) Challenges for organizing along the garment value chain. Experiences from the union network TIE ExChains. https://www.rosalux.de/en/publication/id/39369/challenges-for-organizing-along-the-garment-value-chain/. Accessed 5 Apr 2022

Gereffi G (1994) The organization of buyer-driven global commodity chains: How U.S. retailers shape overseas production network. In: Gereffi G, Korseniewicz M (eds) Commodity chains and global capitalism. Praeger Publishers, Westport, pp 95–122

Gereffi G (1999) International trade and industrial upgrading in the apparel commodity chain. J Int Econ 48:37–70

Gereffi G (2005) The global economy: organization, governance and development. In: Smelser NJ, Swedberg R (eds) The handbook of economic sociology, 2nd edn. Princeton University Press, pp 160–182

Gereffi G, Korseniewicz M (eds) (1994) Commodity chains and global capitalism. Praeger Publishers, Westport

Gereffi G, Humphrey J, Sturgeon T (2005) The governance of global value chains. Rev Int Polit Econ 12:78–104. https://doi.org/10.1080/09692290500049805

Gumbrell-McCormick R, Hyman R (2013) Trade unions in Western Europe: hard times, hard choices. Oxford University Press, Oxford

Haidinger B, Flecker J (2015) Positioning labour in service value chains and networks: the case of parcel delivery. In: Newsome K, Taylor P, Bair J, Rainnie A (eds) Putting labour in its place: labour process analysis and global value chains. Palgrave Macmillan, London and New York, pp 64–82

Hale A, Wills J (eds) (2005) Threads of labour: garment industry supply chains from the workers' perspective. Antipode book series. Blackwell, Malden, Mass, Oxford

Hammer N, Riisgaard L (2015) Labour segmentation in value chains. In: Newsome K, Taylor P, Bair J, Rainnie A (eds) Putting labour in its place: labour process analysis and global value chains. Palgrave Macmillan, London and New York, pp 83–116

Harvey D (1982) The limits to capital. Basil Blackwell, Oxford

Harvey D (2001) Globalization and the "Spatial Fix". Geographische Revue 3:23–30

Hastings T (2019) Leveraging Nordic links: South African labour's role in regulating labour standards in wine global production networks. J Econ Geogr 19:921–942. https://doi.org/10.1093/jeg/lbz010

Hastings T, MacKinnon D (2017) Re-embedding agency at the workplace scale: workers and labour control in Glasgow call centres. Environ Plann A 49:104–120. https://doi.org/10.1177/030851 8X16663206

Hauf F (2017) Paradoxes of transnational labour rights campaigns: the case of play fair in Indonesia. Dev Chang 48:987–1006. https://doi.org/10.1111/dech.12321

Henderson J, Dicken P, Hess M, Coe N, Yeung HW-C (2002) Global production networks and the analysis of economic development. Rev Int Polit Econ 9:436–464. https://doi.org/10.1080/096 92290210150842

Herod A (1997) From a geography of labor to a labor geography: labor's spatial fix and the geography of capitalism. Antipode 29:1–31. https://doi.org/10.1111/1467-8330.00033

Herod A (2001a) Labor geographies: workers and the landscapes of capitalism. Perspectives on economic change. Guilford Press, New York

Herod A (2001b) Labor internationalism and the contradictions of globalization: or, why the local is sometimes still important in a global economy. Antipode 33:407–426. https://doi.org/10.1111/ 1467-8330.00191

Herod A (2007) The agency of labour in global change: reimagining the spaces and scales of trade union praxis within a global economy. In: Hobson JM, Seabrooke L (eds) Everyday politics of the world economy. Cambridge University Press, Cambridge, pp 27–44

Hudson R (2004) Conceptualizing economies and their geographies: spaces, flows and circuits. Prog Hum Geogr 28:447–471. https://doi.org/10.1191/0309132504ph497oa

ILO (1999) Decent work: report of the director-general. Accessed 28 Feb 2020

Jenkins J (2015) The significance of grass-roots organizing in the garment and electrical value chains of Southern India. In: Newsome K, Taylor P, Bair J, Rainnie A (eds) Putting labour in its place: labour process analysis and global value chains. Palgrave Macmillan, London and New York, pp 195–212

Jenkins J, Blyton P (2017) In debt to the time-bank: the manipulation of working time in Indian garment factories and 'working dead horse.' Work Employ Soc 31:90–105. https://doi.org/10. 1177/0950017016664679

Jonas AEG (1996) Local labour control regimes: uneven development and the social regulation of production. Reg Stud 30:323–338. https://doi.org/10.1080/00343409612331349688

Jonas AEG (2009) Labor control regime. In: Kitchin R (ed) International encyclopedia of human geography. Elsevier, Amsterdam, pp 59–65

Kelly PF (2001) The political economy of local labor control in the Philippines. Econ Geogr 77:1–22. https://doi.org/10.2307/3594084

Kelly PF (2002) Spaces of labour control: comparative perspectives from Southeast Asia. Trans Inst Br Geogr 27:395–411. https://doi.org/10.1111/1475-5661.00062

Knight J (1992) Institutions and social conflict. The political economy of institutions and decisions. Cambridge University Press, Cambridge

Kumar A (2014) Interwoven threads: building a labour countermovement in Bangalore's export-oriented garment industry. City 18:789–807. https://doi.org/10.1080/13604813.2014.962894

Kumar A (2019a) A race from the bottom? Lessons from a workers' struggle at a Bangalore warehouse. Compet Change 23:346–377. https://doi.org/10.1177/1024529418815640

Kumar A (2019b) Oligopolistic suppliers, symbiotic value chains and workers' bargaining power: labour contestation in South China at an ascendant global footwear firm. Global Netw 19:394–422. https://doi.org/10.1111/glob.12236

Latham A (2002) Retheorizing the scale of globalization: topologies, actor-networks, and cosmopolitanism. In: Herod A, Wright MW (eds) Geographies of power: placing scale. Blackwell, Malden, MA, pp 115–144

Lerche J (2012) Labour regulations and labour standards in India: decent work? Global Labour J 3. https://doi.org/10.15173/glj.v3i1.1111

Lévesque C, Murray G (2002) Local versus global: activating local union power in the global economy. Labor Stud J 27:39–65. https://doi.org/10.1177/0160449X0202700304

Lévesque C, Murray G (2010) Understanding union power: resources and capabilities for renewing union capacity. Transf Eur Rev Labour Res 16:333–350. https://doi.org/10.1177/102425891037 3867

Levy DL (2008) Political contestation in global production networks. Acad Manag Rev 33:943–963. https://doi.org/10.2307/20159454

Lier DC (2007) Places of work, scales of organising: a review of labour geography. Geogr Compass 1:814–833. https://doi.org/10.1111/j.1749-8198.2007.00047.x

Locke RM, Kochan T, Romis M, Qin F (2007) Beyond corporate codes of conduct: work organization and labour standards at Nike's suppliers. Int Labour Rev 146:21–40. https://doi.org/10.1111/j.1564-913X.2007.00003.x

Lohmeyer N, Schüßler E, Helfen M (2018) Can solidarity be organized "from below" in global supply chains? The case of ExChains. Industrielle Beziehungen 25:400–424

López T (2021) A practice ontology approach to labor control regimes in GPNs: connecting 'sites of labor control' in the Bangalore export garment cluster. Environ Plann A 53:1012–1030. https://doi.org/10.1177/0308518X20987563

López T, Fütterer M (2019) Herausforderungen und Strategien für den Aufbau gewerkschaftlicher Verhandlungsmacht in der Bekleidungswertschöpfungskette: Erfahrungen aus dem TIE-ExChains-Netzwerk. In: Ludwig C, Simon H, Wagner A (eds) Bedingungen und Strategien gewerkschaftlichen Handelns im flexiblen Kapitalismus. Westfälisches Dampfboot, Münster, pp 175–191

López T, Riedler T, Köhnen H, Fütterer M (2021) Digital value chain restructuring and labour process transformations in the fast-fashion sector: evidence from the value chains of Zara & H&M: Online First. Global Netw, 1–17. https://doi.org/10.1111/glob.12353

López Ayala T (2018) Multi-level production of the local labour control regime in the Bangalore readymade garment cluster. In: Butsch C, Follmann A, Müller J (eds) Aktuelle Forschungsbeiträge zu Südasien: (Geographien Südasiens, Band 10). xasia eBooks, Berlin, pp 20–23

Lund-Thomsen P, Coe NM (2013) Corporate social responsibility and labour agency: the case of Nike in Pakistan. J Econ Geogr 15:275–296. https://doi.org/10.1093/jeg/lbt041

MacKinnon D, Cumbers A (2011) Introduction to economic geography: globalization, uneven development and place. Routledge Taylor & Francis Group, London, New York

Marston SA, Jones JP, Woodward K (2005) Human geography without scale. Trans Inst Br Geogr 30:416–432. https://doi.org/10.1111/j.1475-5661.2005.00180.x

Massey D (1984) Spatial divisions of labour: social structures and the geography of production. Macmillan Education UK, London

Massey D (1992) Politics and space/time. New Left Rev 196:65–88

Massey D (1993) Questions of locality. Geography 78:142–149

Mayer FW, Pickles J (2010) Re-embedding governance: global apparel value chains and decent work. Working Paper 2010/01. Capturing the Gains 2010.

McGrath S (2013) Fuelling global production networks with slave labour? Migrant sugar cane workers in the Brazilian ethanol GPN. Geoforum 44:32–43. https://doi.org/10.1016/j.geoforum.2012.06.011

Merk J (2009) Jumping scale and bridging space in the era of corporate social responsibility: cross-border labour struggles in the global garment industry. Third World Q 30:599–615. https://doi.org/10.1080/01436590902742354

Merk J (2014) The rise of tier 1 firms in the global garment industry: challenges for labour rights advocates. Oxf Dev Stud 42:277–295. https://doi.org/10.1080/13600818.2014.908177

Mezzadri A (2008) The rise of neo-liberal globalisation and the 'new old' social regulation of labour: a case of Delhi garment sector. Indian J Labour Econ 51:603–618

Mezzadri A (2010) Globalisation, informalisation and the state in the Indian garment industry. Int Rev Sociol 20:491–511. https://doi.org/10.1080/03906701.2010.511910

Mezzadri A (2016) Class, gender and the sweatshop: on the nexus between labour commodification and exploitation. Third World Q 37:1877–1900. https://doi.org/10.1080/01436597.2016.1180239

Mezzadri A (2017) Sweatshop regimes in the Indian garment industry. Cambridge University Press

Milberg W, Winkler D (2011) Economic and social upgrading in global production networks: problems of theory and measurement. Int Labour Rev 150:341–365. https://doi.org/10.1111/j.1564-913X.2011.00120.x

Moody K (1997) Workers in a lean world: unions in the international economy. Haymarket series. Verso, London, New York

Neethi P (2012) Globalization lived locally: investigating Kerala's local labour control regimes. Dev Chang 43:1239–1263. https://doi.org/10.1111/j.1467-7660.2012.01802.x

Newsome K, Taylor P, Bair J, Rainnie A (eds) (2015) Putting labour in its place: labour process analysis and global value chains. Palgrave Macmillan, London and New York

Ngai P, Smith C (2007) Putting transnational labour process in its place: the dormitory labour regime in post-socialist China. Work Employ Soc 21:27–45. https://doi.org/10.1177/0950017007073611

Nicholls W (2009) Place, networks, space: theorising the geographies of social movements. Trans Inst Br Geog 34:78–93. https://doi.org/10.1111/j.1475-5661.2009.00331.x

Nowak J (2017) Mass strikes in India and Brazil as the terrain for a new social movement unionism. Dev Chang 48:965–986. https://doi.org/10.1111/dech.12320

Padmanabhan N (2012) Globalisation lived locally: a labour geography perspective on control, conflict and response among workers in Kerala. Antipode 44:971–992. https://doi.org/10.1111/j.1467-8330.2011.00918.x

Pattenden J (2016) Working at the margins of global production networks: local labour control regimes and rural-based labourers in South India. Third World Q 37:1809–1833. https://doi.org/10.1080/01436597.2016.1191939

Peck J (1992) Labor and agglomeration: control and flexibility in local labor markets. Econ Geogr 68:325–347. https://doi.org/10.2307/144023

Peck J (1996) Work place: the social regulation of labor markets. Perspectives on economic change. Guilford Press, New York

Peck JA (1989) Reconceptualizing the local labour market. Prog Hum Geogr 13:42–61. https://doi.org/10.1177/030913258901300102

Pye O (2017) A plantation precariat: fragmentation and organizing potential in the palm oil global production network. Dev Chang 48:942–964. https://doi.org/10.1111/dech.12334

Pyke F, Lund-Thomsen P (2016) Social upgrading in developing country industrial clusters: a reflection on the literature. Compet Chang 20:53–68. https://doi.org/10.1177/1024529415611265

Rainnie A, Herod A, McGrath-Champ S (2011) Review and positions: global production networks and labour. Compet Chang 15:155–169. https://doi.org/10.1179/102452911X13025292603714

Riisgaard L, Hammer N (2008) Organised labour and the social regulation of global value chains. DIIS Working Paper. Standards and Agro-Food Exports (SAFE) subseries, 2008:09. DIIS, Copenhagen

Riisgaard L, Hammer N (2011) Prospects for labour in global value chains: labour standards in the cut flower and banana industries. Br J Ind Relat 49:168–190. https://doi.org/10.1111/j.1467-8543.2009.00744.x

Rossi A (2013) Does economic upgrading lead to social upgrading in global production networks? Evidence from Morocco. World Dev 46:223–233. https://doi.org/10.1016/j.worlddev.2013.02.002

Ruwanpura KN (2015) The weakest link? Unions, freedom of association and ethical codes: a case study from a factory setting in Sri Lanka. Ethnography 16:118–141. https://doi.org/10.1177/1466138113520373

Scheper C (2017) Labour networks under supply chain capitalism: the politics of the Bangladesh Accord. Dev Chang 48:1069–1088. https://doi.org/10.1111/dech.12328

Schmalz S, Dörre K (2014) Der Machtressourcenansatz: Ein Instrument zur Analyse gewerkschaftlichen Handlungsvermögens: (The power resource approach: an instrument to analyze trade union action capabilities). Industrielle Beziehungen 21:217–237

Schmalz S, Ludwig C, Webster E (2018) The power resources approach: developments and challenges. Global Labour J 9:113–134. https://doi.org/10.15173/glj.v9i2.3569

Selwyn B (2012) Workers, state and development in Brazil: powers of labour, chains of value. Manchester University Press, Manchester

Selwyn B (2013) Social upgrading and labour in global production networks: a critique and an alternative conception. Compet Chang 17:75–90. https://doi.org/10.1179/1024529412Z.000000 00026

Selwyn B (2015) Commodity chains, creative destruction and global inequality: a class analysis. J Econ Geogr 15:253–274. https://doi.org/10.1093/jeg/lbu014

Selwyn B (2016) Elite development theory: a labour-centred critique. Third World Q 37:781–799. https://doi.org/10.1080/01436597.2015.1120156

Silver BJ (2003) Forces of labor: workers' movements and globalization since 1870. Cambridge studies in comparative politics. Cambridge University Press, Cambridge

Smith C, Pun N (2006) The dormitory labour regime in China as a site for control and resistance. Int J Hum Resour Manag 17:1456–1470. https://doi.org/10.1080/09585190600804762

Smith A, Barbu M, Campling L, Harrison J, Richardson B (2018) Labor regimes, global production networks, and European Union trade policy: labor standards and export production in the Moldovan clothing industry. Econ Geogr 94:550–574. https://doi.org/10.1080/00130095.2018. 1434410

Standing G (2008) The ILO: an agency for globalization? Dev Chang 39:355–384. https://doi.org/ 10.1111/j.1467-7660.2008.00484.x

Swyngedouw E (2003) The Marxian alternative: historical-geographical materialism and the political economy of capitalism. In: Sheppard ES, Barnes TJ (eds) A companion to economic geography. Blackwell, Oxford, Malden, MA, pp 41–59

Taylor P, Newsome K, Bair J, Rainnie A (2015) Putting labour in its place: labour process analysis and global value chains. In: Newsome K, Taylor P, Bair J, Rainnie A (eds) Putting labour in its place: labour process analysis and global value chains. Palgrave Macmillan, London and New York, pp 1–26

Thompson P (2010) The capitalist labour process: concepts and connections. Cap Class 34:7–14. https://doi.org/10.1177/0309816809353475

Thompson P, Smith C (2009) Labour power and labour process: contesting the marginality of the sociology of work. Sociology 43:913–930. https://doi.org/10.1177/0038038509340728

Tsing A (2009) Supply chains and the human condition. Rethink Marx 21:148–176. https://doi.org/ 10.1080/08935690902743088

Tufts S (2007) Emerging labour strategies in Toronto's hotel sector: toward a spatial circuit of union renewal. Environ Plann A 39:2383–2404. https://doi.org/10.1068/a38195

Valentine G (2007) Theorizing and researching intersectionality: a challenge for feminist geography. Prof Geogr 59:10–21. https://doi.org/10.1111/j.1467-9272.2007.00587.x

Webster E (2015) Labour after globalisation: old and new sources of power: ISER Working Paper No. 2015/1. Institute of Social and Economic Research (ISER), Rhodes University, Grahamstown

Webster E, Lambert R, Bezuidenhout A (2008) Grounding globalization: labour in the age of insecurity. Antipode book series. Blackwell, Malden, Mass.

Wells D (2009) Local worker struggles in the global south: reconsidering Northern impacts on international labour standards. Third World Q 30:567–579. https://doi.org/10.1080/014365909 02742339

Werner M (2012) Beyond upgrading: gendered labor and the restructuring of firms in the Dominican Republic. Econ Geogr 88:403–422. https://doi.org/10.1111/j.1944-8287.2012.01163.x

Wickramasingha S, Coe N (2021) Conceptualizing labor regimes in global production networks: uneven outcomes across the Bangladeshi and Sri Lankan apparel industries. Econ Geog, 1–23. https://doi.org/10.1080/00130095.2021.1987879

Wills J (2002) Bargaining for the space to organize in the global economy: a review of the Accor-IUF trade union rights agreement. Rev Int Polit Econ 9:675–700. https://doi.org/10.1080/096922 9022000021853

Wills J (2005) The geography of union organising in low-paid service industries in the UK: lessons from the T&G's campaign to unionise the Dorchester Hotel, London. Antipode 37:139–159. https://doi.org/10.1111/j.0066-4812.2005.00477.x

Wright EO (2000) Working-class power, capitalist-class interests, and class compromise. Am J Sociol 105:957–1002

Zajak S (2017) International allies, institutional layering and power in the making of labour in Bangladesh. Dev Chang 48:1007–1030. https://doi.org/10.1111/dech.12327

Zajak S, Egels-Zandén N, Piper N (2017) Networks of labour activism: collective action across Asia and beyond. An introduction to the debate. Dev Chang 48:899–921. https://doi.org/10.1111/dech.12336

Chapter 3
Towards a Relational Approach for Analysing Labour Control Regimes and Union Agency in GPNs

Abstract This chapter introduces central tenets of relational thinking in economic geography and then develops a relational approach for analysing labour control regimes and union agency in GPNs. It conceptualises place-specific labour control regimes at specific nodes of a GPN as emerging from the articulation of six horizontal (i.e. territorially embedded) and vertical (i.e. network embedded) processual relations: the labour process and workplace, wage, labour market, employment and industrial relations at the horizontal dimension, which in turn intersect with sourcing relations at the vertical, 'network' dimension of the GPN. Moreover, it develops a relational heuristic framework for analysing union agency in GPNs through the lens of three interrelated spaces of labour agency that unions construct through practices of building relations: (1) spaces of organising comprising internal union relations as well as unions organising practices; (2) spaces of collaboration constructed by unions through building relationships of collaboration with other labour and non-labour actors at various levels; and (3) spaces of contestation constructed by unions around specific labour struggles through building antagonistic relationships with employers, lead firms and state actors as well as through practices of drawing other allied actors into spaces of contestation to activate moral power resources.

Keywords Relational economic geography · Practice-oriented approaches · Global production networks · Labour control regimes · Union agency · Labour agency

Since the 2000's, relational analytical approaches have become popular in economic geography as a conceptual alternative to scalar approaches for analysing the global economy (Amin 2002; Dicken et al. 2001; Hudson 2004; Jones and Murphy 2010a, 2010b). Relational approaches have been developed and advanced by scholars such as Ash Amin (2002, 2004), Nigel Thrift (1996, 2008) and Doreen Massey (1994, 1999). These scholars have emphasised 'network practices' and 'relational connectivity' as key features of economic geographies emerging under globalised capitalism (c.f. Amin 2002: 389ff.). Due to their focus on practices as central constituents of social and economic phenomena, relational approaches to the global economy insert themselves within a broader body of practice-oriented work in economic geography.

Practice-oriented work in economic geography is unified by the general concern to connect context, structures and individual agency. This is done through a focus on practices, which are defined as "stabilised, routinised, or improvised social actions that constitute and reproduce economic space" (Jones and Murphy 2010b: 366). In this light, two central meta-theoretical assumptions can be identified that are common to relational approaches in economic geography. Ontologically, relational approaches can be characterised as broadly underpinned by a constructivist paradigm. They do not take economic structures, such as markets, institutions or class relations as conceptual pre-givens, but instead regard them as actively constructed, continuously reproduced and potentially contested in and through socio-economic practices (Hudson 2004: 451; Jones and Murphy 2010b: 372). From this ontological assumption follows the epistemological belief that "to understand higher-order (i.e., local, regional, national, or global scale) economic and social outcomes (e.g., performance, innovation, integration, inequality, exploitation, markets) it is necessary to […] closely observe and understand the micro-social activities (i.e., practices) carried out and performed by people living, labo[u]ring, and creating in the everyday economy" (Jones and Murphy 2010b: 376).

In this chapter, I adopt the focus on practices as constituents of socio-economic structural phenomena as well as a conceptual emphasis on networks and connectivity to develop a relational, practice-oriented approach for analysing labour control regimes and labour agency in GPNs. By doing so, I address the limitations of hitherto dominant scalar approaches, which have not sufficiently explored the interconnections and interdependencies between processes, relationships and practices of labour control and labour agency at different levels (see Sects. 2.3.4 and 2.4.4).

The remainder of this chapter is structured as follows: In the next section, I first outline central conceptual contributions of relational approaches regarding the analytical categories networks, place/space, scale and territories. Thereafter, I develop heuristic frameworks for analysing labour control regimes and union agency in GPNs from a relational perspective.

3.1 A Relational Perspective on Networks, Space/Place, Scale and Territories

A central aim of relational approaches in economic geography is to develop a conceptual alternative to scalar interpretations of the socio-spatialities of globalisation (Amin 2002, 2004; Dicken et al. 2001; Massey 1994, 1999; Thrift 1996). These scalar interpretations emphasise, how under globalisation socio-spatial orders are being transformed through the 're-scaling' of specific socio-economic processes and practices, which are moved from 'higher' to 'lower' scales or vice versa (see e.g. Brenner 1999; Cox 1998; Smith 1993). Scales tend to be understood in the context of scalar interpretations of globalisation as separate, bounded territorial entities that precede and contain social activities and provide "an already partitioned geography" (Smith 1993: 101; see also MacKinnon 2011: 24). Relational thinkers in economic geography, such as Amin (2004: 33), oppose such a 're-scaling' imaginary for ignoring the intertwined flows and connections that characterise contemporary economic geographies, where local territories—such as cities and regions—form constitutive parts of broader global economic networks. In addition, relational approaches have criticised scalar approaches to globalisation for privileging one particular scale or a bifurcation of scales (e.g. global–local) and hence to develop analyses that "preclude alternatives and that obscure subtle variations within, and interconnections between, different scales" (Dicken et al. 2001: 90).

Against this backdrop, relational approaches in economic geography propose to replace the imaginary of hierarchically nested territorial scales as central ordering features of the global economy with the imaginary of multiple horizontally intertwined *networks* that connect actors, practices and places across various distances (Dicken et al. 2001). From a relational perspective, networked relationships represent the central socio-spatial ordering features of the globalised economy. Relational approaches conceptualise the networks of actors and practices constituting the economy as "relational processes, which, when realised empirically within distinct and time- and space-specific contexts, produce observable patterns in the global economy" (Dicken et al. 2001: 91). From a relational perspective, networks as central socio-spatial ordering features are both relational and structural: Networks are relational, since they are constituted through practices of routinised interactions between various actors. These interactions are driven by actors' interests and intentions (Dicken et al. 2001: 96). At the same time, networks are structural in that the specific composition of intertwined relationships between variously powerful actors "constitute structural power relations in which exclusions and inequalities exist" (Dicken et al. 2001: 95). Hence, while networks are constituted through routinised interaction practices in the first place, they also provide the structural context for these interactions. Within networks, spaces for the agency of specific actors are determined by actors' power to shape the nature of interactions and to include or exclude other actors (Dicken et al. 2001: 94f.).

The conceptual imaginary of structural/relational networks as central ordering features of the global economy has specific implications for interpreting key geographical analytical concepts such as space, place and scale and territories (c.f. Jessop et al. 2008). Following this imaginary, economic *space* can be understood as constituted through a mesh of horizontally intertwined and spatially stretched networks, within which money, resources and power flow between actors and places (Amin 2004: 34). Such a relational notion of space rejects the assumption that space "exists as an entity in and of itself, over and above material objects [or actors] and their spatiotemporal relations and extensions" (Jones 2009: 491). Rather, space emerges from and is constituted through networks and event relations that connect actors, material objects and places. As such, space has no a priori hierarchical order, but such order may emerge from power flows within the networks that constitute space (c.f. Schmid 2020: 108).

From this point of view, *places,* in turn, can be conceptualised as nodes within broader networks, where multiple relationships of different lengths intersect and become spatially immanent in specific moments of articulation (Amin 2002: 391). In this sense, Schmid (2020: 71) argues that "the notion of place contextualise[s] (global) power relations that are always produced in concrete sites". Whereas this understanding of 'place' refers to 'the local' as a specific part of broader 'space', a second reading of 'place' is possible from a relational perspective. This reading comprehends place in the sense of concrete sites for human interaction that are constituted through the links of "bodies, artefacts, things, meanings and practices that meet in time and space" (Schmid 2020: 69). In both readings, however, places inevitably need to be understood as constitutive elements of those network relations constituting broader 'space'.

This dialectical understanding of space and place has important implications for the concept of *scale*. Dicken et al. (2001: 95) point out that from a relational perspective, "it becomes meaningless to talk of [ontologically distinct] local versus global processes as in much of the global–local literature". Consequently, relational approaches have proposed an alternative conception of the scales of social and economic life as "practices and relations of different spatial stretch and duration" (Amin 2002: 389). Relational approaches are congruent with scalar approaches in that they recognise the spatiality of social and economic relationships, which give rise to specific socio-spatial orderings. However, relational approaches reject a priori assumptions about specific, reified 'architectures of scale' (c.f. MacKinnon 2011: 22). Instead, the socio-spatial order of specific networks needs to be carved out empirically by mapping its constitutive social relations (c.f. Marston et al. 2005: 426). In this sense, relational analytical approaches are compatible with multi-scalar heuristics, as long as scales are not understood as hierarchical, discrete entities, but rather as spatial stretches that are interwoven in specific networks. Such a relational understanding of scale allows us to understand actors, processes, relationships and practices at various scales—including the workplace, the neighbourhood, the nation-state and the globe—as equally and mutually constitutive of the globalising economy (Dicken et al. 2001: 95).

Whereas in global networks, relationships at various scales are interwoven, networks themselves may also exist at multiple levels. Some networks are boundless, connecting actors and places across various countries and territorial boundaries; other networks are relatively localised in the sense that the relationships that constitute them are bound to specific territories (Henderson et al. 2002; Hess 2004 see also Sect.2.1). In this sense, relational approaches are not opposed to the concepts of 'territory' and 'territoriality'. However, in contrast to 'territorial' spatial approaches which tend to treat territories as conceptual givens, relational approaches emphasise the social construction of *territories* through practices and relationships (Jones 2009: 494; Paasi 2003). In a relational understanding, territories are constructed through practices of 'classifying by area', i.e. of categorising people and things located in space, and enacted through practice relationships between actors within that area (Paasi 2003; Jones 2009). Consequently, from a relational standpoint, territories are not "frozen frameworks" but "typically contested and actively negotiated" (Paasi 2003: 110).

In summary, relational approaches emphasise the interweaving of practices, actors, places and territories within economic space. Therefore, networks—or, more accurately, networked relationships—become the central socio-spatial ordering feature of the economy. These networks are both structural and relational; they represent contested fields within which actors exercise power by including or excluding other actors and places from these networks. Networks may go beyond territorial boundaries or they may be tied to specific territories, whereas, in the latter case, networks and territories may also be considered as mutually constitutive.

As laid out in Sect. 2.1, this type of relational thinking informed the inception of the GPN framework as a "relational framework for analysing the global economy" in the early 2000s (Dicken et al. 2001; Coe et al. 2008: 272f.). However, as illustrated in the preceding literature review, particularly within literature on labour in GPNs, relational perspectives have been overshadowed by scalar analytical perspectives. Against this background, with this book, I aim to bring the relational perspective that informed the early GPN framework back into debates on labour control and labour agency in GPNs. To this end, I highlight the networked character of the 'architectures of labour control' underpinning GPNs and of local unions' agency strategies in GPNs. With the two heuristic frameworks developed in the remainder of this chapter, I aim to reveal how everyday practices of labour control and labour agency in GPNs are intertwined in multiple processual relationships stretching across various distances and territories. I propose that it is only by focussing on the dialectic and mutually constitutive relationship between 'local' and 'global' processes and practice relations, and between territorially embedded and network dynamics that we can really comprehend the relational constitution and workings of labour control and labour agency in GPNs. Such comprehension is in turn crucial to understand the conditions that constrain and enable the building of local union power in garment producing countries, as will be shown in the empirical analysis in Chapters 6 and 7.

In the following two sections, I develop two heuristic frameworks for studying labour control and union agency in GPNs through the lens of a relational perspective.

3.2   A Relational Approach to Labour Control Regimes in GPNs: Intertwining Processual Relations of Labour Control

Gaining a refined understanding of the 'labour control architectures' underpinning a particular GPN is crucial for understanding the structural contexts and constraining conditions that shape and potentially limit spaces for the agency of local unions. In this section, I develop a relational, practice-oriented approach for analysing these labour control architectures. I start from the conceptual assumption that we need to understand GPNs not primarily as networks of firms but rather as networks of *territorially embedded labour processes*, which lie "at the heart at the heart of all systems of commodity production" under capitalism (Cumbers et al. 2008). Labour processes are necessarily territorially embedded because they are tied to the practices of labouring bodies who are located in space and who—as David Harvey (1989: 19) famously stated—'need to go home every night'. Moreover, many labour processes depend on specific material infrastructure that is (at least temporarily) fixed in specific places (Harvey 1989). This is the case particularly in the labour-intensive export industries 'at the bottom' of GPNs, which produce mass consumer goods in Fordist production arrangements. Labour processes in these industries are usually tied to material infrastructure in the form of large factories hosting workers, machines and production facilities (see e.g. Kumar 2014; Ngai and Smith 2007; Smith and Pun 2006).

To ensure the extraction of surplus value *in* and the unobstructed reproduction *of* territorially embedded labour processes at specific nodes of a GPN, labour control regimes are necessary (see Sect. 2.3.2). In this light, I propose that we can think of the 'labour control architectures' underpinning GPNs as a mesh of intertwined, place-specific *labour control regimes*. These place-specific labour control regimes emerge around labour processes at specific nodes of the GPN from the intertwining of various institutionalised, *processual relations,* which, together, ensure the reproduction of the labour process in its profit-maximising form. All these processual relations are constituted through networks of routinised practices of labour control that link diverse actors across various distances. Drawing on Neethi's (2012) extended notion of labour control (see Sect. 3.2), *labour control practices* are defined here as encompassing exploiting and disciplining practices, on the one hand, and all practices that directly or indirectly ensure the smooth reproduction of the labour process on the other hand. Therefore, I consider the following three types of practices as labour control practices: (1) *exploiting practices* directed at minimising labour costs while maximising labour productivity; (2) *disciplining practices* directed at undermining or preventing collective labour organisation; (3) *practices that contribute to (re-) producing the broader conditions for capitalist production.*

As a result, the processual relations intertwined in labour control regimes are linked to the labour control process in various ways and fulfil different functions in relation to the capitalist accumulation process. More specifically, processual relations that are interwoven in labour control regimes either (1) directly *shape the labour process* through exploiting practices directed at maximising surplus value, (2) *ensure the subordination of labour to the labour process* through disciplining practices or (3) *secure the broader conditions for the reproduction of the labour process*, such as securing adequate labour supply.

In doing so, processual relations that are interwoven in labour control regimes also *shape the terrain for labour agency* at a specific node of the GPN in two regards. On the one hand, these processual relations represent 'contested fields' (c.f. Levy 2008), that are the *subject of potential disputes* between capital and labour and that may be challenged and transformed when the capital-labour balance is shifted in favour of workers. But, on the other hand, the processual relations that constitute the labour control regime *constrain the opportunities for workers and unions to build and/or activate the power resources* that would enable them to shift the capital-labour power balance.

Without claiming this to be an exhaustive list, in this study, I identify six processual relations that intersect with the labour process at specific nodes of a GPN and constitute labour control regimes. These six relations are: sourcing relations at the vertical 'network' dimension of the GPN, and territorially embedded wage relations, work-place relations, industrial relations, employment relations and labour market relations at the horizontal dimension of the GPN. These processual relations show distinct socio-spatial features and linkages to the labour process. The following list gives a short characterisation of each type of relations, including their spatial extension, the actors and practices that hang together in them, their function within the broader labour control regime (i.e. exploiting, disciplining or securing the broader conditions for capital accumulation), the intersections of these labour control relations with the labour process and their implications for labour agency.

- *Sourcing relations* link global lead firms within a GPN with local suppliers at the various nodes of the GPN; they are therefore situated at the vertical 'network' dimension of the GPN. Since lead firms have the power to set up and control geographically dispersed supplier networks, sourcing relations are primarily constituted through lead firms' purchasing practices as well as through lead firm practices of managing supplier pools and organising the sourcing process. Various studies have highlighted how lead firms' exploitative sourcing practices shape localised labour processes at specific nodes of a GPN, for example through 'squeezing prices'. As a result, suppliers rely on a range of exploitation practices to ensure value capture for both themselves and the lead firm (see e.g. Anner 2019, 2020). Therefore, these studies have argued that to transform exploitative practices in the workplace, unions must also tackle the sourcing practices of lead firms through networked agency strategies (López 2021).
- *Wage relations* link workers, employers and state actors within a specific region, state or country. Wage relations are hence territorially embedded and usually structured through specific legal-institutional frameworks that fix, for example, statutory minimum wages or regulations for collective wage bargaining within a particular territory. Practices that constitute wage relations may either enact these frameworks or seek to circumvent or undermine these frameworks to minimise labour costs. Wage relations intersect with the labour process, because wage levels directly influence the amount of surplus value that employers are able to generate from the labour process. At the same time, wage relations are shaped by capital-labour power relations: Where labour power is weak, wage relations tend to be shaped predominantly by exploitative employer practices that ensure maximum surplus extraction through keeping wages low. Where labour power is high, workers may, however, be able to contest employers' exploiting practices and hence to achieve a more equitable redistribution of the surplus value produced in the labour process through higher wages. Wage relations therefore do not only represent a domain of exploitation within labour control regimes but also a field of contestation between capital and labour.
- *Workplace relations* are constituted through the interaction practices between workers and management in a specific site of production. Workplace relations are usually localised and territorially embedded. Workplace relations tend to

emerge around place-specific legal-institutional frameworks for worker representation, worker-management dialogue or co-determination. The way these legal-institutional frameworks are enacted in a specific site of production is highly shaped by capital-labour power relations in the workplace. In workplaces where workers have low bargaining power, workplace relations potentially represent a domain of labour control since managers are then able to construct workplace relations through disciplining practices directed at suppressing collective worker organisation. In contrast, when workers possess high workplace bargaining power due to strong collective organisation and/or their strategic position in the GPN (see Sect. 2.4.1), they may be able to construct workplace relations through practices of genuine worker management dialogue that can help to improve workers' conditions.

- *Industrial relations* are constituted through territorially embedded relationships between employers and their organisations, workers and their organisations, and the state in a specific region, sector and/or country. Industrial relations are usually constructed around legal frameworks of collective bargaining and industrial dispute settlement. In this sense, industrial relations intersect with workplace relations as well as with wage relations. Due to the special role of collective worker representation by trade unions vis-à-vis employers and state actors, which exceeds interactions at the workplace and addresses topics beyond wages, I however treat industrial relations as a separate set of relations. Industrial relations are usually the domain that may allow workers to activate institutional power resources in the form of legal frameworks for ensuring labour rights and settling industrial disputes. However, studies have shown that in many industrial export sectors 'at the bottom' of GPNs, where workers possess low structural and associational power, state actors and employers frequently undermine the implementation of these legal frameworks through a variety of disciplining practices (see e.g. Anner 2015a; Ruwanpura 2015). In these cases, employers' and state actors' disciplining practices significantly constrain the terrain for worker organising and collective bargaining.

- *Employment relations* are constituted through the interrelated practices that form the relationship between employers and workers, in which workers sell their labour power to an employer. Consequently, employment relations link employers with individual workers. Employment relations may also be mediated through third parties such as temporary work agencies. Employment relations are usually also territorially embedded since they are generally structured through legal frameworks that apply to a specific territory and that define the rights and obligations of employers and employees. Studies have shown that employers in export sectors 'at the bottom of GPNs' rely on various exploiting and disciplining practices when constructing employment relations. First, employers construct employment relationships that allow them to outsource economic risks to workers through flexible, informal and/or mediated employment models such as piece-rate work or contract work (see e.g. Anner 2019; Mezzadri 2017). Second, employers construct employment relationships that increase workforce segmentation to save labour costs and to hamper collective worker organisation (e.g. Flecker 2009; Flecker and Meil 2011; Flecker et al. 2013). As a result, employment relations represent a potential field of contestation for workers and unions, who may seek to challenge employers' practices of constructing exploitative employment relations. However, employers' employment practices may also constrain workers' and unions' capacities to build associational power resources.

- *Labour market relations* are defined here as the practices and relationships that secure adequate labour supply and thereby contribute to ensuring the broader conditions for capital accumulation at a specific node of a GPN. Practices that constitute labour market relations include recruiting practices, linking employers, workers and potentially third-party actors such as head hunters or recruiting agencies. Moreover, labour market relations are constituted through training and skilling practices, which link workers and employers with educational and training organisations. Compared to wage, workplace, industrial and employment relations, which are usually spatially delimited, the spatial extension of labour market relations is flexible. Labour market relations may be rather localised, when they are constructed around local or regional (often informal) networks (see e.g. Kelly 2001; Padmanabhan 2012). Labour market relations may, however, also be rather unbounded when they are constructed around national or international migration regimes that link workers and employers from different regions within a country (see e.g. Ngai and Smith 2007) or even across countries (see e.g. Azmeh 2014; Pye 2017). Labour market relations represent an important structural context for the agency of labour since the nature of the relations and practices that constitute labour markets may enhance or constrain workers' marketplace bargaining power at specific nodes of the GPN (see Sect.2.4.1). In situations of limited local labour supply, workers' marketplace bargaining power increases, enabling workers to demand higher wages. In these cases, to hedge labour's increased marketplace bargaining power, employers need to expand the territory of the labour market, e.g. through setting up migration regimes to channel surplus labour force from geographically more or less distant places into localised labour processes.

In a nutshell, labour control regimes at specific nodes of a GPN emerge from the intertwining of the six processual relations listed above with the localised labour process. All processual relations of labour control either shape the labour process or secure its reproduction. As illustrated, each set of processual relations has its own socio-spatiality: Sourcing relations are territorially unbounded and usually stretch across various countries, they are therefore characterised by *network embeddedness* (see Sect. 2.1). On the contrary, wage relations, workplace relations, industrial relations and employment relations are characterised by *territorial embeddedness* (see Sect. 2.1). They link actors only within the workplace or within a specific region or country, where they also intersect with place-specific social and power relations constructed around categories of age, gender or ethnicity. Moreover, wage relations, workplace relations, industrial relations and employment relations are usually constructed around and (to varying extents) shaped by institutional-legal frameworks

linked to specific local, state or national administrative territories. Labour market relations, in turn, vary regarding their territorial extension and type of embeddedness,
since they may link workers and employers only within one region or country and
across countries. Hence, the place-specific nature of 'local' labour control regimes
at specific nodes of a GPN does not result from the localised nature of the practices
and relationships that constitute the labour control regime. Instead, the place-specific
nature of the labour control regime at a specific node of a GPN results from the particular articulations of multiple processual relations of labour control stretching across
various distances with the labour processes at that particular node of the GPN. The
various labour control regimes at different nodes of a GPN—that together constitute
the 'architectures of labour control' underpinning a GPN—are hence characterised
by distinctive socio-spatialities that cannot be assumed a priori but that need to be
deduced (c.f. Schmid 2020).

Figure 3.1 illustrates how labour control regimes at specific nodes of the GPN
emerge from the articulation of the six sets of processual relations at the vertical and
horizontal dimensions that intersect with localised labour processes.

Linking the labour control regime back to the central question of this study, we can
summarise that the labour control regime at a specific node of the GPN represents an
important structural context for local union agency for two reasons. First, it shapes the
terrain for the agency of local unions since it is constituted through various exploiting
practices by capital and state actors that unions need to challenge and transform to

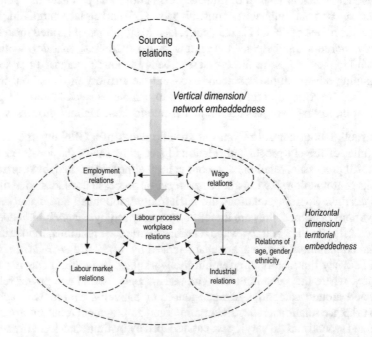

Fig. 3.1 Labour control regime at a specific node of the GPN as constituted through place-specific
articulations of processual relations of labour control. *Source* Author

improve workers' conditions. Second, the labour control regime constrains workers' opportunities to build and/or activate structural, associational and institutional power resources through the disciplining practices that are interwoven in it. Consequently, at nodes of a GPN with tight labour control regimes, workers and unions need to construct their own spaces for agency through constructing processual relations that are not dominated by capital and pro-capital state actors. These spaces can then provide structural contexts, within which workers and unions can develop strategic capacities and build power resources. In the next section, I develop a relational heuristic framework for analysing how unions construct such 'spaces for labour agency'.

3.3 A Relational Approach to Union Agency in GPNs: Linking Spaces of Labour Agency

As labour geographers have reiterated, the agency of workers and unions in GPNs is not limited to contesting the networked processual relations of labour control that emerge from the intertwined exploiting and disciplining practices of capital and state actors. In addition, workers and unions forge their own relational networks across various distances and thereby add their own relational layers to the GPN (see e.g. Cumbers 2015; Hastings 2019; Pye 2017). The relational networks constructed by workers and unions within GPNs are of great analytical importance since they represent structural contexts (besides labour control regimes) that may potentially have *enabling* effects on workers and unions.

In this light, I propose that we can conceptualise the relational networks constructed by labour actors through the lens of different '*spaces of labour agency*'. As opposed to labour control relations, in which firm and state actors monopolise planning and decision-making practices, spaces of labour agency are primarily constructed through planning, analysing and solidarity-building practices performed by workers, unions and/or their allies (see also Sect. 2.4.4). Within the routinised interactions that constitute spaces of labour agency workers and unions can develop the individual and collective capacities that are fundamental to shifting the capital-labour power balance (see also Gindin 1998). Consequently, spaces of labour agency can provide *enabling* contexts, within which local unions in GPNs can develop and/or activate different types of power resources and thereby bring about sustained improvements for workers. At the same time, it is essential to note that the processual relations that constitute spaces of labour agency are in themselves structured through flows of resources and power, which may not be distributed equally among all actors. As a result, where power and resources are monopolised, for example by labour's allies (e.g. NGOs or consumer organisations), the local unions' strategic capacities and hence their associational and organisational power may be limited.

I distinguish between three *spaces of labour agency* that are constituted through different types of processual relationships constructed by workers and local unions at specific nodes of a GPN: (1) spaces of organising, (2) spaces of collaboration and (3) spaces of contestation. In the following, I will briefly characterise each space, giving an overview of the practices and actors that are intertwined within these spaces, their spatiality, and of the potential capacities and power resources that workers and unions may build within these spaces:

Unions construct *spaces of organising* through practices of building solidarity among workers around common interests with the aim to act collectively vis-à-vis capital and state actors. Three types of practices create and constitute spaces of organising: practices of membership recruiting, practices of internal solidarity-building within the union and practices of training and capacity-building. Practices of membership recruiting are directed at building relationships with workers in a specific factory, sector or community to make them become a part of the union. Practices of internal solidarity-building include building relationships among workers as rank-and-file-members, and between rank-and-file-members and union leadership. Practices of training and capacity-building, lastly, are directed at developing workers' 'oppositional consciousness' (Cumbers et al. 2010) and enabling workers to participate in union life. Practices of training and capacity-building may include teaching knowledge on relevant legal frameworks or training workers to participate in collective strategy-building and decision-making processes. Hence, within spaces of organising unions build associational power resources but also organisational power resources.

Spaces of collaboration, in turn, are constructed by local unions through building relationships of collaboration with other labour and non-labour actors. These actors may, for example, be NGOs, consumer campaigning networks, international organisations, community organisations or worker organisations. Unions may construct relationships with these actors for different purposes—from acquiring financial resources through funded project collaborations to activating consumers' moral power in specific labour struggles. Whereas past studies have highlighted network building by Global South unions with actors from the Global North (see e.g. Anner 2015b; Kumar 2014; Zajak 2017), it is important to note that relationships of collaboration are not limited to transnational relationships. Local unions at a specific node of the GPN may, for example, also build solidary relationships with local community organisations to organise workers outside of formal employment relationships (see Sect. 2.4). In a nutshell, spaces of collaboration serve unions to build 'coalitional' or 'societal' power resources in the form of moral power or access to financial and informational resources. When collaborative relations with allies involve processes of joint strategy-building and decision-making, these interactions may enhance the strategic capacities of union leaders and members, and thereby contribute to strengthening the union's organisational and associational power resources.

Lastly, *spaces of contestation* are those spaces that unions construct around specific labour struggles. Spaces of contestation differ from spaces of organising and spaces of collaboration in that they are constructed primarily through antagonistic relationships with employers, lead firms and, in some cases, state actors as well. At the same time, to exercise leverage against these actors, unions frequently 'draw' other allied actors into spaces of contestation to activate moral power resources. As a result, the specific practices through which unions construct spaces of contestation in struggles targeting capital and/or state are interrelated with unions' practices of building solidary relations within spaces of organising and spaces of collaboration. Only if unions have previously forged solidarity relations within these spaces and thereby built associational and coalitional power resources, will they be able to activate these power resources when constructing spaces of contestation around specific labour struggles. Spaces of contestation are therefore constituted by the intertwining of antagonistic and solidary relationships at various levels. Spaces of contestation may be constructed as territorially embedded spaces when only local capital and state actors, workers and allies are drawn into a conflict. However, spaces of contestation may also intertwine territorial and network spaces when actors in other countries, such as lead firms or consumer groups, are drawn into the conflict. The extent to which union leadership and rank-and-file members are able to develop strategic capacities within spaces of contestation depends, in turn, on the extent to which union leaders and workers actively participate in planning and executing the antagonistic interactions with capital actors that are at the core of spaces of contestation.

It is important to note that the three spaces outlined above are not discrete containers for social action. Instead, they need to be understood as dynamic networks of relationships that variously fold into each other: The collaborative relationships that unions maintain with external actors, for example, influence how unions construct intra-union relations and relations with workers as potential members. At the same time, the practices through which unions construct antagonistic relations with capital actors in specific labour struggles are shaped by unions' internal relationships and external collaborations. Therefore, to evaluate a union's strategic approach regarding its capacity for building lasting bargaining power, we need to analyse the practices through which unions construct all three spaces of agency and how these influence each other.

Figure 3.2 provides an exemplary (and non-exhaustive) graphical representation of the different processual relations that potentially constitute each space of labour agency. It is important to note that the actors in the graphical representation do not represent an exhaustive list of all actors with which local unions may potentially build relations but rather an exemplary selection.

In summary, each space of labour agency is constituted through a mesh of networked relationships constructed by unions, with each space linking unions to different actors across varying spatial stretches. Neither space can, however, be equated a priori with a specific scale of labour agency. Different unions may construct spaces of organising, collaboration and contestation across different distances. Therefore, the distinctive socio-spatialities of the spaces of labour agency created by a specific union need to be deduced empirically.

The following section summarises the relational approaches to labour control regimes and union agency in GPNs developed so far in this chapter and discusses their analytical benefits.

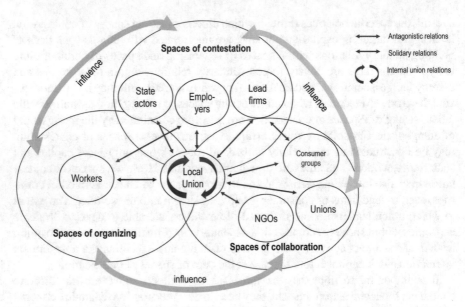

Fig. 3.2 Spaces of labour agency in GPNs. *Source* Author

3.4   Interim Conclusion: Benefits of a Practice-Oriented, Relational Approach to Labour Control and Labour Agency in GPN

In this chapter, I have developed a relational approach for studying labour control and labour agency—or, more specifically, union agency—in GPNs. In doing so, I have sought to develop a conceptual alternative to dominant scalar approaches for analysing labour control and labour agency in GPNs. My relational approach to labour control in GPNs proposes to conceptualise place-specific labour control regimes at specific nodes of a GPN as emerging from the articulation of various horizontal (i.e. territorially embedded) and vertical (i.e. 'network') processual relations. Together, these relations ensure the reproduction of the localised labour process in its profit-maximising form. Horizontal, territorially embedded processual relations include, for example, workplace relations, wage relations and industrial relations, which are usually constructed around specific locales of production, or within specific administrative territories to which legal-institutional frameworks for wages or collective bargaining are applicable. Vertical, 'network' processual relations, in turn, include sourcing relations, which link globally acting lead firms with local suppliers at particular nodes of a GPN. In this light, the 'local' character of labour control regimes at specific nodes of a GPN does not result in the first place from the fact that the practices and actors that constitute it are 'local' ones. Instead, place-specific labour control regimes at specific nodes of a GPN emerge from the unique articulation of territorially embedded processual relations that link actors within a specific workplace, city, region or country and cross-border processual relationships embedded with the broader production networks.

I argue that understanding the interrelations and interdependencies between the various relations constituting labour control regimes at particular nodes of a GPN is paramount to understand the structural context for local union agency in GPNs in two regards. First, a relational perspective focussing on the interconnections between different practices, processes and relations of labour control can advance our understanding of the conditions that constrain unions' capacities for building and leveraging associational and institutional power resources. I argue that the great challenge for local unions lies specifically in the complex, intertwined nature of labour control regimes in GPNs, in which manifold exploiting and disciplining practices performed by actors in more or less distant places enable and shape each other. As a result, a more nuanced understanding of the relational constitution of the labour control regime as structural context for labour agency is, second, crucial for developing networked labour agency strategies that simultaneously address multiple actors and practices (see Sect. 2.4.3).

To develop and carry out networked agency strategies, local unions need to build their own relational networks independent of those constructed by capital. This is necessary to create spaces in which workers and unions can build power resources and strategic capacities. The relational framework for analysing union agency in GPNs developed in this chapter has proposed to conceptualise unions' agency strategies as emerging at the intersection of three relational spaces of labour agency: (1) spaces of organising, (2) spaces of collaboration and (3) spaces of contestation. *Spaces of organising* are constituted through practices and relations that link unionists and union members, on the one hand, and unionists and workers who are not (yet) union members, on the other. Spaces of organising are hence spaces where workers and unions can develop strategic capacities for communicating and mobilising, as well as an active internal union life to build associational and organisational power resources (see Sect. 2.4.2). *Spaces of collaboration,* in turn, are constructed through unions' practices of building alliances with other societal actors. By constructing spaces of collaborations, unions and workers may build coalitional power resources granting unions access to financial or informational resources and moral power resources that can be leveraged vis-à-vis capital or state actors. Lastly, *spaces of contestation* are constructed by unions around specific labour struggles targeting capital and/or state actors. Spaces of contestation consequently differ from spaces of organising and spaces of collaboration in the sense that they are constructed primarily through antagonistic relationships. However, solidarity relations also play a role in spaces of contestation: workers and unions may draw allies into spaces of contestation to activate coalitional and moral power resources. It is important to note that the three spaces are not discrete entities or vessels for social actions, but that the relations they constitute variously shape and fold into each other.

What are the benefits of such a relational, practice-oriented approach for studying labour control regimes and spaces of labour agency at specific nodes of a GPN? I argue that the relational approach to labour control and labour agency in GPNs developed in this book can overcome the limitations of dominant scalar approaches to labour control and labour agency in GPNs. It can provide novel insights that enhance our understanding of the conditions that constrain and enable local union power at particular nodes of a GPN. The *relational approach to labour control regimes* can enhance our understanding of the constitution of labour control regimes as complex, structural contexts for the agency of local unions at specific nodes of a GPN in three regards: First, the relational approach to labour control regimes developed in this chapter is able to reveal the manifold intertwined practices and relations through which labour control regimes as structural contexts for labour agency are constituted and reproduced. In doing so, the relational approach allows us to grasp the complex, networked structures of labour control, within which various exploiting and disciplining practices enable and shape each other, and to evaluate resulting constraints for union agency.

Second, a relational approach is more sensitive towards the complex socio-spatialities of labour control regimes at different nodes of a GPN. Whereas scalar approaches have tried to fit the spatialities of labour control regimes into universal, pre-defined scalar categories (see Sect. 2.3.4), a relational, practice-oriented approach follows the empirical connections between practices and actors that stretch across various distances. Therefore, a relational analytical approach is able to reveal the distinct socio-spatialities of empirically existing labour control regimes at different nodes of the GPN. This, in turn, allows us to produce more refined analyses of the spatial labour control practices and dynamics that constrain workers' and unions' abilities of building power resources.

Third, I argue that the relational analytical perspective developed here allows for an enhanced conceptualisation of the dialectical relationships between labour control and labour agency in GPNs. From a relational perspective, labour control structures are constituted through spatially situated practices and processual (power) relations that may be contested and transformed by labour. A practice approach to labour control structures, thus, gives visibility to the 'small transformations' (Latham 2002) through which workers and unions may bring about important improvements for workers.

The *relational concept of spaces of labour agency* developed in this book is, in turn, well-equipped to generate a more fine-grained understanding of the conditions, contexts and networks that *enable* workers and unions to challenge and transform practices or even broader relations of labour control. I argue that, particularly in comparison to scalar approaches, the relational approach introduced in this chapter presents three benefits: First, scalar approaches have mostly neglected intra-union relations as an important scale for the agency of workers and unions. However, as laid out in Sect. 2.4.1, unions internal organising practices matter since building democratic intra-union relations and actively involving members in union life is a central condition for building associational and organisational power. By shifting the focus to the networked practices and relations through which unions construct spaces of organising, the here-developed analytical approach is well-equipped to identify organising practices and relations that contribute to building sustained associational and organisational power.

Second, the relational heuristic framework of 'spaces of labour agency' developed here allows for a higher sensibility towards variations in the practices through which workers and unions may construct different types of solidary relations at the same scale. Scalar approaches have conflated very different sets of transnational relations forged by workers and unions under the same notion of 'up-scaling'. By focussing on the practices and power flows that constitute unions' collaborative networks, the relational analytical approach developed in this chapter sheds light on the distinct constitution of each collaborative network. Moreover, it reveals the distinct capacities and power resources that unions can access in different networks—independent from the scale at which these networks are forged.

Third, the here-developed relational approach allows for a better understanding of how the different types of relationships that workers and unions construct influence each other. The proposed concept of 'spaces of labour agency' understands all sets of relations as power-laden and, therefore, structural. Consequently, uneven power flows may create uneven opportunities for capacity development for various powerful actors. By highlighting the structural effects of the networks constructed by workers and unions, the concept of 'spaces of labour agency' can grasp the inter-relations between external and internal union relations. More specifically, it is able to assess which *external* relations enable workers and unions to develop *internal* strategic capacities for building sustained associational and organisational power (see Sect. 2.4.2).

Before applying the relational approach for studying labour control regimes and union agency in GPNs to the Bangalore export-garment cluster, in the next chapter, I first lay out the research design and methodology underpinning this study.

References

Amin A (2002) Spatialities of globalisation. Environ Plan A 34:385–399. https://doi.org/10.1068/a3439

Amin A (2004) Regions unbound: towards a new politics of place. Geogr Ann: Ser B, Hum Geogr 86:33–44. https://doi.org/10.1111/j.0435-3684.2004.00152.x

Anner M (2015a) Labor control regimes and worker resistance in global supply chains. Labor History 56:292–307. https://doi.org/10.1080/0023656X.2015.1042771

Anner M (2015b) Social downgrading and worker resistance in apparel global value chains. In: Newsome K, Taylor P, Bair J, Rainnie A (eds) Putting labour in its place: labour process analysis and global value chains. Palgrave Macmillan, London and New York, pp 152–170

Anner M (2019) Predatory purchasing practices in global apparel supply chains and the employment relations squeeze in the Indian garment export industry. Int Labour Rev 158:705–727. https://doi.org/10.1111/ilr.12149

Anner M (2020) Squeezing workers' rights in global supply chains: purchasing practices in the Bangladesh garment export sector in comparative perspective. Rev Int Polit Econ 27:320–347. https://doi.org/10.1080/09692290.2019.1625426

Azmeh S (2014) Labour in global production networks: workers in the qualifying industrial zones (QIZs) of Egypt and Jordan. Global Netw 14:495–513. https://doi.org/10.1111/glob.12047

Brenner N (1999) Globalisation as reterritorialisation: the re-scaling of Urban Governance in the European Union. Urban Stud 36:431–451. https://doi.org/10.1080/0042098993466

Coe NM, Dicken P, Hess M (2008) Global production networks: realizing the potential. J Econ Geogr 8:271–295. https://doi.org/10.1093/jeg/lbn002

Cox KR (1998) Spaces of dependence, spaces of engagement and the politics of scale, or: looking for local politics. Polit Geogr 17:1–23. https://doi.org/10.1016/S0962-6298(97)00048-6

Cumbers A, Nativel C, Routledge P (2008) Labour agency and union positionalities in global production networks. J Econ Geogr 8:369–387. https://doi.org/10.1093/jeg/lbn008

Cumbers A (2015) Understanding labour's agency under globalization; embedding GPNs within an open political economy. In: Newsome K, Taylor P, Bair J, Rainnie A (eds) Putting labour in its place: labour process analysis and global value chains. Palgrave Macmillan, London and New York, pp 135–151

Cumbers A, Helms G, Swanson K (2010) Class, agency and resistance in the old industrial city. Antipode 42:46–73. https://doi.org/10.1111/j.1467-8330.2009.00731.x

Dicken P, Kelly PF, Olds K, Yeung HW-C (2001) Chains and networks, territories and scales: towards a relational framework for analysing the global economy. Global Netw 1:89–112

Flecker J (2009) Outsourcing, spatial relocation and the fragmentation of employment. Compet Chang 13:251–266. https://doi.org/10.1179/102452909X451369

Flecker J, Meil P (2011) Organisational restructuring and emerging service value chains—implications for work and employment. Work Employ Soc 24:1–19

Flecker J, Haidinger B, Schönauer A (2013) Divide and serve: the labour process in service value chains and networks. Compet Chang 17:6–23. https://doi.org/10.1179/1024529412Z.000 00000022

Gindin S (1998) Socialism with sober senses: developing worker's capacities. Social Regist:75–101

Harvey D (1989) The urban experience. Blackwell, Oxford

Hastings T (2019) Leveraging Nordic links: South African labour's role in regulating labour standards in wine global production networks. J Econ Geogr 19:921–942. https://doi.org/10.1093/jeg/lbz010

Henderson J, Dicken P, Hess M, Coe N, Yeung HW-C (2002) Global production networks and the analysis of economic development. Rev Int Polit Econ 9:436–464. https://doi.org/10.1080/096 92290210150842

Hess M (2004) 'Spatial' relationships? Towards a reconceptualization of embeddedness. Prog Hum Geogr 28:165–186. https://doi.org/10.1191/0309132504ph479oa

Hudson R (2004) Conceptualizing economies and their geographies: spaces, flows and circuits. Prog Hum Geogr 28:447–471. https://doi.org/10.1191/0309132504ph497oa

Jessop B, Brenner N, Jones M (2008) Theorizing sociospatial relations. Environ Plan D 26:389–401. https://doi.org/10.1068/d9107

Jones M (2009) Phase space: geography, relational thinking, and beyond. Prog Hum Geogr 33:487–506. https://doi.org/10.1177/0309132508101599

Jones A, Murphy JT (2010a) Practice and economic geography. Geogr Compass 4:303–319. https://doi.org/10.1111/j.1749-8198.2009.00315.x

Jones A, Murphy JT (2010b) Theorizing practice in economic geography: foundations, challenges, and possibilities. Prog Hum Geogr 35:366–392. https://doi.org/10.1177/0309132510375585

Kelly PF (2001) The political economy of local labor control in the Philippines. Econ Geogr 77:1–22. https://doi.org/10.2307/3594084

Kumar A (2014) Interwoven threads: building a labour countermovement in Bangalore's export-oriented garment industry. City 18:789–807. https://doi.org/10.1080/13604813.2014.962894

Latham A (2002) Retheorizing the scale of globalization: topologies, actor-networks, and cosmopolitanism. In: Herod A, Wright MW (eds) Geographies of power: placing scale. Blackwell, Malden, MA, pp 115–144

Levy DL (2008) Political contestation in global production networks. Acad Manag Rev 33:943–963. https://doi.org/10.2307/20159454

López T (2021) A practice ontology approach to labor control regimes in GPNs: connecting 'sites of labor control' in the Bangalore export garment cluster. Environ Plan A 53:1012–1030. https://doi.org/10.1177/0308518X20987563

MacKinnon D (2011) Reconstructing scale: towards a new scalar politics. Prog Hum Geogr 35:21–36. https://doi.org/10.1177/0309132510367841

Marston SA, Jones JP, Woodward K (2005) Human geography without scale. Trans Inst Br Geogr 30:416–432. https://doi.org/10.1111/j.1475-5661.2005.00180.x

Massey D (1999) Imagining globalization: power Geometries of time-space. In: Brah A, Hickman MJ, Ghaill MM (eds) Global futures: migration, environment and globalization. Palgrave Macmillan UK; Imprint: Palgrave Macmillan, London, pp 27–44

Massey DB (1994) Space, place and gender. Polity Press, Cambridge

Mezzadri A (2017) Sweatshop regimes in the Indian garment industry. Cambridge University Press

Neethi P (2012) Globalization lived locally: investigating Kerala's local labour control regimes. Dev Change 43:1239–1263. https://doi.org/10.1111/j.1467-7660.2012.01802.x

Ngai P, Smith C (2007) Putting transnational labour process in its place. Work Employ Soc 21:27–45. https://doi.org/10.1177/0950017007073611

Paasi A (2003) Territory. In: Agnew JA, Mitchell K, Toal G (eds) A companion to political geography. John Wiley & Sons Ltd, Hoboken, pp 109–122

Padmanabhan N (2012) Globalisation lived locally: a labour geography perspective on control, conflict and response among workers in Kerala. Antipode 44:971–992. https://doi.org/10.1111/j.1467-8330.2011.00918.x

Pye O (2017) A plantation precariat: fragmentation and organizing potential in the palm oil global production network. Dev Chang 48:942–964. https://doi.org/10.1111/dech.12334

Ruwanpura KN (2015) The weakest link?: unions, freedom of association and ethical codes: a case study from a factory setting in Sri Lanka. Ethnography 16:118–141. https://doi.org/10.1177/1466138113520373

Schmid B (2020) Making transformative geographies: lessons from Stuttgart's community economy. Social and cultural geography = Sozial- und Kulturgeographie. transcript, Bielefeld

Smith N (1993) Homeless/global: scaling places. In: Bird J (ed) Mapping the futures: local cultures, global change. Routledge, London and New York, pp 87–119

Smith C, Pun N (2006) The dormitory labour regime in China as a site for control and resistance. Int J Hum Resour Manag 17:1456–1470. https://doi.org/10.1080/09585190600804762

Thrift NJ (1996) Spatial formations. Theory, culture & society. Sage, London

Thrift NJ (2008) Non-representational theory: space, politics, affect. International library of sociology. Routledge, London

Zajak S (2017) International allies, institutional layering and power in the making of labour in Bangladesh. Dev Chang 48:1007–1030. https://doi.org/10.1111/dech.12327

Part III
Research Design and Methodology

Chapter 4
Grounding Dynamics of Labour Control and Labour Agency in GPNs Through an 'Extended Single Embedded Case Study Design'

Abstract This chapter introduces the research design and methodology of this study. It starts by setting out the key philosophical assumptions underpinning this study, characterised by a constructivist or reflexive research approach. Drawing on Burawoy's extended case study method and Yin's single embedded case study model, the chapter then develops an 'extended single embedded case study' design for studying the interrelations between place-specific dynamics of labour control and labour agency, and broader governance dynamics in the garment GPN. The chapter further illustrates how, for this study, the Bangalore export-garment cluster was constructed as a single case with three local garment unions representing embedded sub-units of analysis. Thereafter, the data collection process through participant observations and in-depth interviews is described. In this context, the chapter discusses challenges and strategies for interviewing managers and state actors as well as workers and unions in light of the power relations, which structure interactions between the researcher and the research subjects. The chapter concludes by outlining the data preparation, analysis and interpretation process.

Keywords Extended case study method · Single embedded case study · Positionality · Power relations in the field · Participant observations · In-depth interviews

This chapter introduces the research design and methodology of this study. The research objective of this study is two-fold. Firstly, it aims to assess the extent to which different agency strategies of local grassroots unions in the garment GPN enhance or constrain the building of labour bargaining power, following an understanding of labour's agency as embedded within wider structural conditions (Coe and Jordhus-Lier 2010). Second, this study seeks to better understand the mechanisms and practices that (re-)produce the 'labour control architectures underpinning GPNs' (Baglioni 2018). To this end, this study employs a qualitative research approach that seeks to 'empirically ground' dynamics of labour agency and labour control within GPNs by studying how these dynamics materialise at a specific node of the garment GPN (c.f. Yeung 2020). The node of the garment GPN that this study focusses on is

© The Author(s) 2023
T. López, *Labour Control and Union Agency in Global Production Networks*, Economic Geography, https://doi.org/10.1007/978-3-031-27387-2_4

the Bangalore export-garment cluster. Following Yin's (2014) embedded case study design, I construct the Bangalore export-garment cluster as the primary unit of analysis, with three local garment unions as embedded sub-units of analysis. To analyse the dynamics of labour control and labour agency in the Bangalore export-garment cluster as constituted through territorially embedded practices and relationships and at the same time shaped by broader 'network forces', I combine the single embedded case study design with Burawoy's (1998, 2009) extended case method.

Section 4.1 introduces the broader research design of this study. After that, Sect. 4.2 discusses the field research and data collection process. Lastly, Sect. 4.3 outlines the data preparation, analysis and interpretation process.

4.1   Research Design: The 'Extended Single Embedded Case Study'

A study's research design provides the broader perspective with which a specific topic is approached and thereby requires specific methods of data collection (Leavy 2017: 14). The research design of this study combines a qualitative single embedded case study design (Yin 2014) with the ethnographic 'extended case method' developed by Burawoy (1998). The following section illustrates the scientific-philosophical assumptions underpinning the research design of this study. Thereafter, I demonstrate how I constructed the Bangalore export-garment cluster with the presence of three local garment unions as a single embedded case study (Sect. 4.1.2). Last, I introduce the 'extended case method' representing the second tenet of the research design underpinning this study (Sect. 4.1.3).

4.1.1   Scientific-Philosophical Positioning of This Study

"Philosophy is to research as grammar is to language, whether we immediately recognise it or not. Just as we cannot speak a language successfully without following certain grammatical rules, so we cannot conduct a piece of research without making certain philosophical choices. Philosophy, like grammar, is always there." (Graham 2005)

As the quote by Graham exemplifies, philosophical assumptions are a central element of every research process. They significantly influence all decisions of the research process, "from topic selection all the way down to the final representation and dissemination of research findings" (Leavy 2017: 11). In line with the practice-oriented, relational perspective adopted in this book, the philosophical assumptions underpinning this study can be characterised as broadly informed by a *'reflexive' or 'constructivist' paradigm* (Burawoy 1998; Flick et al. 2004: 88ff.). Whereas these two paradigms are further differentiated and not entirely congruent with each other, several common assumptions can be identified. From an ontological point of view, the reflexive paradigm and (some positions within) the constructivist paradigm hold that no 'objective' social reality independent from human perception and agency exists 'out there'. Instead, social reality is constructed through human interaction. Hence, different people experience and attribute meaning to the social world in different ways (Burawoy 1998: 14f.). This ontological assumption has two significant epistemological consequences. First, the central research interest of studies adopting a reflexive or interpretivist perspective consists in generating insights into the social processes involved in creating certain social phenomena (Flick et al. 2004). Second, from a constructivist or interpretivist perspective, the research process itself is constituted through various social interactions, within which knowledge and interpretations are constructed (Flick et al. 2004: 6). As a result, knowledge and interpretations generated in the research process are always situated, i.e. produced by specific people and under specific circumstances (Burawoy 1998: 14).

Given the situatedness of knowledge, feminist and critical scholars have stressed the need for the researcher to reflect upon their positionality in the research process (Rose 1997; Sultana 2007). In this context, the term 'positionality' is used to refer to the researcher's pre-study beliefs and values (Holmes 2020: 2) and to his or her position within the 'multidimensional geography of power' (Rose 1997: 308). This position is shaped by the researcher's embeddedness in the grid of social relations of gender, class, race, educational status and geographical provenience inter alia. Given the assumption that a researcher's position has a significant impact on all stages of the research process—from study design, over data collection to the interpretation of research findings—critical and feminist scholars have reiterated the importance for researchers to adopt a 'reflexive research approach' (McDowell 2001; Sultana 2007). Adopting a 'reflexive research approach' means that researchers should critically reflect and be transparent about the impact of their positionality on research-related decisions and relationships with research participants (Rose 1997). In this light, reflections on my own positionality and how it has shaped decisions concerning research design, data collection and interpretation will be included in all remaining sections of this chapter.

Flick et al. (2004: 6) propose that studies within the reflexive/constructivist paradigm may be well suited with ethnographic research designs, since these are particularly apt to produce "descriptions of processes of creation of social situations". Accordingly, the research design adopted in this study builds on the ethnographic methodological approach of Burawoy's (1998, 2009) 'extended case method', in combination with Yin's (2014) 'single embedded case study' design. In the next section, the construction of the Bangalore export-garment cluster with the presence of three local garment unions as a single embedded case study is outlined.

4.1.2 Constructing the Bangalore Export-garment Cluster and Local Garment Unions as a Single Embedded Case

The single embedded case study is a specific form of case study, in which a selected main case contains various sub-units of analysis (Yin 2014: 265). This study uses the single embedded case study design since it allows us to study and compare the strategies of three Bangalore-based local labour unions *and* the structural context within which all three unions are embedded. The three garment unions represent the sub-units of analysis, which are, in turn, embedded within the same main case, i.e. the Bangalore export-garment cluster. Figure 4.1 provides a graphic representation of the main case and the embedded sub-units of analysis. The varying sizes of the circles representing the four case study unions indicate their different sizes measured by the number of their members.

The Bangalore export-garment cluster, as the main unit of analysis, was selected based on a purposeful sampling approach. Purposeful sampling is "a strategic approach to sampling in which 'information-rich cases' are sought out to best address the research purpose and questions" (Leavy 2017: 79). Against this backdrop, the Bangalore export-garment cluster was selected as the main case for two reasons: First, it represents an important node within the garment GPN and accounts for a significant share of global garment production. India ranks among the world's top five garment exporting countries (WTO 2020), and the Bangalore garment cluster accounts for 15–20% of India's garment exports (López Ayala 2018). Second, Bangalore is the garment cluster with the highest union activity in India, hence making it an apt case for a grounded study of labour agency within the garment GPN. A total of three local trade union organisations are currently active in organising garment workers in the Bangalore export-garment cluster. Over the past decade, these three

Fig. 4.1 Main case and sub-units of analysis in this study. *Source* Author

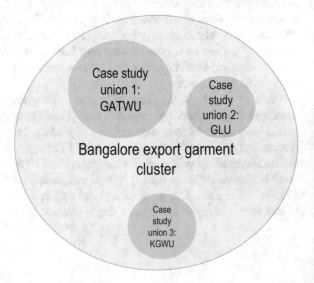

unions have achieved important improvements in wages and working conditions (CWM 2014, 2015; Jenkins 2013, 2015; Kumar 2014). In light of these features, the Bangalore export-garment cluster can be considered an information-rich case that is particularly apt to generate comprehensive data for a grounded study of interlinked dynamics of labour control and labour agency within the garment GPN.

To define the geographical study area, I drew on a relational notion of the cluster as a network of local and regional economic relations (c.f. Porter 2000). Traditionally, garment factories have been concentrated within the administrative boundaries of the Bangalore urban territory. However, over the past decade, large export-garment manufacturing companies have gradually been shifting their factories towards the outskirts of the city or to rural areas up to 150 km distance from Bangalore (Kumar 2014). Therefore, the geographical study area for this case study was expanded to a radius of 150 km around the Bangalore urban area. Figure 4.2 shows the geographical distribution of garment factories in the Bangalore export-garment cluster.

The rationale underlying this decision was twofold: First, most of the production units in rural areas around Bangalore belong to larger export-garment companies originating from Bangalore, where their headquarters are still based. As a result, the organisation of the labour process and working conditions in 'rural' units are similar to those in 'urban' units. Second, at least two of the three case study unions organise workers in 'rural' factory units in up to 150 km distance from Bangalore. Hence, from a relational point of view, even though these relocated production units

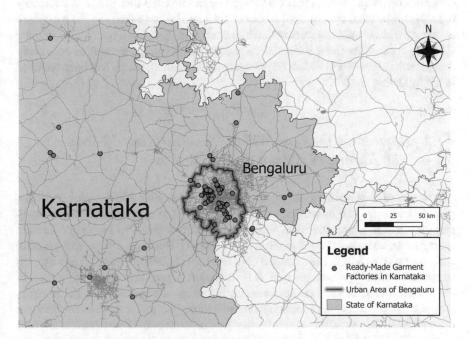

Fig. 4.2 Geographical distribution of export-garment factory units in the Bangalore export-garment cluster. *Data source* Own data/QGis; Map elaboration: Wilhelm Felk/Tatiana López

are located beyond the administrative boundaries of the city of Bangalore, they are embedded within the same networks of social and economic relations that constitute the Bangalore export-garment cluster.

Within the Bangalore export-garment cluster, three Bangalore-based local trade unions were identified as sub-units of analysis following a purposeful sampling approach (Leavy 2017: 79). To be included in the sample, unions had to meet two criteria: Firstly, unions had to actively address labour issues in the Bangalore garment cluster. Second, unions had to actively organise garment workers. Three local trade unions fulfilled these criteria and were hence defined as sub-units of analysis: the Garment and Textile Workers Union (GATWU), the Garment Labour Union (GLU) and the Karnataka Garment Workers Union (KGWU). To identify these unions, I relied on a snowballing approach (c.f. Leavy 2017: 80). Pre-existing contacts with GATWU provided the starting point for a sequential process, in which I asked inter-viewed union leaders whether they were aware of any other trade unions currently organising garment workers in and around Bangalore. Besides the three selected unions, two other unions had been pointed out by interviewees in the snowballing process. However, upon contacting representatives from those unions, it turned out that while they had some members in garment factories, they had no strategic approach for organising workers. Consequently, these two unions were not included in the sample.

As stated before, this study's main aim is to provide a grounded analysis of broader dynamics of labour control and labour agency in GPN. To this end, it is necessary to analyse how local dynamics of labour control and labour agency in the Bangalore export-garment cluster intersect with wider structural 'network dynamics' within the GPN. To account for these intersections, I combine the single embedded case study design with Burawoy's (1998, 2009) extended case method, which will be introduced in the following section.

4.1.3   Embedding the Case Within Wider Structural Forces with Burawoy's Extended Case Method

The extended case method developed by Burawoy (1998, 2009) employs ethno-graphic methods such as participant observation and in-depth interviews to study everyday life in a specific local setting while also seeking to locate the studied setting in its 'extralocal' context (Burawoy 1998: 4). Burawoy defines main central analytical phases within the extended case method: (1) intervention, (2) process, (3) structuration and (4) reconstruction. It is important to note that these four phases in the extended case method do not necessarily follow one another in a linear struc-ture. Instead, the research process is structured through circular movements between these phases. The remainder of this section first introduces the four central analytical phases of the extended case method. Thereafter, I describe how these phases have structured my own research process.

The first phase of *intervention* refers to the fact that the researcher intervenes in the life world of research participants through ethnographic methods. It is important to note that 'intervention', in this case, does not carry the meaning of an intentional political intervention as proposed by participatory action research methodologies (c.f. Rowe 2014). Instead, 'intervention' refers to the researcher creating situations of social interaction with research participants, interactions that cause disruptions to participants' usual routines (Burawoy 1998). It is in the 'mutual reaction' caused by the intervention that researcher and research participants together 'discover' the properties of the social order (Burawoy 1998). However, since the knowledge constructed through each intervention is necessarily situated and, therefore, partial, situated knowledge needs to be aggregated by the researcher into broader social processes that are significant for the case or phenomenon under study.

Process, as the second central phase in the extended case method, hence represents a first moment of extension, in which the researcher extends from the individual interactions with research participants to broader social processes that go beyond the immediate experiences and perspectives of individual research participants (Burawoy 1998: 15).

In the third phase, *structuration*, the researcher goes yet another step further by seeking to delineate the broader 'external field of forces' that shapes the social processes within the principal ethnographic locale. The external field of forces is constituted through social processes that lie outside of the local study setting; it has "systemic features of its own, operating with its own principles of coordination and contradiction, and its own dynamics" (Burawoy 1998: 15). In this light, the phase of structuration represents a second moment of 'extending' from the social relations of the primary case to the social relations constituting the external field of forces. It is important to note that different empirical methods may be applied when reconstructing the social relations that constitute the ethnographic locale and the external field of forces, respectively. Whereas reconstructing the social processes within the ethnographic locale requires qualitative ethnographic methods, the social processes that constitute the external field of forces may be reconstructed by studying secondary sources or through quantitative methods (Burawoy 1998: 29).

The fourth phase, *theoretical reconstruction*,[1] represents the last moment of extension in the extended case method. In this phase, the researcher extends from the empirical into the theoretical realm. It shall be stressed here again that theoretical reconstruction does not follow empirical data collection and analysis linearly. Instead, theory is reconstructed in dialogue with the empirical findings in the course of the research process.

In this study, I have applied the extended case method to reconstruct how processual relations and practice dynamics within the Bangalore export-garment cluster are shaped by broader network forces within the garment GPN (c.f. Yeung 2020). The

---

[1] Burawoy (1998) only speaks of 'reconstruction' as the fourth principle and last moment of extension in the extended case method. However, to distinguish the process of reconstructing theory from the process of reconstructing empirical social relations based on the gathered data, I chose to denominate the process of reconstructing theory as 'theoretical reconstruction'.

theoretical perspective that provided the point of departure for this research project and informed the empirical data collection process was a Marxist political economy approach to labour control and labour agency in GPNs. This theoretical perspective informed data collection during my first field research trip to Bangalore in October 2016. During this trip, I acted as a translator for a German union delegation during meetings with local garment workers and unionists and during two factory visits. The aggregated situational knowledge from this first field trip allowed me to reconstruct some central social *processes* shaping the labour control regime and union strategies in the Bangalore garment industry. At the same time, I was able to engage in *structuration* by identifying several external forces shaping the relations of production in the Bangalore garment cluster, namely brands' sourcing and CSR practices.

This first research stay was followed by a period of '*theory deconstruction*', in which I identified several blind spots within the theoretical concepts and frameworks that had informed my first observations in the field. As a result, the aim to 'improve' existing theories (c.f. Burawoy 1998: 28) guided my interventions during the following two field trips to India in March/April 2017 and in September/October 2017. On each field trip, I spent around four weeks in Bangalore conducting participant observations and in-depth interviews with union representatives, workers, factory managers, and industry, state and NGO representatives (for more details on the data collection process, see Sect. 4.2). In light of my aim to reconstruct the external structural forces that shape the 'grounded' dynamics of labour control and labour agency in the Bangalore export-garment cluster, I also undertook trips to New Delhi. There, I conducted additional interviews with representatives of national industry associations and of national and global union federations.

Research trips were again followed by an extensive period of *theoretical reconstruction* during which I experimented with giving existing theories of labour control regimes and union agency in GPN a "novel angle […] of vision" (Burawoy 1998: 16) by re-thinking them from a relational perspective. Finally, in February/ March 2019, I undertook a fourth and last research field trip to Bangalore to conduct follow-up interviews with the three case study unions, followed by a last phase of intertwined empirical data analysis, theory reconstruction and writing. Table 4.1 provides an overview of the phases of the research process and of the moments of intervention, process reconstruction, structuration and theory reconstruction during each phase.

The following section discusses the field access and data collection process in more detail.

4.2 Field Access and Data Collection

This section gives insights into the field access and empirical data collection process. It starts by discussing the applied strategies for 'opening up' the field (Sect. 4.2.1) and then gives insights into the two main methods of empirical data collection employed: participant observations (Sect. 4.2.2) and in-depth interviews (Sect. 4.2.3).

Table 4.1 Moments of interventions, process reconstruction, structuration and theory reconstruction in the research process

Time period	Place(s)	Phase	Moments of extended case method	Methods of data collection/ analysis
January–September 2016	Cologne, Germany	Explorative review of literature on labour control and labour agency in GPN	–	Literature analysis
October 2016	Bangalore, India Dhaka, Bangladesh	1st field trip with German union delegation	Intervention (mediated), process reconstruction, structuration	Participant observation
November 2016–February 2017	Cologne, Germany	Focussed review of literature on labour control and labour agency in GPN	Theory deconstruction	Literature analysis in dialogue with empirical data
March/April 2017	Bangalore, India Chennai, India New Delhi, India	2nd field research stay (6 weeks)	Intervention (direct), process reconstruction, structuration	Participant observation, in-depth interviews
Apri–August 2017	Cologne, Germany	Focussed review of literature on labour control and labour agency in GPN	Theory reconstruction	Literature analysis in dialogue with empirical data
September 2017	Bangalore, India New Delhi, India	3rd field research stay (5 weeks)	Intervention (direct), process reconstruction, structuration	Participant observation, semi-structured interviews
October 2017–February 2019	Cologne, Germany	Exploration of alternative theoretical perspectives on labour control and labour agency in GPN	Theory reconstruction	Literature analysis in dialogue with empirical data
March 2019	Bangalore, India	4th field research stay (4 weeks)	Intervention (direct), process reconstruction,	In-depth interviews, participant observation

(continued)

Table 4.1 (continued)

Time period	Place(s)	Phase	Moments of extended case method	Methods of data collection/ analysis
April 2019–June 2021	Cologne, Germany	Development of heuristic framework for data analysis and interpretation; writing process	Theory reconstruction (final phase)	Qualitative content analysis of empirical data in dialogue with literature

4.2.1 'Opening Up' the Field: Dealing with Gatekeepers and Multiple Researcher Positionalities

Literature on qualitative field research stresses as the main challenge for 'opening up' the field the need to build rapport with research participants, particularly when studying a sensitive topic and/or working with disadvantaged groups (see, e.g., Chaudhry 2017; Huisman 2008; Weiner-Levy 2008). Taking this into consideration, methodological literature has proposed two strategies for 'opening up' the field. On the one hand, literature has stressed the importance of gaining the trust and confidence of gatekeepers that may facilitate (or otherwise constrain) the researcher's access to research participants (see, e.g., Lata 2020; Mandel 2003). On the other hand, various scholars have observed that to gain the trust of potential research participants researchers may accentuate different aspects of their positionality—that is different roles and parts of their identity (Bachmann 2011; Herod 1999). In the following, I outline how I confronted the challenge of gaining access to and building trust and rapport with three types of actors: labour actors, capital actors and state actors.

To gain access to *labour actors*, i.e. local unionists and workers, I relied on the support of two primary gatekeepers. The first gatekeeper was the leadership of GATWU, the oldest and biggest garment union in Bangalore. I first met GATWU leaders during my first trip to Bangalore in October 2016 as a translator for the German delegation of the TIE global union network. Whereas I openly communicated my role as a PhD student and researcher, the salient role was that of an active member and supporter of the TIE union network. Since GATWU had long-established working relationships with the TIE union network, my 'insider position' as a member of the network allowed me to gain the trust and confidence of GATWU's leadership and of their local network of supporters (c.f. Herod 1999: 320). During the following three research trips, GAWTU leaders allowed me to join union meetings and trainings, and kindly facilitated contacts with worker activists. The second important gatekeeper was a labour rights NGO called Cividep, which works closely with GLU—the second biggest garment union in Bangalore. In light of Cividep's interest in making their work in the garment industry internationally visible, accentuating my position as a German researcher with contacts to German labour unions and consumer activist groups facilitated rapport building. Cividep's director arranged a meeting with GLU leaders and facilitated the contact of a Cividep staff member to provide translation. Through GLU, I, in turn, learned about the existence of KGWU, the third union engaged in organising workers in the Bangalore export-garment sector. To maintain a good rapport with all three unions, I made a conscious effort to maintain a position as a 'sympathetic and informed outsider' in my relations with GLU and KGWU (c.f. Herod 1999: 322f.). Accentuating my outsider positionality as an independent researcher was crucial for navigating a research field where inter-union relationships are marked by personal and political tensions.

Compared to building rapport with labour actors, gaining access to *capital actors* proved to be far more challenging. My initial intent to 'open up' the field of capital actors focussed on gaining access to local factory managements. However, my attempts to get in touch with factory managers through official email addresses and phone numbers resulted in little success. In the only interview obtained through this method with the HR manager of a leading Bangalore based garment exporter, it became evident that my positionality as Western researcher combined with my research interest in industrial relations presented a severe barrier to building trust with local managers. In this light, gaining the support of a local industry association representative proved to be crucial for gaining access to factory managers. To gain the trust of this representative, I followed a strategy employed by Herod (1999), who found it helpful to use the terminology used in HR management when interacting with managers. As a result, when introducing the aim of my research project, I consciously avoided politically charged terms, such as 'working conditions', 'industrial relations' or 'collective bargaining'. The trust that I had built in my personal interaction with the local industry association representative, in turn, served to provide factory managers with 'a personal sense of security' and to convince them of my integrity (c.f. Lata 2020). Nevertheless, interviews with local factory managers remained limited to a number of five. Whereas this might be considered a limitation of this study, I found that the five conducted interviews with factory managers provided enough information to reach a point of saturation, where additional interviews did not yield additional insights regarding Bangalore garment manufacturers' labour control practices (c.f. Leavy 2017: 78).

Besides contacting local factory managers, I also undertook several unsuccessful efforts to get access to multinational garment retailers sourcing from Bangalore-based factories. I was, however, able to (at least partially) compensate for this limitation by gathering comprehensive information on retailers' sourcing practices and CSR practices from several other sources. These sources included participant observation of a meeting between German unionists and Bangalore garment factory managers, publicly accessible company and NGO reports, and existing academic research on this topic (see, e.g., Anner 2019; Mezzadri 2017).

Lastly, my process of opening up the field included efforts to identify and gain access to relevant *state actors*. Given the fact that many state websites only exist in Kannada, the local language of the state of Karnataka, the first challenge consisted in identifying relevant departments and finding contact details online. I overcame this first challenge by identifying relevant state departments in my interviews with union leaders, who helped me to obtain contact details for these departments. When contacting state departments, I strategically accentuated my role as researcher from a German university. With this introduction, I could obtain interviews with representatives of two state departments involved in industrial land development and in promoting the garment industry. The fact that accentuating my position as a foreign researcher proved successful when contacting state actors can most probably be attributed to the dynamics described by Sabot (1999). Sabot (1999) argues that a foreign researchers' visit always has a flattering element for local political actors, since it heralds international recognition and promises to project their achievements abroad.

In summary, this section has exemplified the various challenges I encountered in my process of 'opening up' the field. It has further illustrated how building trust with gatekeepers and accentuating different aspects of my positionality has helped me to overcome these challenges. The following section outlines the two main methods for data collection: participant observations and in-depth interviews.

4.2.2 Collecting Data Through Participant Observations

Participant observation is a method for data collection most prominent in ethnography, but which is also frequently used in combination with other methods in human geography (Mattissek et al. 2013: 142ff.). As the name indicates, participant observation consists of the combination of two moments: participation and observation. The 'participation' moment in participant observation refers to the researcher getting involved in the everyday lifeworld of a specific context. Even if this involvement is minimal, "observing social activity is [always] predicated on participating, even in the most minimal ways […]; we change the spaces we are present in to greater or lesser degrees, even when we are seemingly passive" (Laurier 2016: 171f.). The 'observation' moment in participant observation, in turn, refers to a systematically planned, documented and analysed observation (as opposed to unsystematic everyday observations). Observations may be structured, meaning that the documentation process follows pre-defined categories or they may be conducted in an unstructured manner, meaning that the researcher stays open to new aspects and developments that may be relevant to the research question (Mattissek et al. 2013: 150). Structured observations are common in quantitative research design, whereas unstructured observations tend to prevail in qualitative research designs.

The participant observations conducted for this study can therefore be characterised as systematic yet unstructured. Observations were guided by the aim to generate empirical data on three aspects: first, the spatial organisation of the export-garment industry; second, the organisation of the labour process in Bangalore export-garment factories; and third, the studied unions' organisational practices.

To generate insights into the *spatial organisation of the export-garment industry*, I spent several mornings and afternoons strolling through three main garment industry hubs in Bangalore: Mysore Road, Peenya and Yelahanka. During these strolls, I observed, among other aspects, the spatial layout of the roads and factory buildings, the built features and practices through which everyday life inside the factories was shielded from outsiders, how workers left the factories after their shifts, and which mode of transport they took home. Since garment factories in these areas usually consist of large multi-storey buildings with high walls and often no windows, one challenge during these participant observations consisted in singling out garment factories among the multitude of factories present in each area. On two occasions, I was accompanied during my observations by local unionists, who indicated garment factories to me and also gave me some background information on each factory. When setting out alone for my participant observations in industrial areas, I usually carried a list of factory names in the respective area retrieved from H&M's (2021) publicly available supplier list to identify factories from the list by their names when passing by. I recorded my observations during these strolls in the form of audio notes that I captured with a voice recorder. Moreover, I took several photos of factory buildings.

Besides undertaking participant observations in industrial areas to observe the everyday life and spatial settings in industrial areas, I secondly undertook participant observations inside two garment factories to gain insights into the *organisation of the labour process in Bangalore export-garment factories*. The most extensive insights resulted from my first 'covert' participant observation (c.f. Cook 2005) during a two-hour tour through an export-garment factory in Peenya, in which I officially participated during my first field trip as part of a German union delegation. My role as a translator during this tour allowed me to take detailed notes of all explanations given by the production manager, who guided us through the different stages of the production process. Moreover, members of the German union delegation took multitudinous photos and were kind enough to share them with me. The second participant observation that I was able to conduct inside a garment factory was in a factory in Yelahanka. Compared to the first tour, this second was much shorter, with a duration of approximately 30 minutes.

Lastly, I employed participant observation to gain insight into *unions' organisational practices* and how these construct and shape internal and external union relations. To this end, I participated in a total of eleven union strategy meetings, worker assemblies, and organiser trainings and protests. In addition, I spent several days in union offices observing everyday office life. When spending time at a union office, I usually occupied myself by writing down notes on my observations. Whenever possible, I also engaged with union staff to assist with whatever I could be useful for. During union meetings and trainings, my involvement was usually limited to introducing myself at the beginning and thereafter taking notes and photos (of course, after getting participants' consent). Therefore, my role during my participant observations in union meetings and offices can be best described with the term coined by Lave and Wenger (1991) as 'legitimate peripheral participation'.

In some cases, this state of 'peripheral' participation was, however, actively challenged, when unionists or other present actors asked me to leave my role as an uninvolved researcher and to take on the role of an international supporter. This involvement was rather unproblematic when union meetings and training sessions were held in private. In my dealings with local garment unions, I was very upfront about my own union engagement in Germany and my general 'pro-worker' stance (c.f. Castree 2007: 856). Therefore, switching roles from independent research to supporter of the international labour movement was consistent with my own positionality. However, on one occasion, I joined a public union protest against the state government, when a camera team from a local TV station asked me to give a message in support of workers' demands. While workers around me were cheering, I felt that publicly challenging the government of Karnataka on local TV might compromise my ability to conduct interviews with government officials or local managers. Therefore, I declined the camera team's request, much to the disappointment of the workers surrounding me. My experience during this protest highlights the complexity of constructing and negotiating multiple roles in the field, particularly when working with participant observations that involve close working and personal relationships with research participants over a prolonged time period (c.f. Bachmann 2011; Chereni 2014; Sultana 2007).

Besides the need to negotiate my role during participant observations, a second challenge for conducting participant observations of unions' organisational practices was facilitating communication. As Mattissek et al. (2013) highlight, documenting conversations and other forms of communication is usually a central part of conducting a participant observation. However, my abilities to document and engage in verbal communication during my participant observations with the three case study unions were limited by the fact that communication in union offices and during union meetings and training sessions usually took place in Kannada. As a result, during my observations in the union offices, I had to rely on union leaders or office staff who spoke both English and Kannada to provide me with short accounts of the issues discussed with workers who came by the office.

In summary, conducting participant observations in union offices, trainings and meetings was not only beneficial to generate insights into unions' internal organisational practices; participating in union training sessions and meetings also proved to be a good way to build rapport with union leaders and organisers during breaks. Building rapport at union meetings, in turn, allowed me to conduct subsequent interviews with union leaders and organisers in a relaxed atmosphere, with interviewees trusting me enough to speak freely and also talk about sensitive issues.

4.2.3 Collecting Data Through In-Depth Interviews

Following the dominant research approach in human geography to combine data collection from participant observation with other methods of data collection (c.f. Mattissek et al. 2013: 148), I employed in-depth interviews as a second central method for data collection (c.f. Legard et al. 2011). In total, I conducted 53 in-depth interviews with a variety of actors, including leaders, organisers and workers from the three case study unions as well as factory managers, industry representatives, state officials, NGOs and researchers. Frequently, interviews were conducted with more than one interview partner. Representatives of local case study unions were interviewed multiple times throughout the research project. Table 4.2 provides an overview of the number of interviews and interviewees for each actor category (for a detailed list of interview partners see Annex I).

In-depth interviews "combine structure with flexibility" (Legard et al. 2011: 141). While the researcher usually uses an interview guide that sets out key topics and issues for the interview, topics may be covered in a flexible order to allow for a natural flow of conversation. The in-depth interview provides openness for new issues and aspects that the researcher had not previously thought of to emerge during the conversation

Table 4.2 Conducted in-depth interviews

Category	Interviewed actors	Number of interviewees	Number of conducted interviews
Labour actors	Local case study garment union leaders and organisers	11	16
	Garment workers	9	3
	Other local unions	1	1
	National trade union federations	2	2
	Global union federations/ networks	2	3
Total labour actors		**25**	**25**
Industry actors	National garment industry associations	3	3
	Training organisations for the garment industry	2	1
	Factory managers	5	5
	Sourcing company	2	3
Total industry actors		**11**	**10**
Civil society actors	Local NGOs / Community organisations	6	4
	International labour rights NGOs/ Multi-stakeholder-initiatives	3	3
	Labour researchers	6	7
	Labour lawyer	1	1
Total civil society actors		**16**	**15**
State actors	Karnataka State Department of Handlooms and Textiles	2	1
	Karnataka State Infrastructure Development Cooperation	1	1
Total state actors		**3**	**2**
Total		**55**	**53**

(Legard et al. 2011: 146). As the name suggests, the in-depth interview aims to produce detailed accounts of specific issues while exploring "all the factors that underpin participants' answers: reasons, feelings, opinions and beliefs" (Legard et al. 2011: 141). Hence, during the interview, the researcher usually seeks to create more 'depth' in participants' answers through follow-up questions and probes (Legard et al. 2011: 146). When combined with participant observation, in-depth interviews are frequently used to gather additional expert knowledge on studied setting or to collect the subjective views of specific actors (Hopf 2004: 204).

In this study, I employed in-depth interviews with different actor groups for different purposes. When interviewing union leaders, organisers and workers from the four case study unions, my objective was twofold. Firstly, I sought to get insights into their subjective experiences in the union and/ or in the workplace and into how they attribute meaning to their role within the union and to the union as an organisation. Secondly, when interviewing union leaders, I did not only address them as individual subjects but also as experts, who possess specialised knowledge about the history and organisational structure of the union, and whose actions and opinions significantly shape the unions' strategies (c.f. Bogner et al. 2014).

In contrast, further labour, industry, civil society and state actors were exclusively interviewed as experts to access their knowledge on the Bangalore and Indian export-garment industry and/or the garment GPN (c.f. Gläser and Laudel 2010: 10). In most cases, interviewees' expert knowledge stemmed from the fact that they held relatively high ranks within their organisation. Consequently, interviewing these actors involved dealing with several methodological challenges associated with interviewing social elites (see, e.g., Herod 1999; Mikecz 2012; Sabot 1999). In the following section, I will provide more insights into the interview process with union leaders, organisers and workers from the three case study unions, and with other labour, civil society, industry and state actors who were interviewed in their role as experts.

4.2.3.1   Interviewing Unionists, Organisers and Workers from Local Unions

In total, I conducted 20 interviews with leaders, organisers and workers from local unions. Interviews usually took place at their respective union offices. The first introductory interview was commonly conducted as a group interview, in which several union leaders and full-time organisers were present. Introductory interviews followed a semi-structured approach and covered questions regarding the union's collective action frame, strategic repertoire and internal organisational practices (c.f. Ross 2008). Given their nature as group interviews, the broad range of topics covered and the significant amount of time spent on personal introductions, introductory interviews with case study unions lasted between 120 and 180 min. In cases where the introductory interview represented the first contact with the union, I chose not to record the interview and instead to rely on handwritten notes as a mode of documentation to create an easy and informal atmosphere and, therefore, facilitate trust and rapport building (c.f. Lata 2020).

Subsequent follow-up interviews with union leaders were usually recorded. Compared to the introductory interviews, follow-up interviews with union leaders were conducted in a more open manner, with my role being limited to providing an initial stimulus for union leaders to narrate the development of a specific labour struggle. After this initial prompt, I only interrupted union leaders' narrations to ask for more details on specific issues and to ask for clarifications or point out apparent contradictions in their narrations (c.f. Legard et al. 2011: 146). Given the more focussed nature of follow-up interviews, each follow-up interview usually lasted between 60 and 90 min. Generally, interviews with union leaders were conducted in English. Only follow-up interviews with one union leader were conducted with the help of a translator.

Interviews with workers also took place at the unions' offices. Only one interview took place at the home of a worker. Since union leaders facilitated all contacts with workers, all interviewed workers were union members. Out of the nine interviewed workers, seven workers were women, and two were men. Given the limited time frame that particularly women workers had for an interview due to their double burden of wage work and care work, two or three workers were interviewed simultaneously. Interviews with workers usually lasted about 30 min and covered questions about working conditions and work organisation, capital-labour relations in the workplace and workers' strategies to address problems at the workplace. To communicate with workers, I relied on the help of union leaders or of a translator to translate my questions from English to Kannada and workers' responses from Kannada to English. Building trust and creating a reassuring environment in which workers feel safe to talk about working conditions and work-related problems is crucial when interviewing workers (c.f. Hale and Hurley 2005: 72). For this reason, I decided not to record interviews with workers and instead to rely on taking detailed written notes, for which I had plenty of time due to the consecutive translation. Moreover, in the interview documentation and throughout this book, workers' identities are protected through anonymisation. Any information that might make workers identifiable has been omitted.

A significant challenge when interviewing workers also consisted in overcoming barriers for confidential conversations due to my Western-European background and perceived power asymmetries in terms of social status and education. Especially in one interview situation where I met a group of workers in their living area accompanied by a translator but without a union representative, workers were reluctant to say anything negative about their experiences in the garment factory. Only towards the end of the interview, when I encouraged workers to ask me any questions they might have, it became clear that workers had thought I was a social auditor coming from a Western NGO, since it is a common practice of social auditors to interview workers in their neighbourhoods. Hence, workers were very cautious about mentioning any problem in the factory for fear that providing such information may have negative consequences if it became known to the management. In other worker interviews, such barriers could be mitigated when I was introduced to workers by a union representative whom workers knew well and trusted and who encouraged workers to talk to me about their problems at the workplace. Hence, this experience highlights once more the importance of generating interpersonal trust through gatekeepers when interviewing potentially vulnerable social groups on sensitive issues (c.f. Lata 2020).

4.2.3.2   Interviewing Other Labour, Civil Society, Industry and State Actors as Experts

Besides interviewing unionists and workers from case study unions, I conducted 33 semi-structured in-depth interviews with representatives of other local unions, national and global union federations, local and international labour rights NGOs, labour researchers, factory managers and state officials as experts. Experts are "persons who – based on specific practical or experiential knowledge that relates to a clearly definable problem area – have created the possibility of structuring the concrete field of action in a meaningful and action-guiding way for others" (Bogner et al. 2014: 13, own translation). Who qualifies as an expert is defined by the researcher in relation to the specific topic under study (Bogner et al. 2014: 11f.). Experts interviewed for this study were selected due to their special context knowledge of the Bangalore and Indian export-garment industry (c.f. Przyborski and Wohlrab-Sahr 2014: 119f.). Almost all experts held leading or senior positions within their organisations and can therefore be considered to form part of a 'functional elite' (Meuser and Nagel 1994). As a result, conducting expert interviews also brought along several challenges linked to interviewing elites (c.f. Mikecz 2012; Nader 1972; Sabot 1999).

Expert interviews usually took place at their respective offices and lasted between 60 and 90 min on average. At the beginning of each interview, I introduced myself and my research interest. In this introduction, I usually sought to gain credibility and trust by accentuating my position as a researcher affiliated to a German university, bound to academic research's ethics (c.f. Gläser and Laudel 2010: 178f.). After this initial introduction, expert interviews were carried out in the form of a relatively open conversation around central topics listed in an interview guide. Careful preparation of interview questions is particularly important when interviewing experts for several reasons: On the one hand, since experts usually only have limited time for the interview, careful preparation enables the researcher to ask the right questions that will elicit useful and in-depth information from interviewees within a limited time period (Mattissek et al. 2013: 176). On the other hand, to engage in a conversation with experts that can reveal new and interesting aspects, the researcher needs to present himself or herself as competent in the respective topic. Thereby, the researcher avoids spending too much time on general information that could also be accessed publicly (c.f. Mikecz 2012: 489). Consequently, preparing the interview guide for each interview usually required several hours of background research about the experts' organisation, the organisation's work and—if publicly available—the expert's person.

Given the fact that most interviewed experts belonged to a functional elite (Meuser and Nagel 1994: 181), experts "usually [had] the ability and power to protect themselves from exposure to criticisms" (Mikecz 2012: 484). Therefore, I employed several interviewing techniques to elicit answers beyond 'public relations' versions (ibid.). For example, Laurila (1997: 411) suggests that "managers' vanity may be exploited by emphasising the view that they now have a chance to teach the researcher". Following this technique, when interviewing factory managers, I maintained a careful balance between displaying my theoretical knowledge on the garment industry, on the one hand, and stressing my position as a novice researcher in the field, on the other hand. In this line, I also employed an interviewing technique often referred to as "being naïve" in methodological literature. This technique involves asking multitudinous follow-up questions to let managers explain seemingly 'obvious' procedures, such as measuring workers' productivity, in detail (c.f. Hermanns 2004: 13). This strategy successfully generated detailed accounts of the workplace practices that structure the labour process in the Bangalore export-garment industry.

After this detailed discussion of my field access and data collection process, I now turn to the data analysis and interpretation process.

4.3 Data Analysis and Interpretation

In this section, I lay out the data analysis and interpretation process. I first provide insights into the process of data preparation (Sect. 4.3.1). Thereafter, I introduce the 'qualitative content analysis' as the method of choice for data analysis (Sect. 4.3.2).

4.3.1 Preparing the Data: Protocols and Transcriptions

To analyse data collected through participant observations and interviews, the data needed to be converted into text. This conversion of visual and audio data from participant observations and interviews was undertaken through the preparation of written protocols and transcripts (c.f. Mattissek et al. 2013: 191). The main source of information for protocols was audio notes that I recorded during and after participant observations and directly after each interview. These audio memos (c.f. Bogner et al. 2014: 61) comprised detailed accounts of the place of the interview or of the observed event, of the atmosphere before, during and after the interview or event, and of informal conversations.

Based on these memos and notes taken during the interview or observed event, I prepared a detailed interview or observation protocol as soon as possible. For participant observations and interviews that had not been recorded, I prepared detailed reconstructions of the interview or of observed conversations. Whereas these reconstructions may be considered less truthful' it is important to note that, according to the reflexive/constructivist research paradigm, any posterior documentation of an observed event or interview represents a selective and impartial reconstruction, shaped by the researcher's subjective interpretation (Kowal and O'Connell 2004: 249). The subjectivity of protocols and transcripts is manifested, for example, in the fact that the researcher decides alone which observed details or parts of interviewees' responses are important enough to become part of the protocol or transcript (c.f. Mattissek et al. 2013: 195).

In addition to interview and observation protocols, for 30 recorded interviews, transcripts were prepared. Transcripts were prepared by a team of people comprising student assistants, myself, and a local contractor from Bangalore. In light of the chosen method for data analysis—the qualitative content analysis (Kuckartz 2016)—interviews were transcribed in standard orthography (Kowal and O'Connell 2004: 250). Since the primary goal of the qualitative content analysis was to extract and systematise relevant information from interviews, the specific characteristics of the spoken language were therefore largely irrelevant for the analysis process (c.f. Schmidt 2004). To enhance legibility, interviewees' responses were smoothed in the transcription process by omitting unfinished beginnings of phrases and interjections and by correcting grammatical errors (c.f. Mattissek et al. 2013: 193f.).

4.3.2 Analysing Data with Qualitative Content Analysis

The analysis of semi-structured interviews requires a methodological approach that reflects the relatively open form of this interview type. Since in semi-structured interviews new issues and topics may emerge during the interview process, analytical categories need to be developed, not only a priori based on theoretical reflections but also in dialogue with the empirical material (Schmidt 2004: 253). Against this background, a qualitative content analysis approach (Kuckartz 2016) was chosen for analysing the empirical data collected for this study.

Multiple variants of qualitative content analysis exist, setting out different procedures (see, e.g., Kuckartz 2016; Mayring 2004; Schmidt 2004). Generally, four general phases can be identified as common to all variants of qualitative content analysis: (1) developing main analytical categories, (2) developing sub-categories, (3) coding the entire empirical data material (4) and comparing relevant text segments and developing interpretations guided by the central research question(s) (Fig. 4.3).

In the first phase, a relatively limited number of rather broad main analytical categories (also called 'codes') are developed. Main analytical categories may be developed deductively from theoretical concepts that have guided the research process or from topics set out in the interview guide. Alternatively, main analytical categories may be developed inductively from the empirical material itself (Schmidt 2004: 254f.).

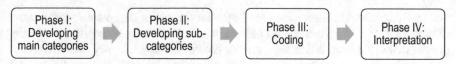

Fig. 4.3 Four phases of qualitative content analysis. *Source* Own elaboration drawing on Schmidt (2004), Mayring (2004), and Kuckartz (2016)

In the second phase, these main analytical categories are then further differentiated by developing sub-categories in dialogue with the empirical material (Kuckartz 2016: 97). Frequently, this second phase is used to enrich main categories derived from analytical concepts or broader topics successively with empirical content. As a result, a hierarchical category system is obtained that sets out the different main categories with their respective sub-categories, and which serves as a heuristic framework for the subsequent coding and interpretation process (Kuckartz 2016: 97).

In the third phase, this category system is then systematically applied to the entire data material through the coding process. Coding means that selected text segments are assigned to specific categories. Afterwards, they are marked and colour coded according to their information (Rädiker and Kuckartz 2020: 54ff.). When dealing with large data, literature on qualitative data analysis recommends using computer-assisted methods for the coding process. These allow quick retrieval of all text segments coded with the same category (Kelle 2004: 279).

This easy and quick retrieval of text segments, in turn, facilitates the comparison of text segments in the fourth and last phase. Here, coded text segments are compared and interpreted in light of the research questions and according to the specific research purpose (Schmidt 2004: 357). Whereas phases one to three are rather standardised, in the fourth phase, the researcher engages in the 'imaginative work of interpretation' (Coffey and Atkinson 2013: 6f.). In the remainder of this section, I lay out the data analysis process underpinning this study according to the just described four phases.

The data analysis process started with developing main analytical categories from both the theoretical concepts that guided data collection and the collected empirical material. In a first step, I distinguished between 'labour agency' and 'labour control regime' as two overarching analytical categories (c.f. Kuckartz 2016: 34). For the category of 'labour agency', I derived three further sub-categories based on the concepts of 'spaces of organising', 'spaces of collaboration' and 'spaces of contestation'. For the category 'labour control regime', I identified the labour processes and six further processual relations in the empirical material as constituting parts of the labour control regime in the Bangalore export-garment industry. These processes were then established as sub-categories for the main category 'labour control regime'.

In phase two, the empirical data—that is interview transcripts and observation protocols—was coded with these initial main categories with the help of the software MAXQDA (c.f. Rädiker and Kuckartz 2020). In this process, the main categories established in the first phase were further classified by generating sub-categories for each main category from the empirical data. These sub-categories were predominantly factual categories (i.e. categories referring to empirical facts, places or events) and in vivo codes, i.e. expressions used by interviewees to designate specific issues (Kuckartz 2016: 34f.). As a result of phase two, I obtained two comprehensive category systems (Kuckartz 2016: 38)—one for the overarching category 'labour agency' and one for the overarching category 'labour control regime'. Figures 4.4 and 4.5 provide a stylised graphical representation of the two category systems.

In the third phase, all protocols and transcripts were coded using these category systems. During this process, I created several further thematic categories to code text passages with additional relevant context information on the Indian or Bangalore garment industry that, however, did not fit within the categories developed in phase two.

Lastly, I retrieved all text extracts for each category in the fourth phase and compared them to develop interpretations. To provide insights into the structural context for the agency of local unions in the garment GPN, I firstly compared extracts for the overarching category 'labour control regime' to develop thick descriptions (Cousin 2005) of the different processual relations that constitute the labour control

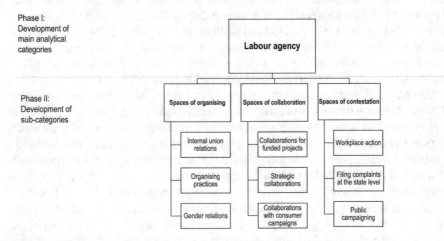

Fig. 4.4 Category system for overarching category 'labour agency'. *Source* Author

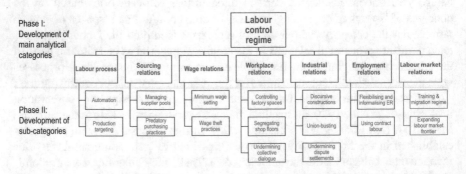

Fig. 4.5 Category system for overarching category 'labour control regime'

regime in the Bangalore export-garment industry. In addition, I sought to detect the specific mechanisms through which these different relations are interlinked by looking for categories that frequently appeared together. To ensure credibility, I employed different strategies of data triangulation when reconstructing the relations, processes and practices that constitute the labour control regime from the retrieved text extracts (c.f. Schuermans 2017: 7; Bogner et al. 2014: 95). Comparing information on specific dynamics or practices from interviews with different actors allowed me to identify potential contradictions or lacunas in the empirical data. In these cases, empirical data was further triangulated and complemented with data from additional sources, such as public company reports, internal company documents provided by interviewees or policy documents.

To assess different unions' strategic approaches regarding their potential and limits for building local bargaining power, I compared the retrieved text passages for the overarching category 'labour agency'. In this process, I identified interrelations between the three main categories: 'spaces of organising', 'spaces of collaboration' and 'spaces of contestation'. Identifying these interrelations, in turn, allowed me to develop an interpretation of how the relations and practices constituting each space influence one another. Thereafter, I compared the three case study unions' strategic approaches regarding their limits and potentials for building local union power with to develop more generalised conclusions about the enabling and constraining factors for building local bargaining power in the garment GPN.

After these detailed insights into the research design and into the data collection and analysis process, in the following section, I introduce the Bangalore export-garment cluster representing the empirical case in the focus of this study.

References

Anner M (2019) Predatory purchasing practices in global apparel supply chains and the employment relations squeeze in the Indian garment export industry. Int Labour Rev 158:705–727. https://doi. org/10.1111/ilr.12149

Bachmann V (2011) Participating and observing: positionality and fieldwork relations during Kenya's post-election crisis. Area 43:362–368

Baglioni E (2018) Labour control and the labour question in global production networks: exploitation and disciplining in Senegalese export horticulture. J Econ Geogr 18:111–137. https://doi. org/10.1093/jeg/lbx013

Bogner A, Littig B, Menz W (2014) Interviews mit experten: eine praxisorientierte Einführung. Lehrbuch. Springer VS, Wiesbaden

Burawoy M (1998) The extended case method. Soc Theory 16:4–33. https://doi.org/10.1111/0735-2751.00040

Burawoy M (2009) The extended case method: four countries, four decades, four great transformations, and one theoretical tradition. University of California Press, Berkeley

Castree N (2007) Labour geography: a work in progress. Int J Urban Reg Res 31. https://doi.org/ 10.1111/j.1468-2427.2007.00761.x

Chaudhry V (2017) Knowing through tripping: a performative praxis for co-constructing knowledge as a didsabled halfie. Qual Inq 24:70–82. https://doi.org/10.1177/1077800417728961

Chereni A (2014) Positionality and collaboration during fieldwork: insights from research with co-nationals living abroad. Forum Qual Soc Res 15:1–21. https://doi.org/10.17169/FQS-15.3. 2058

Coe NM, Jordhus-Lier DC (2010) Re-embedding the agency of labour. In: Bergene AC, Endresen SB, Knutsen HM (eds) Missing links in labour geography. Ashgate Pub, Farnham, pp 29–42

Coffey A, Atkinson P (2013) Making sense of qualitative data: complementary research strategies, 10th edn. Sage, Thousand Oaks, CA

Cook I (2005) Participant observation. In: Martin D, Flowerdew R (eds) Methods in human geography: a guide for students doing a research project. Routledge, London

Cousin G (2005) Case study research. J Geogr High Educ 29:421–427. https://doi.org/10.1080/030 98260500290967

CWM, Centre for Workers' Management (2014) State of garment workers in Bangalore. CWM, New Delhi

CWM, Centre for Workers' Management (2015) Wage and work intensity: study of the garment industry in greater Bangalore. Research Paper Series (vol. 1). CWM, New Delhi

Flick U, von Kardorff E, Steinke I (eds) (2004) A companion to qualitative research. Sage, London

Gläser J, Laudel G (2010) Experteninterviews und qualitative Inhaltsanalyse als Instrumente rekonstruierender Untersuchungen, 4th edn. VS Verlag, Wiesbaden, Lehrbuch

Graham E (2005) Philosophies underlying human geography research. In: Martin D, Flowerdew R (eds) Methods in human geography: a guide for students doing a research project. Routledge, London, pp 8–33

Hale A, Hurley J (2005) Action research: tracing the threads of labour in the global garment industry. In: Hale A, Wills J (eds) Threads of labour: garment industry supply chains from the workers' perspective. Blackwell, Malden, MA, and Oxford, pp 69–94

Hermanns H (2004) Interviewing as an activity. In: Flick U, von Kardorff E, Steinke I (eds) A companion to qualitative research. Sage, London, pp 209–213

Herod A (1999) Reflections on interviewing foreign elites: praxis, positionality, validity, and the cult of the insider. Geoforum 30:313–327. https://doi.org/10.1016/S0016-7185(99)00024-X

Holmes AGD (2020) Researcher positionality—a consideration of its influence and place in qualitative research—a new researcher Guide. Educ 8:1–10. https://doi.org/10.34293/education.v8i4. 3232

Hopf C (2004) Qualitative interviews: an overview. In: Flick U, von Kardorff E, Steinke I (eds) A companion to qualitative research. Sage, London, pp 203–208

Huisman K (2008) Does this mean you're not going to come visit me anymore? An inquiry into an ethics of reciprocity and positionality in feminist ethnographic research. Sociol Inq 78:372–396. https://doi.org/10.1111/j.1475-682X.2008.00244.x

Jenkins J (2013) Organizing 'spaces of hope': union formation by Indian garment workers. Br J Ind Relat 51:623–643. https://doi.org/10.1111/j.1467-8543.2012.00917.x

Jenkins J (2015) The significance of grass-roots organizing in the garment and electrical value chains of Southern India. In: Newsome K, Taylor P, Bair J, Rainnie A (eds) Putting labour in its place: labour process analysis and global value chains. Palgrave Macmillan, London/New York, pp 195–212

Kelle U (2004) Computer-assisted analysis of qualitative data. In: Flick U, von Kardorff E, Steinke I (eds) A companion to qualitative research. Sage, London, pp 276–283

Kowal S, O'Connell DC (2004) The transcription of conversations. In: Flick U, von Kardorff E, Steinke I (eds) A companion to qualitative research. Sage, London, pp 248–252

Kuckartz U (2016) Qualitative Inhaltsanalyse: Methoden, Praxis, Computerunterstützung, 3rd edn. Grundlagentexte Methoden. Beltz Juventa, Weinheim, Basel

Kumar A (2014) Interwoven threads: building a labour countermovement in Bangalore's export-oriented garment industry. City 18:789–807. https://doi.org/10.1080/13604813.2014.962894

Lata LN (2020) Negotiating gatekeepers and positionality in building trust for accessing the urban poor in the Global South. QRJ Ahead-of-Print. https://doi.org/10.1108/QRJ-03-2020-0017

Laurier E (2016) Participant and non-participant observation. In: Clifford NJ, Cope M, Gillespie T, French S (eds) Key methods in geography. Sage, Los Angeles, London, New Delhi, and Singapore, pp 169–181

Laurila J (1997) Promoting research access and informant rapport in corporate settings: notes from research on a crisis company. Scand J Manag 13:407–418. https://doi.org/10.1016/S0956-522 1(97)00026-2

Lave J, Wenger E (1991) Situated learning: legitimate peripheral participation. Learning in doing. Cambridge University Press, Cambridge

Leavy P (2017) Research design: quantitative, qualitative, mixed methods, arts-based, and community-based participatory research approaches. The Guilford Press, New York and London

Legard R, Keegan J, Ward K (2011) In-depth interviews. In: Ritchie J (ed) Qualitative research practice: a guide for social science students and researchers. Sage, Los Angeles, CA, pp 139–169

López Ayala T (2018) Multi-level production of the local labour control regime in the Bangalore readymade garment cluster. In: Butsch C, Follmann A, Müller J (eds) Aktuelle Forschungsbeiträge zu Südasien (Geographien Südasiens, Band 10). xasia eBooks, Berlin, pp 20–23

Mandel JL (2003) Negotiating expectations in the field: gatekeepers, research fatigue and cultural biases. Singapore J Trop Geo 24:198–210. https://doi.org/10.1111/1467-9493.00152

Mattissek A, Pfaffenbach C, Reuber P (2013) Methoden der empirischen Humangeographie, 2nd edn. Das Geographische Seminar. Westermann, Braunschweig

Mayring P (2004) Qualitative content analysis. In: Flick U, von Kardorff E, Steinke I (eds) A companion to qualitative research. Sage, London, pp 266–275

McDowell L (2001) "It's that Linda again": ethical, practical and political issues involved in longitudinal research with young men. Ethics, Place Environ 4:87–100. https://doi.org/10.1080/136 68790124226

Meuser M, Nagel U (1994) Expertenwissen und Experteninterview. In: Hitzler R, Honer A, Maeder C (eds) Expertenwissen: Die institutionalisierte Kompetenz zur Konstruktion von Wirklichkeit. Vieweg+Teubner Verlag, Wiesbaden, pp 180–192

Mezzadri A (2017) Sweatshop regimes in the Indian garment industry. Cambridge University Press

Mikecz R (2012) Interviewing elites. Qual Inq 18:482–493. https://doi.org/10.1177/107780041244 2818

Nader L (1972) Up the anthropologist: perspectives gained from studying up. https://files.eric.ed. gov/fulltext/ED065375.pdf. Accessed 5 Apr 2022

Porter ME (2000) Location, competition, and economic development: local clusters in a global economy. Econ Dev Q 14:15–34. https://doi.org/10.1177/089124240001400105

Przyborski A, Wohlrab-Sahr M (2014) Qualitative Sozialforschung: Ein Arbeitsbuch, 4th edn. Oldenbourg Verlag, München, Lehr- und Handbücher der Soziologie

Rädiker S, Kuckartz U (2020) Focused analysis of qualitative interviews with MAXQDA, 1st edn. MAXQDA Press

Rose G (1997) Situating knowledges: positionality, reflexivities and other tactics. Prog Hum Geogr 21:305–320. https://doi.org/10.1191/030913297673302122

Ross S (2008) Social unionism and membership participation: what role for union democracy? Studies in Political Economy 81:129–157. https://doi.org/10.1080/19187033.2008.11675075

Rowe WE (2014) Positionality. In: Coghlan D, Brydon-Miller M (eds) The Sage encyclopedia of action research. Sage, Thousand Oaks, CA, p 628

Sabot EC (1999) Dr Jekyl, Mr H(i)de: the contrasting face of elites at interview. Geoforum 30:329–335. https://doi.org/10.1016/S0016-7185(99)00023-8

Schmidt C (2004) The analysis of semi-structured interviews. In: Flick U, von Kardorff E, Steinke I (eds) A companion to qualitative research. Sage, London, pp 253–258

Schuermans N (2017) Towards rigour in qualitative research: paper to be presented at Torino, 29 June. https://www.dist.polito.it/news/allegato/(idnews)/9384/(ord)/0/(idc. Accessed 21 Jan 2021

Sultana F (2007) Reflexivity, positionality and participatory ethics: negotiating fieldwork dilemmas in international research. ACME: Int E-J CritAl Geogr 6:374–385

Weiner-Levy N (2008) When the hegemony studies the minority—an Israeli Jewish researcher studies Druze women: transformations of power, alienation, and affinity in the field. Qual Inq 15:721–739. https://doi.org/10.1177/1077800408330343
WTO (2020) World trade statistical review 2020. https://www.wto.org/english/res_e/statis_e/wts 2020_e/wts2020_e.pdf. Accessed 25 Mar 2021
Yeung HW-C (2020) The trouble with global production networks. Environ Plan A:1–11
Yin RK (2014) Case study research: design and methods. Sage, Los Angeles

Part IV
Introduction of Empirical Case

Chapter 5
Situating the Bangalore Export-garment Cluster Within the Garment GPN

Abstract This chapter introduces the Bangalore export-garment cluster as the main case of this study and situates it with the broader structural context of the garment GPN. To this end, the chapter first outlines the historical and geographical development of the garment GPN as well as the power relations structuring it. In this context, the chapter identifies three subsequent trends that have characterised the garment GPN since the early 2000s: (1) the geographical consolidation of garment retailers' sourcing networks with a particularly strong growth of the industry in China and India during the 2000s; (2) thereafter, the emergence of new low-wage sourcing destinations in South and South-East Asia, and in Africa; and, most recently, (3) a selective shift towards 'near sourcing' by fast fashion retailers for higher value-added, time critical fashion garments. Thereafter, the chapter lays out the historical and geographical development of the export-garment industry in India and in Bangalore and gives an overview of the industrial relations in the cluster.

Keywords Garment production network · Garment industry · India · Bangalore · Geography · Power relations · Industrial relations

This chapter introduces the main case of this study: the Bangalore export-garment cluster. Given the relational perspective adopted in this study, the Bangalore export-garment cluster is conceptualised as one specific node within the broader garment GPN. The next section (5.1) first characterises the garment GPN focussing on its vertical dimension value chain dimension. Thereafter, I shift the focus to the horizontal dimension of the GPN and give an overview of the historical and geographical evolution of the Indian garment industry, the Bangalore export-garment cluster and industrial relations in the Bangalore export-garment industry (Sect. 5.2).

5.1 The Vertical Dimension: The Garment GVC

For the purpose of this study, the garment GPN is conceptualised as the relational network that emerges around garment GVCs, representing the vertical dimension of the GPN. Garment GVCs have traditionally been set up, coordinated and

© The Author(s) 2023
T. López, *Labour Control and Union Agency in Global Production Networks*,
Economic Geography, https://doi.org/10.1007/978-3-031-27387-2_5

controlled by large garment retailers or branded manufacturers with headquarters in the Global North. These retailers or branded manufacturers control the higher value-added processes of research and development, design and retail, and outsource the lower value-added, labour-intensive manufacturing process to independent suppliers located predominantly in the Global South. For this reason, the garment GVC has also been labelled as the prototype of a buyer-driven GVC, in which retailers or branded manufacturers act as 'lead firms' (Gereffi 1994).

5.1.1   Historical Development of the Garment GPN

GVCs in the garment sector first emerged in the 1970s, when US and EU garment retailers started to source garments from overseas manufacturers. During this time, Asia emerged as a central garment manufacturing hub for the global market (Gereffi 1999). Between the 1970s and the 1990s, the geographies of garment GVCs became increasingly fragmented due to two factors: the Multi-Fibre-Agreement and a dynamic of organisational succession among buyers (Gereffi 1999). The Multi-Fibre-Agreement, which was in place from 1974 to 1994, fixed quotas for the importation of ready-made garments produced in 'developed countries' into EU countries and into the US to protect the domestic garment industry in these countries. These import quotas led retailers to establish geographically fragmented supplier networks across Asia and the Caribbean, since the amount of garments that could be imported from each country was limited (Abernathy et al. 2006).

This dynamic of distributing sourcing activities over a range of countries was in turn characterised by a logic of organisational succession refers to the historical fact that usually retailers with lower quality and style demands—such as department stores with their own store brands—started to source from a specific country, with suppliers taking on the role of assemblers of inputs provided by retailers (Gereffi 1999). By carrying out assembly production, garment manufacturers learned the necessary skills of production planning, managing quality and dealing with overseas retailers. Thereby these garment manufacturers became attractive as suppliers for other retailers with higher quality and style demands, such as branded marketers and fashion retailers. Fashion retailers trained assembly manufacturers to also assume responsibility for sourcing production inputs and become Original Equipment Manufacturers (OEM). In Asia, the so-called tiger states, i.e. Hong Kong, Taiwan, South Korea and Singapore, were the first countries to assume the role of OEM suppliers for European and US branded retailers and marketers (Gereffi 1999).

'Upgrading' of suppliers from mere assembly production to OEM production involved several shifts in the practices that construct and structure the relationships between retailers and lead firms as suppliers. As opposed to department store retailers, who usually relate to suppliers through buying houses or trading companies, branded fashion retailers (such as GAP, Inditex or H&M) or branded marketers (such as Nike or Adidas) established direct sourcing relations with their suppliers. In these relations, branded retailers and marketers provide the design as well as specifications regarding production inputs and quality while suppliers take over the tasks of sourcing of inputs, dying, cutting and trimming the fabric, and producing the finished garment. Suppliers' upgrading process from assembly to OEM producers therefore enabled manufacturers to demand higher prices and thus increase their value capture. At the same time, however, suppliers' need for more qualified and skilled workers led to wage increases. As a result, suppliers acting as OEM producers became unattractive to department store retailers who compete mainly by price with other retailers as opposed to design or quality. Hence, department store retailers such as Walmart shifted their sourcing activities to countries where wages were still lower (Gereffi 1994).

5.1.2 Geography of the Garment GPN

As a result, the garment GPN is geographically highly diversified and fragmented (Dicken 2015). Since the phasing out of the Multi-Fibre Agreement in 2005, three trends have shaped the geographies and nature of lead firm-supplier relationships in the garment GPN: As a *first trend*, sourcing locations for garment retailers have consolidated with a particularly strong growth of the industry in China and India (Frederick and Gereffi 2011; Rasiah and Ofreneo 2009). This geographical consolidation of garment retailers' sourcing networks has also been driven by retailers' shift towards a fast-fashion business model since the beginning of the 2000s. Under the fast-fashion model, retailers transform the latest catwalk styles into affordable garments for mainstream consumers, introducing up to 24 new collections per year (Tokatli 2008). The need for shorter time-to-market and higher cost-efficiency under the fast-fashion business model has led garment retailers to introduce lean and agile supply chain strategies, in which flexible and demand-oriented production allows retailers to minimise stocks (Christopher et al. 2004). In this context, retailers have increasingly sought to consolidate their production networks and to concentrate business on a smaller number of larger and more capable suppliers. As a result, over the past two decades, large tier one suppliers have emerged in established garment producing countries such as China or India (Appelbaum 2008; Merk 2014). These tier one suppliers usually maintain networks of a large number of company-owned or contracted factories in one or more countries and often take over key logistical tasks for retailers, such as managing inventories (Azmeh and Nadvi 2013, 2014).

With suppliers in consolidated garment sourcing destinations such as China and India increasingly taking on higher value-added tasks, wage levels have also risen in these countries. Therefore, as a *second trend*, over the past decade, we can observe the emergence of new low-wage sourcing destinations in South and South-East Asia, such as Myanmar or Vietnam, but also in Africa (e.g. Ethiopia) as new nodes in the garment GPN (Bae et al. 2021; Whitfield et al. 2020). Given the competition by these new players, garment manufacturers in established sourcing destinations are under increasing pressure to introduce more capital-intensive, semi-automated production models to remain competitive (see also López et al. 2021).

Third and last, most recently, the emergence of new 'ultra-fast' pure online fashion retailers using regional supplier networks close to consumer markets has further increased pressures on garment retailers to increase the time and cost-efficiencies of their supply chains. Against this backdrop, traditional fast-fashion retailers have recently also increased their 'near sourcing' activities from destinations closer to consumer markets for higher value-added, time critical fashion garments (Berg et al. 2017).

Despite the tendencies of the garment industry to diversify again in geographical terms after the initial trend for concentration in a few Asian countries, up to date, Asia remains the most important macro-regional hub for garment OEM production: China, Bangladesh, Vietnam, India, Indonesia and Cambodia alone account for roughly 50% of global garment exports (WTO 2020).

After outlining the geographical evolution of the garment GPN, in the next section, I discuss the various power flows that structure lead firm-supplier relations in the garment GPN.

5.1.3 Power Relations in the Garment GPN

In light of the relational approach adopted in this study that understands GPNs as networks of relational processes structured by power flows (c.f. Dicken et al. 2001), in this section, I characterise the power flows inherent to lead firm-supplier relations within the garment GPN. To understand these power flows, we need to take a closer look at the specific practices through which retailers as lead firms set up, coordinate and control relationships with suppliers. In the garment value chain, retailers and brands as lead firms have the power to include or exclude suppliers from their networks of suppliers. To be included into the supplier pool of a specific retailer or brand, manufacturers need to undergo several technical and social audits. In these audits, auditors employed or hired by lead firms evaluate the production capacities of the manufacturers and assess whether the specific factory will be able to meet the required quality standards. Moreover, auditors assess whether suppliers meet specific minimum social standards that retailers define in their company-specific codes of conduct (Sum and Ngai 2005).

Relations between retailers as lead firms and manufacturers as suppliers are therefore characterised by largely unilateral power flows from retailers towards their suppliers, with retailers imposing a variety of technical and social production standards. Retailers can do so for two reasons: On the one hand, retailers concentrate and control essential resources—such as financial resources, access to consumer markets in the Global North, and specialised knowledge in international law, markets and management—enabling retailers to set up and coordinate global supplier networks. On the other hand, the widespread adoption of industrialisation strategies focussing on export-garment production by governments of newly industrialising countries allows retailers to shift their sourcing activities to lower wage countries, when wages in a specific country raise above a specific threshold and increased costs outweigh the benefits linked to the reliability of long-established supplier relations (c.f. Berg et al. 2017).

In light of the above-mentioned emergence of large tier one suppliers, several scholars have argued that we can observe changes in the power flows between retailers and suppliers. As large tier one suppliers increasingly occupy strategic positions in retailers' lean and agile value chains and take over strategic tasks (e.g. in design and logistics), researchers have proposed that we can observe a shift from largely asymmetrical power relations characterised by unilateral dependence of suppliers on retailers towards more balanced power relationships characterised by mutual dependence (see e.g. Azmeh and Nadvi 2014; Kumar 2019; Merk 2014). I maintain, however, that while this observation may apply to the relationships of lead firms with a limited number of core strategic suppliers, we need to be cautious not to over-generalise this observation for three reasons. First, there are indications that overall, power relations between garment retailers and their suppliers remain highly asymmetrical. Supplier relations in the garment GPN are still characterised by a state of hyper-competition in light of the general over-production in global garment markets where production usually exceeds demand (Anner 2019). This state of hyper-competition is further fuelled by the fact that—despite trends for consolidation—retailers' supplier networks usually still encompass several hundred factories in various countries, allowing retailers to play suppliers off against each other based on price.

Second, the shift towards more balanced or 'symbiotic' lead firm-supplier relationships in the garment GPN needs to be understood as geographically differentiated. As mentioned earlier, in recent years, retailers have re-discovered 'near sourcing' locations for higher value-added, more time-sensitive fashion items. At the same time, however, retailers maintain distant sourcing networks for less time-sensitive basic items. In light of this 'dual sourcing' strategy (Andersson et al. 2018), it is likely that the degree of coordination and mutual dependence in lead firm-supplier relations varies between near and distant sourcing networks. Whereas for more fashionable items time-to-market and flexible production are the predominant sourcing parameters, for basic items price remains the most important parameter. Therefore, large tier one suppliers in established distant sourcing countries, such as India or China, compete with the emerging garment sectors in lower wage countries such as Myanmar or Ethiopia. Symbiotic relationships thus seem to be more likely to emerge between large, strategic suppliers in near-sourcing networks, where close cooperation and coordination between lead firms and suppliers is necessary to ensure flexible adaptation of designs according to customer demand and on-time delivery.

Third and last, even where relationships between retailers and strategic tier one suppliers are shifting from suppliers' unilateral dependence on retailers towards mutual dependence, retailers as lead firms continue to dictate the terms and conditions that structure retailer-supplier relations. Retailers' continued position to unilaterally define terms and conditions has been exemplified most recently during the COVID-19 crisis: In reaction to nationwide lockdowns in consumer countries, fashion retailers unilaterally cancelled orders under referral to a 'force majeure' clause in contracts with suppliers (Brydges and Hanlon 2020). As a result, suppliers had to bear the costs for already purchased production inputs and wages for already produced orders, in many cases without receiving any compensation from retailers (Anner 2020).

In summary, I argue that the emergence of strategic suppliers and the related shifts in power and dependence relations between retailers and suppliers need to be understood as shaped by two rather contradictory dynamics. On the one hand, large tier one suppliers benefit from their new position as strategic partners of garment retailers since establishing long-term business relations grants them economic stability in terms of continuous order flows. Moreover, due to their capacity to process larger orders, these tier one suppliers can enhance their position within retailers' supply chains, e.g. by adopting new production technologies allowing for increased time and cost-efficiency. On the other hand, however, retailers largely maintain a position of power due to their ability to maintain large geographically dispersed supplier networks allowing them to unilaterally dictate the terms and conditions of exchange by keeping suppliers in a state of constant competition. Power imbalances between retailers as lead firms and suppliers persist for those suppliers concentrating on basic, less fashionable and lower value-added products, as is the case for manufacturers in the Bangalore export-garment cluster. In the Bangalore export-garment cluster, power relations between manufacturers and suppliers are hence shaped by two partially contradicting tendencies. On the one hand, the industry is dominated by large export-garment companies, which maintain factory networks of between 10 and 50 factories and act as strategic OEM suppliers for branded fashion and garment retailers. As a result, most companies in the Bangalore export-garment cluster have achieved relatively stable business relationships with retailers. On the other hand, power asymmetries, however. remain due to the specialisation of the cluster in casual men's wear, which generally comprises less time-intensive and lower value-added products. Accordingly, barriers to shifting production to lower wage locations are relatively low, allowing retailers to exercise significant price pressure on Bangalore garment manufacturers.

5.2   The Horizontal Dimension: The Indian Garment Industry and the Bangalore Export-garment Cluster

5.2.1   Situating the Bangalore Export-garment Cluster within the Indian Garment Industry

India has become integrated into the production networks of major transnational retailers as a supplier of apparel since the 1980s (Ramaswamy and Gereffi 2000). Coming from a long tradition of weaving industries and possessing a large rural labour reserve, India was among the first Asian countries to be included in Northern retailers' production networks after the initial East Asian Tiger States. Up until the

2000s, the export-garment industry's growth was still rather limited due to the industrial quota regime under the Multi-Fibre Agreement. Under this regime, the government reserved production quota especially for small-scale industries to preserve the labour-intensive character of the industry and thereby promote employment generation (Mezzadri and Srivasta 2015; Mezzadri 2017). The quota regime and a series of further government policies favouring small factories, led to a development pattern in the Indian garment industry that is characterised in India with the term 'unorganised' (Singh Yadav 2020: 15f.). As an 'unorganised' industry, the Indian garment industry has been traditionally characterised by a large number of predominantly small, often workshop-like factories with less than 100 or even less than 10 workers. Working and employment conditions in these factories were largely unregulated due to various exemptions for the applicability of general labour laws to factories with less than 10 or 100 workers, respectively. According to the Indian Factories Act of 1948, production establishments with less than 10 workers have, for example, traditionally been exempted from labour regulations regarding legal minimum wages and bonus payments, working hours and health and safety provisions. Factories with more than 10 but less than 100 workers, in turn, have traditionally been bound by these provisions but still enjoyed certain exemptions such as retrenching or laying off workers and closing their operations without prior government permission.[1] As a result, employment in the garment industry under the quota regime tended to be largely informal, unstable and insecure, with factories frequently closing down when they had fulfilled their allocated quotas (Kumar 2014).

Since the 2000s, the Indian garment industry has, however, undergone a range of transformations and strong growth stimulated by a shift in national policies. With the introduction of the National Textile Policy in 2000, the Indian government abolished the reservation of production quota for small-scale industries and allowed 100 per cent foreign investment. In subsequent policies, the government further introduced several incentives for capital investments in the garment sector, such as the 'Technological Upgradation Fund' and a 'Capital Subsidy Scheme' (Kalhan 2008). As a consequence of these policy changes, the Indian garment industry experienced a steep growth. Whereas in the year 2000, India still ranked 9th on the list of the world's top garment export countries, accounting for 2.8% of garments on the world market (WTO 2001: 154), in 2019, India was the 5th biggest garment exporting country in the world. Today, the Indian garment industry accounts for 3.5% of global exports (WTO 2020: 10) and employs over a million workers (ILO 2015). Besides having undergone significant growth over the past two decades, the Indian garment industry has also undergone a qualitative transformation. Since the end of the Multi-Fibre-Agreement, the industry has overall become more 'organised' due to increasing market consolidation and concentration. In this process, smaller factories have dwindled or shifted their business strategy to sub-contracted or domestic production and

---

[1] Under India's most recent labour law reform that has merged 29 labour laws into four so-called 'Codes', this threshold has been further elevated. According to the most recent version of the Code on Industrial Relations, which has been passed by the Indian Parliament in 2020, only factories with more than 300 workers now need to obtain permission by the Labour Department to retrench workers (Indian Ministry of Law and Justice 2020).

large business conglomerates owning several factories have concentrated increasing market shares in the export-garment industry. This increasing market concentration has also been driven by buyers' efforts to consolidate their supplier networks in the face of increased demands for social and technical production standards (Merk 2014). Responding to these demands, garment manufacturers have increasingly reorganised production and labour processes inside factories. Whereas during the quota regime, one worker was usually in charge of assembling a whole garment piece, over the past two decades, most tier one garment suppliers have introduced assembly line production with lean or flexible production systems.

Nevertheless, the Indian garment industry remains highly segmented, and there are large differences between 'organised' tier one suppliers and often still workshop-like tier two or three sub-contractors. Large tier one suppliers commonly outsource particularly labour-intensive tasks such as embroidery and embellishment, and surplus orders that exceed a factory's production capacity to networks of tier two and three sub-contractors (Mezzadri 2017: 34; AEPC 2009). As a result, the Indian Apparel Export Promotion Council (AEPC) found in a survey with garment manufacturers in ten garment export clusters that more than three quarters of all garment manufacturing units had less than 40 machines, usually equalling employment of less than 100 workers (AEPC 2009).

Furthermore, the Indian garment industry is geographically fragmented, with high levels of regional specialisation that also coincide with levels of 'organisation' of the industry. India's export-garment industry is concentrated in various local clusters, with the four biggest and most significant clusters being located in the New Capital Region (NCR) in and around New Delhi as well as in and around the cities of Chennai, Tiruppur and Bangalore located in the South Indian states of Tamil Nadu and Karnataka. Together, these clusters account for over half of India's garment exports (AEPC 2009: 8f.). Each cluster specialises in a different product type and shows distinctive industry organisation and workforce. Whereas the NCR garment cluster is specialised in embroidered women's clothes (Mezzadri 2008, 2012), the Tiruppur cluster in the state of Tamil Nadu is specialised in knitwear (Chari 2000; Neve 2012). In contrast, production in Chennai and Bangalore focusses predominantly on casual and formal men's wear, such as shirts, trousers and denim products (Mezzadri 2017: 80ff.). Product variations are also linked to different models of industrial organisation. In the NCR region, a smaller number of large tier-one factories co-exists with a large number of small, often informal workshops and home-based piece-rate workers carrying out manual embroidery and embellishment work for tier one suppliers. In Tirupur and Chennai, conversely, production is carried out predominantly in small and medium-sized factories, of which 80% have less than 100 workers and can therefore be characterised as unorganised (AEPC 2009: 12).

In comparison to the other three major export-garment clusters, Bangalore therefore stands out with a particularly high share of large tier one factories, which has earned it the reputation as India's most 'organised' garment cluster (RoyChowdhury 2005). According to a survey by Anner (2019), tier one suppliers in Bangalore employ, on average, 881 workers—a much higher number compared to tier one factories in the NCR region, which had an average of 361 workers according to the survey. Workers employed in Bangalore's tier one export-garment factories usually receive formal employment contracts and are paid according to the legally fixed minimum wage. In total, around 450,000 workers are directly employed in the Bangalore export-garment industry. As opposed to the NCR cluster where most workers are male migrant workers, in Bangalore, 85% of garment workers are women (Mezzadri 2017). Garment companies in Bangalore produce for all major US and EU branded fashion and garment retailers from the mid-price segment, such as American Eagle, G.A.P., H&M or Zara, but also for some higher priced brands such as Tommy Hilfiger, Hugo Boss or Calvin Klein (Kumar 2014). Given its specialisation in casual and formal men's wear, which are relatively low value-added products, the Bangalore export-garment cluster has been facing increasing competition from other Asian countries such as China and Bangladesh over the past years (AEPC 2009: 37).

5.2.2 Historical and Geographical Development of the Bangalore Export-garment Cluster

Bangalore has traditionally been a centre of large textile mills for silk and fabrics, making the city an attractive sourcing location for US and EU retailers in the emerging stages of the global garment industry in the 1980s. Besides the presence of a skilled workforce, its mild climate and convenient location with an international airport and relative proximity to the container shipping harbour of Madras (today Chennai) contributed to Bangalore's early inclusion into the global garment GPN. With rising globalisation and the first round of outsourcing in the garment industry driven by large retailers in the 1970s and 1980s, a ready-made garment sector started to evolve in Bangalore. The rapid development of the ready-made garment sector in subsequent decades was further catalysed by the decline of the textile mill industry: Due to the beginning economic liberalisation, which also included the liberalisation of import regulations, garment manufacturers were able to import cheaper fabrics and yarns from China, where automation had already progressed. In light of this new competition from China, many of the bigger branded textile mill companies started to shift their business towards garment production while outsourcing fabric production to small production centres run by formally self-employed workers (Kumar 2014). In the early years of the Bangalore garment industry, garment factories were generally still rather small—with usually less than 100 workers—and located in the central area of Bangalore. The workforce in these early garment factories was composed predominantly of men from the Bangalore urban area since industrial work was still a predominantly male domain (RoyChowdhury 2005).

With the economic liberalisation in the 1990s, which involved India's opening towards foreign direct investments, the Bangalore export-garment industry, however, underwent a first round of significant organisational and geographical restructuring. With the lift of the quota regime and buyers' increasing demands for shorter lead times and lower prices in the context of the shift towards a fast-fashion business model, garment production shifted from smaller workshops to larger factory settings with several hundred workers. The shift towards larger factory settings was also linked to a shift in production organisation towards semi-automated and assembly-line production models. In Bangalore's large tier one supplier factories, the pre-production steps of design, marker making and cutting have been increasingly automated and digitised through computer-aided technologies. At the same time, the sewing process is organised in a tightly controlled assembly line production model (Kalhan 2008). This organisational shift of the industry also went hand in hand with an increasing market concentration within the Bangalore export-garment sector, giving rise to a dozen large garment export company conglomerates dominating the market in the Bangalore export-garment cluster.

This restructuring of the market and of production organisation also had important implications for the required skill profile and hence for the social and geographical composition of the workforce. In the 'unorganised' workshops, workers usually assembled the whole garment and therefore required high levels of dexterity, skills and experience. In contrast, in the assembly line production system, each worker carries out one only specific stitch, such as sewing a sleeve or collar of a shirt to the body. Whereas skilled workforce is still needed in the pre-production process, e.g. to program and maintain computer-aided cutting machines, the largest share of the workforce employed in the assembly-line sewing process is now classified as unskilled or semi-skilled. This skill segmentation has also created new gender divisions in Bangalore garment factories. While higher-skilled tasks in the pre-production process are predominantly carried out by men, the larger share of unskilled or semi-skilled tasks in the sewing process are usually carried out by women (López et al. 2021).

Simultaneously with the shift towards large factory set-ups, the Bangalore garment industry underwent a geographical restructuring process. Due to the general expansion of the city and the bigger spaces needed for larger factory set-ups employing several hundred workers, the export-garment industry started to concentrate along major traffic axes leading out of Bangalore and in sub-urban industrial areas that were left vacant with the decline of public sector manufacturing (Kumar 2014). Therefore, today, factory units in the Bangalore export-garment clusters are located predominantly at the outskirts of the city along the major traffic axes Mysore Road and Hosur Road and in the industrial areas Peenya and Yelahanka. Figure 5.1 illustrates the geographical agglomerations of factories within the Bangalore export-garment cluster. The figure distinguishes between manufacturing factories that carry out all production steps from fabric cutting to assembly and processing factories that concentrate on dyeing, printing, washing or bleaching fabrics or finished garments.

As a result of this organisational and geographical restructuring, the workforce composition in the Bangalore export-garment cluster gradually shifted from skilled workers from the Bangalore urban area to include an increasing number of unskilled or semi-skilled women workers from rural areas. Today, most garment workers labouring in Bangalore's export-garment factories are women originating from rural areas in Karnataka, who have moved to Bangalore alone or with their families in search of employment. Most of these women are first-generation industrial workers

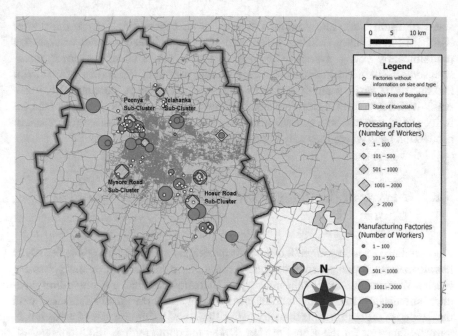

Fig. 5.1 Geographical sub-clusters of the garment industry in the Bangalore urban area. *Source* Data from own research/ H&M 2021; elaboration: Wilhelm Felk/ Tatiana López/ QGis

with limited formal school education—usually no further than the 8th or 9th grade. Age-wise, the average age span of women workers is 18 to 40 years since the high work pressure and repetitive movements usually make it impossible to work in the industry for a very long time.

Over the past decade, the Bangalore export-garment cluster has entered a second round of geographical and workforce restructuring characterised by two trends. First, increasing land and rent prices in the greater Bangalore area and the rapid growth of the service sector absorbing unskilled workers, have led garment manufacturers to move production facilities further out of the city. Garment factories have thus been increasingly shifted to semi-urban or rural areas along the major transport axes connecting Bangalore with smaller towns in rural Karnataka. Furthermore, various state subsidies aiming to foster industrial development and generate employment in the rural, so-called 'backward areas' of Karnataka have incentivised export-garment companies to set up large factories in various rural towns within the State of Karnataka or in newly created Textile and Apparel parks in up to 150 km distance from Bangalore. Seizing economies of scale and cheap land prices, these factories in rural towns and apparel parks are usually very large, employing more than 5000 workers (see Sect. 6.7).

As a second trend within the most recent round of industry restructuring, the Bangalore export-garment industry has seen an increasing influx from migrant workers not only from rural Karnataka and from the neighbouring Southern Indian states of Tamil Nadu and Andhra Pradesh but also increasingly from the poorer Northern Indian states Odisha, Assam and Jharkhand. These migrant workers are predominantly young women who migrate to Bangalore under the government-sponsored Skill India Program. As part of this program, the Indian government has since 2011 set up training centres in rural areas classified as 'economically backward' with a focus on the Northern states. In these training centres, young women are trained as sewing machine operators and then sent to the major urban export-garment production hubs where they stay in hostels (see Sect. 6.7).

5.2.3   Industrial Relations in the Bangalore Export-garment Cluster

The continued restructuring of the cluster's geography and workforce since the 1990s has also had consequences for the industrial relations in the Bangalore export-garment cluster. Whereas the textile mill industry in Bangalore still had a strong union presence, unionisation rates in the textiles and garment sector drastically decreased with the decline of the textile mills industry in the 1980s and the emergence of ready-made garment factories for the global market. Under the quota regime, the 'unorganised' character of the Bangalore export-garment industry posed significant challenges to India's central trade unions. Being affiliated with India's major political parties, central trade unions have traditionally focused on legal mechanisms such as tripartite industrial dispute settlement mechanisms to gain improvements for workers in 'organised' industrial sectors with formal employment and stable workforces, first and foremost in the public sector. However, with the economic liberalisation starting in 1991 and the related deregulation and flexibilisation of labour markets, central trade unions found themselves confronted with a growing number of 'unorganised' industries characterised by rather unstable and informalised employment—such as the textile and garment sector (Kumar 2014).

In Bangalore, the Centre of Indian Trade Unions (CITU) and the All India Trade Union Confederation (AITUC)—two of the more militant trade unions among India's central trade unions, with political affiliation to the Indian communist parties—had undertaken various attempts to negotiate collective bargaining agreements with garment manufacturers in the 1980s and 1990s. However, none of these attempts were met with success for two reasons: First, given the regulatory void for small production units with less than 100 workers, manufacturers could quickly close down production units and re-open them in another location in response to unions' organising attempt. Second, given the lower market concentration in the early decades of the export-garment industry, even larger garment companies with more than 100 workers did not have the financial capacity to survive a prolonged strike by workers. Accordingly, in the collective memory of Bangalore trade unions, even today the case of a month-long strike organised by CITU in a garment company called 'Asoke' is present which ended in the company closing down (RoyChowdhury 2005).

With the consolidation of the industry and the emergence of larger production units starting from the 2000s, companies had increased financial resilience. However, new challenges for organising workers arose due to the geographical sprawl of the industry and the increasing de-skilling and feminisation of the workforce. Firstly, the low skill requirements of assembly line production systems and new labour supply from rural women entering the labour market decreased workers' market-place bargaining power. Second, the social profile of women workers as mostly first-generation industrial workers from rural areas who were unfamiliar with unionisation made it difficult for established, central trade unions to organise women workers. In light of these challenges, central trade unions ceased active organising efforts in the Bangalore garment industry after several unsuccessful attempts in the 1990s (Kumar 2014).

Given the lack of involvement from established trade unions and a deliberate 'laissez-faire' approach by the Indian government towards the garment industry, work in the Bangalore export-garment industry was characterised by large-scale labour rights violations in the 1990s and early 2000s. Basic legally prescribed labour standards, such as payment of minimum and overtime wage, and basic facilities, such as creches, medical centres or drinking water, were non-existent in many facto-ries. These conditions were, however, not unique to the Bangalore export-garment industry but rather prevalent in most Asian garment-export industries. Against this background, by the end of the 1990s, anti-sweatshop movements in consumer coun-tries started to attract international attention to exploitative working conditions in the Asian garment industry, forcing garment retailers and brands to introduce social standards and auditing mechanisms. With this international attention to working conditions in the garment industry, a market was created for local NGOs in produc-tion countries to engage in social activism in the garment sector through projects sponsored by Northern NGOs. In this context, two local NGOs called Cividep and FEDINA started organising garment workers in Bangalore through a community organising approach (see also Jenkins 2013).

The community organising approach, however, had limits for creating better working conditions since, as NGOs, Cividep and FEDINA could not legally repre-sent women workers before the management or state organs in work-related conflicts. Consequently, in 2006, NGO activists and garment workers registered the first trade union in Bangalore with the specific mandate to organise workers in the garment sector: the Garment and Textile Workers Union (GATWU). As a union that grew out of an NGO-led organising project, GATWU is exemplary of a new type of 'indepen-dent' trade union in India that is neither affiliated to nor financed by a political party. GATWU, as a union at the state industry level, is affiliated to the New Trade Union Initiative (NTUI) at the national level—India's first independent trade union federa-tion characterised by a broadly socialist ideology but without ties to any particular political party (Gross 2013). In 2009, and 2012, two further independent trade unions for the Bangalore export-garment sector were created through splits from GATWU: the Karnataka Garment Workers Union (KGWU) and the Garment Labour Union

(GLU), respectively. Whereas KGWU is not affiliated with any trade union organisation at a higher level, GLU has affiliated itself with the central trade union federation Hind Mazdoor Sabha (HMS) in 2017. Initially founded by the Socialists in 1984, today, HMS considers itself politically independent and committed to democratic socialism. It is considered India's most pragmatic central trade union federation (Höllen 2010).

A central issue for the split of the three unions was their dependence on financial means from NGOs to support full-time union organisers: Whereas GATWU decided to cut all financial ties with international and local NGOs in the form of project work in 2011, KGWU and GLU continued maintaining close ties with local and international NGOs. The latter provide the primary financial means for their organising work through project funds. According to their own accounts, at the time of data collection, GATWU, GLU and KGWU together had a membership of about 19,000 workers in the garment industry in Karnataka (see Table 5.1).

Given that around 450,000 workers are directly employed in the Karnataka garment industry, the overall membership of the three garment unions amounts to a unionisation rate of around 4%. Despite this relatively low unionisation rate, the three garment unions have achieved significant improvements for Bangalore garment workers. By exerting pressure on politicians and employers to implement legally prescribed regular minimum wage revisions, trade unions have achieved significant wage increases for workers over the past years. Moreover, by negotiating with management and filing complaints to the labour and factory departments, unions have achieved basic health and safety standards, such as drinking water, ventilation and medical facilities in most factories. Nevertheless, wages in the Bangalore garment sector remain at around 10,000 Rupees per month (approx. 130 US$)—representing the legal minimum wage for the Karnataka garment industry—significantly below a subsistence level.

Table 5.1 Overview of main characteristics of three Bangalore-based garment unions

Name of trade union	Year of foundation	Political affiliation	Affiliation to central trade union federation	Number of members in the garment industry
Karnataka Garment and Textile Workers Union (GATWU)	2006	Independent	New Trade Union Initiative (NTUI)	10,000
Karnataka Garment Workers Unions (KGWU)	2009	Independent	None	3000
Garment Labour Union (GLU)	2012	Independent	Hind Mazdoor Sabha (HMS)	6000
Total members				**19,000**

Source Unions' own reports, 2019

The Indian Trade Union Act of 1948[2] allows trade unions to negotiate individual collective bargaining agreements with managements in which wages beyond the legally prescribed minimum wage and other gratuities beyond the legally mandatory ones may be fixed. To date, the only collective bargaining agreement in the Bangalore export-garment industry has been signed by GATWU with the multinational garment label producer Avery Dennison. Generally, in most workplaces, relationships between unions and management are rather informal, with unions not being recognised as official bargaining partners by management. Moreover, union busting practices such as victimising or dismissing union members or worker leaders at the workplace are widespread in the cluster, hampering unions' abilities to build associational power resources (for more details, see Sect. 6.5). As a result, in the few cases where unions have established (albeit informal) working and bargaining relationships with the management, these achievements have, in most cases, been won with help of additional extra-local pressure from transnational consumer networks and brands.

It is important to note that the three unions in this case study represent a specific type of union within the broader Indian union landscape. Traditionally, the Indian union landscape has been dominated by five large central union federations: the left AITUC and CITU, the social-democrat HMS, the centrist Indian Trade Union Congress (INTUC) and the Hindu-nationalist Bharatiya Mazdoor Sangh (BMS, engl. Indian Workers Union) (Höllen 2010). Except for the HMS, these unions serve as labour wings of specific political parties. Close relations with politicians have historically represented a key source of power for India's biggest trade unions to extract concessions for workers from the state as a provider of social welfare and as an employer in India's large public sector. With the shift to neoliberal policies in India's post-liberalisation period and the privatisation of large parts of India's public industries, the central trade union federations have, however, been confronted with a decline of their historical political power and of their membership base (Ferus-Comelo 2007). The 'laissez-faire' attitude of the state towards employers in the export industries and the overall weakening of public labour institutions have eroded the traditional power sources of India's central trade unions. In the face of their traditional focus on the public sector and often strongly hierarchical and bureaucratic organisational structures that are frequently dominated by male and rent-seeking leaderships, central trade unions were ill-equipped for organising the growing share of first-generation industrial workers from rural backgrounds in India's growing export-garment industry (RoyChowdhury 2005; see also Sherlock 2003).

Against this backdrop, the three local garment unions in this case study were founded by local civil society and worker activists as independent trade unions without political ties and instead with strong ties to local labour rights NGOs. The independent character of the three case study unions opened up new opportunities for

---

[2] The provisions of the Indian Trade Union Act of 1948 have in 2020 been integrated into the Industrial Relation Code, which however kept most provisions from the Trade Union Act intact. The most significant change concerns the organising threshold allowing unions to form a collective bargaining committee at the factory level, which has been raised from 10 to 20%.

developing alternative organising strategies that were not centred on economic and workplace issues but instead addressed garment workers problems more holistically (Jenkins 2013). On the other hand, these unions also face heightened challenges for building bargaining power vis-à-vis employers and the state due to their lack of political influence and limited financial resources. As a result, all three Bangalore-based garment unions have developed networked agency strategies and built alliances with worker organisations and NGOs from the Global North to harness coalitional power resources.

With this taken into consideration, in the next chapter, I analyse the labour control regime in the Bangalore export-garment industry to illustrate the various conditions that constrain unions' capacities to build bargaining power in more detail. I demonstrate how the biggest constraint for unions to build bargaining power vis-à-vis employers results from the complex nature of the labour control regime, which emerges from a web of intersecting social and economic relations that connect actors across various distances.

References

Abernathy FH, Volpe A, Weil D (2006) The future of the apparel and textile industries: prospects and choices for public and private actors. Environ Plan A 38:2207–2232. https://doi.org/10.1068/a38114

AEPC, Apparel Export Promotion Council (2009) Indian apparel clusters: an assessment. AEPC, New Delhi

Andersson J, Berg A, Hedrich S, Ibanez P, Janmark J, Magnus K-H (2018) Is apparel manufacturing coming home?: nearshoring, automation, and sustainability—establishing a demand-focused apparel value chain. file:///C:/Users/wmc205/Downloads/Is-apparel-manufacturing-coming-home_vf.pdf

Anner M (2019) Predatory purchasing practices in global apparel supply chains and the employment relations squeeze in the Indian garment export industry. Int Labour Rev 158:705–727. https://doi.org/10.1111/ilr.12149

Anner M (2020) Squeezing workers' rights in global supply chains: purchasing practices in the Bangladesh garment export sector in comparative perspective. Rev Int Polit Econ 27:320–347. https://doi.org/10.1080/09692290.2019.1625426

Appelbaum RP (2008) Giant transnational contractors in East Asia: emergent trends in global supply chains. Compet Chang 12:69–87. https://doi.org/10.1179/102452908X264539

Azmeh S, Nadvi K (2013) 'Greater Chinese' global production networks in the middle east: the rise of the jordanian garment industry. Dev Chang 44:1317–1340. https://doi.org/10.1111/dech.12065

Azmeh S, Nadvi K (2014) Asian firms and the restructuring of global value chains. Int Bus Rev 23:708–717. https://doi.org/10.1016/j.ibusrev.2014.03.007

Bae J, Lund-Thomsen P, Lindgreen A (2021) Global value chains and supplier perceptions of corporate social responsibility: a case study of garment manufacturers in Myanmar. Global Netw 21:653–680. https://doi.org/10.1111/glob.12298

Berg A, Hedrich S, Lange T, Magnus K-H, Mathews B (2017) The apparel sourcing caravan's next stop: digitization: McKinsey Apparel CPO Survey 2017. https://www.mckinsey.com/~/media/mckinsey/industries/retail/our%20insights/digitization%20the%20next%20stop%20for%20the%20apparel%20sourcing%20caravan/the-next-stop-for-the-apparel-sourcing-caravan-digitization.pdf. Accessed 20 August 2020

Brydges T, Hanlon M (2020) Garment worker rights and the fashion industry's response to COVID-19. Dialogues Hum Geogr 10:195–198. https://doi.org/10.1177/2043820620933851

Chari S (2000) The Agrarian origins of the knitwear industrial cluster in Tiruppur, India. World Dev 28:579–599. https://doi.org/10.1016/S0305-750X(99)00143-6

Christopher M, Lowson R, Peck H (2004) Creating agile supply chains in the fashion industry. Int J Retail Distrib Manag 32:367–376. https://doi.org/10.1108/09590550410546188

de Neve G (2012) Fordism, flexible specialization and CSR: How Indian garment workers critique neoliberal labour regimes. Ethnography 15:184–207. https://doi.org/10.1177/1466138112463801

Dicken P (2015) Global shift: mapping the changing contours of the world economy. Guilford Pr, New York, NY

Dicken P, Kelly PF, Olds K, Yeung HW-C (2001) Chains and networks, territories and scales: towards a relational framework for analysing the global economy. Global Netw 1:89–112

Ferus-Comelo A (2007) Unions in India at critical crossroads. In: Phelan C (ed) Trade union revitalisation: trends and prospects in 34 countries, 1st edn. Peter Lang UK, Bern, pp 475–488

Frederick S, Gereffi G (2011) Upgrading and restructuring in the global apparel value chain: why China and Asia are outperforming Mexico and Central America. Int J Technol Learn, Innov Dev 4:67–95

Gereffi G (1994) The organization of buyer-driven global commodity chains: how U.S. retailers shape overseas production network. In: Gereffi G, Korseniewicz M (eds) Commodity chains and global capitalism. Praeger Publishers, Westport, pp 95–122

Gereffi G (1999) International trade and industrial upgrading in the apparel commodity chain. J Int Econ 48:37–70

Gross T (2013) "Raising the voice of workers in global supply chains": global leverages in the organising strategies of three 'new' labour unions in the Indian garment sector. RLS South Asia Working Paper Series. https://www.rosalux.de/fileadmin/rls_uploads/pdfs/engl/GROSS_Tandiwe_Raising-the-voice_160217.pdf. Accessed 31 August 2016

Höllen F (2010) Die indischen Gewerkschaften der Linken: Ein aktueller Überblick mit thematischen Schwerpunkten. RLS South Asia Working Paper Factories December 2010. Thought Factory, vol 01. Rosa Luxemburg Stiftung, Berlin/Pune

ILO (2015) Insights into working conditions in India's garment industry: International Labour Office, fundamental principles and rights at work (fundamentals). ILO Publications, Geneva

Jenkins J (2013) Organizing 'spaces of hope': union formation by Indian garment workers. Br J Ind Relat 51:623–643. https://doi.org/10.1111/j.1467-8543.2012.00917.x

Kalhan A (2008) Permanently temporary workers in the global ready-made garment hub of Bangalore. Indian J Labour Econ 51:115–128

Kumar A (2014) Interwoven threads: building a labour countermovement in Bangalore's export-oriented garment industry. City 18:789–807. https://doi.org/10.1080/13604813.2014.962894

Kumar A (2019) Oligopolistic suppliers, symbiotic value chains and workers' bargaining power: labour contestation in South China at an ascendant global footwear firm. Global Netw 19:394–422. https://doi.org/10.1111/glob.12236

López T, Riedler T, Köhnen H, Fütterer M (2021) Digital value chain restructuring and labour process transformations in the fast-fashion sector: evidence from the value chains of Zara & H&M: online first. Global Netw:1–17. https://doi.org/10.1111/glob.12353

Merk J (2014) The rise of tier 1 firms in the global garment industry: challenges for labour rights advocates. Oxf Dev Stud 42:277–295. https://doi.org/10.1080/13600818.2014.908177

Mezzadri A (2008) The rise of neo-liberal globalisation and the 'new old' social regulation of labour: a case of Delhi garment sector. Indian J Labour Econ 51:603–618

Mezzadri A (2012) Reflections on globalisation and labour standards in the Indian garment industry: codes of conduct versus 'codes of practice' imposed by the firm. Global Labour J 3:40–62. https://doi.org/10.15173/glj.v3i1.1112

Mezzadri A (2017) Sweatshop regimes in the Indian garment industry. Cambridge University Press

Mezzadri A, Srivasta R (2015) Labour regimes in the Indian garment sector: capital-labour relations, social reproduction and labour standards in the National Capital Region. CDPR SOAS

Ramaswamy KV, Gereffi G (2000) India's apparel exports: the challenge of global markets. Dev Econ 38:186–210

Rasiah R, Ofreneo RE (2009) Introduction: the dynamics of textile and garment manufacturing in Asia. J Contemp Asia 39:501–511. https://doi.org/10.1080/00472330903076727

RoyChowdhury S (2005) Labour activism and women in the unorganised sector: garment export industry in Bangalore. Econ Pol Wkly 40:2250–2255

Sherlock S (2003) Labour and the remaking of Bombay. In: Brown A, Hutchison J (eds) Organising labour in globalising Asia. Taylor and Francis, Hoboken, pp 152–172

Singh Yadav S (2020) Exploring the competitiveness of home-based women workers in Delhi: a case study of the garment and apparel sector. https://kups.ub.uni-koeln.de/11375/1/Doktorarbeit-Satyendra%20Singh%20Yadav-final-KUPS.pdf. Accessed 31 December 2021

Sum N-L, Ngai P (2005) Globalization and paradoxes of ethical transnational production: code of conduct in a Chinese workplace. Compet Chang 9:181–200. https://doi.org/10.1179/102452905 X45427

Tokatli N (2008) Global sourcing: insights from the global clothing industry the case of Zara, a fast fashion retailer. J Econ Geog 8:21–38. https://doi.org/10.1093/jeg/lbm035

Whitfield L, Staritz C, Morris M (2020) Global value chains, industrial policy and economic upgrading in Ethiopia's apparel sector. Dev Chang 51:1018–1043. https://doi.org/10.1111/dech.12590

WTO (2001) International trade statistics 2001: trade by sector. https://www.wto.org/english/res_e/statis_e/its2001_e/chp_4_e.pdf. Accessed 25 March 2021

WTO (2020) World Trade Statistical Review 2020. https://www.wto.org/english/res_e/statis_e/wts 2020_e/wts2020_e.pdf. Accessed 25 March 2021

Part V
Empirical Analysis

Chapter 6
A Relational Analysis of the Labour Control Regime in the Bangalore Export-garment Cluster

Abstract This chapter applies a practice-oriented, relational analytical approach to labour control regimes in GPNs to the empirical case of the Bangalore export-garment cluster. It illustrates how the labour control regime in the Bangalore export-garment cluster emerges from the intersection of six different sets of processual relations with the labour process: sourcing relations, wage relations, workplace relations, industrial relations, employment relations and labour market relations. For each set of relations, the chapter reveals the specific exploiting and disciplining practices performed by actors at various levels, which together constitute structural labour control relations. These practices include inter alia Bangalore garment managers' production targeting, union-busting and wage theft practices, garment retailers' predatory purchasing practices, and employers' and state actors' practices of constructing a complex multi-level training and migration regime to secure adequate labour supply. In the face of this complex mesh of labour control practices, the chapter highlights the various constraints and challenges for local garment unions to build and activate associational and institutional power resources.

Keywords Bangalore · Garment industry · Labour control regime · Labour process · Labour market · Employment relations · Industrial relations

In this chapter, I apply the practice-oriented, relational analytical approach for analysing labour control regimes in GPNs developed in Chapter 3 to the empirical case of the Bangalore export-garment cluster. I illustrate how the labour control regime in the Bangalore export-garment cluster emerges from the intersection of six different sets of processual relations with the labour process. These processual relations are (1) sourcing relations, (2) wage relations, (3) workplace relations, (4) industrial relations, (5) employment relations and (6) labour market relations. As laid out in Sect. 3.2, each relation is constituted through various networked practices that fulfil three functions regarding the labour process: They ensure the generation of surplus value in the labour process (➜ exploiting practices); they secure the smooth reproduction of the labour process by mitigating the class conflict inherent to capitalist production (➜ disciplining practices); and they create the broader social relations and conditions required for the continued reproduction of the labour process, e.g. through ensuring that adequate labour supply is available. In the following, I

T. López, *Labour Control and Union Agency in Global Production Networks*, Economic Geography, https://doi.org/10.1007/978-3-031-27387-2_6

illustrate the practices through which the labour process and each set of processual relations are constructed in the Bangalore export-garment cluster. Moreover, I discuss implications for and interrelations with union agency. On the one hand, I show how the complex interplay of manifold exploiting and disciplining practices constrains the capacities of local garment unions to build and activate associational and institutional power resources and thereby to build sustained bargaining power vis-à-vis employers. On the other hand, I also shed light on the 'small transformations' that Bangalore garment unions have achieved so far through contesting and stopping selected exploiting and disciplining practices.

Section 6.1 introduces the practices through which the *labour process* is constructed in the Bangalore export-garment industry. It is followed by the analysis of the specific practices through which retailers construct *sourcing relations* with Bangalore export-garment manufacturers. These sourcing relations are characterised by asymmetrical power relations that enable retailers to unilaterally dictate terms and conditions of exchange (Sect. 6.2). Section 6.3 then lays out the various practices through which employers and state actors construct wage relations as de facto exploiting relations that ensure maximum surplus generation in the labour process through keeping wages low. Next, in Sect. 6.4, I discuss how Bangalore garment manufacturers construct *workplace relations* as de facto disciplining relations through constructing tightly controlled workplaces and through creating gendered and segregated shop floors. Section 6.5 then turns to *industrial relations* in the Bangalore export-garment cluster and illustrates the interplay of employers' union-busting practices at the workplace level, which are enabled by state actors' pro-business practices when performing the legal-institutional industrial dispute settlement process. After that, Sect. 6.6 illustrates the various practices through which Bangalore export-garment manufacturers construct flexibilised and informalised *employment relations*. Lastly, in Sect. 6.7, I reveal the various practices through which Bangalore employers territorially expand labour market relations to secure adequate labour. The chapter ends with an interim conclusion summarising the main findings and discusses implications from the analysis of the labour control regime for union agency in the Bangalore export-garment cluster.

6.1   Labour Process in the Bangalore Export-garment Cluster

The labour process in the Bangalore export-garment cluster (i.e. the networked practices and relations involved in producing a garment) is highly localised since it is tied to the material setting of predominantly large factory buildings. Commonly, export-garment companies maintain local networks of several factories. Typically, each factory is specialised in a specific step of the production process, such as fabric dyeing and cutting, assembling the final garment, processing (involving, e.g. dyeing techniques to produce a 'used look') or finishing. However, over the past years also several ultra-large factories have emerged that centralise all steps of the production process—from fabric dyeing and cutting to assembling the final garment, washing and packing—within one building complex. In this sense, the spatiality of the labour process in the Bangalore export-garment cluster differs, for example, from the spatiality of the labour process in the NCR region. There, the production process is geographically more fragmented, linking workers labouring in-factories and piece-rate workers carrying out embellishment tasks from home or in informal workshop settings. In contrast, in the Bangalore export-garment industry, the labour process is constituted through practices and relations that link workers, managers and machines predominantly within factory buildings.

Within these factories, the labour process is generally structured according to practices of Taylorist work organisation. Taylorist practices include dividing the production process into various standardised steps. In Bangalore export-garment factories, the production process starts with designing marker patterns and cutting the fabric—which is usually bought from external suppliers—into specific pieces of cloth, such as bodies, sleeves, collars or cuffs. These pieces are then assembled by sewing machine operators (in the same or in a different factory building) in an assembly line system. In this assembly line system, each machine operator usually carries out only one specific task, such as sewing a shirt sleeve to the collar (see Fig. 6.1).

After the final garment is assembled, it may undergo another step of processing depending on the type of garment. For denim products, one main product category

Fig. 6.1 Assembly line in Bangalore export-garment factory. *Source* Photo by Hubert Thiemeyer; stylised by author to ensure anonymity of depicted persons and company

in the Bangalore garment export cluster, a common processing step is treating the garment with automated laser technology or manually with sandpaper to create a 'distressed' or 'used' look. In the final production step, 'checkers' then cut possible hanging threads and conduct a final quality check. From the checking station, approved garments are carried to the finishing section, where they are prepared for shipping. Preparing garments for shipping may involve various steps depending on retailers' specific requirements. Generally, with fast-fashion retailers' introduction of 'lean store' systems, Bangalore manufacturers have taken over increasing tasks to prepare garments for sale, such as attaching price labels and security tags, ironing garments and placing garments on hangers in some cases (INT53). As the last step, garments are 'packed' in boxes by packers and transported to a warehouse from where garments are shipped when a production order has been completed.

Given buyers' price squeeze and demand for ever-shorter lead times, managers must maximise labour process efficiency to ensure the capture of surplus value. In this light, Bangalore manufacturers have, over the years, developed two sets of practices to maximise production efficiency in the labour process while at the same time keeping labour costs down: labour process automating and production targeting. These sets of practices will be introduced in the remainder of this chapter.

6.1.1   Hedging the 'Indeterminacy of Labour' Through Labour Process Automating

In the designing and cutting parts of the labour process, manufacturers have substituted significant parts of human labour through computer-aided design and cutting technology. Computer-aided design is used, for example, for marker making with software that automatically calculates marker patterns, which are then transferred digitally to semi-automated laser cutting machines. With the introduction of these technologies, manufacturers have not only been able to speed up the design and cutting process but to reduce labour costs as well, since marker design and cutting traditionally required relatively skilled workers with higher wages. With the introduction of computer-aided design and manufacturing technologies, the number of workers working in these departments could be significantly reduced. In the sewing process, which concentrates the largest share of workers, a complete substitution of human labour through automation is so far not viable due to the high investment costs. Manufacturers have, however, introduced semi-automated machines for complex stitches that require high levels of experience and dexterity, such as J-stitches for trousers, welt pocket stitches or bottom hemstitches. These machines require human operators to place the fabric correctly in the machine, which is then automatically moved under the needle. To place the fabric in the machine correctly, workers only require a short introduction instead of the extensive on-the-job experience needed to perform the earlier-mentioned complex stitches manually.

In this light, labour process automation fulfils two central functions with regard to ensuring the reproduction of the labour process in its profit-maximising form: First, by automating complex tasks within the labour process that are time-intensive and require high levels of tacit knowledge, garment manufacturers achieve to hedge the 'indeterminacy' of labour (c.f. Thompson 2010) by reducing human error rates and resulting disturbances to the production flow. Hedging the indeterminacy of labour is of particular importance for export-garment manufacturers in light of fashion retailers' growing demands for shorter lead times, flexibility in styles and lower prices. To ensure that production orders are completed within the time frames set by buyers, manufacturers need to minimise fluctuations in productivity levels. In this light, (semi-)automation of complex production steps also helps manufacturers avoid fluctuations in productivity that may arise from workers' learning curves and initially lower productivity when switching to a new style. In addition, (semi-) automation enables manufacturers to increase the speed of the labour process and thereby shorten lead times. Bangalore export-garment manufacturers' practices of automating production processes hence need to be seen as being directly shaped by retailers' sourcing practices (see Sect. 6.2).

Second, (semi-)automating production processes also enable garment manufacturers to reduce labour costs by hiring less skilled workers. Whereas skilled workers are needed to operate and maintain digital cutting machines in the automated cutting process, they are less in numbers. In the sewing process, the introduction of semi-automated sewing machines has enabled manufacturers to recruit more workers without previous experience in factory work, classifying them as unskilled workers (INT53). By reducing the number of skilled or semi-skilled workers in the factory, employers are, on the one hand, able to reduce wage costs and balance out statutory minimum wage increases. On the other hand, manufacturers can ensure continued labour supply through hiring migrant workers from rural areas in face of a tight local labour market. In this sense, the role of manufacturers' automating practices in ensuring the reproduction of the labour process can be best understood when looking at the intersections of the labour process with wage relations (see Sect. 6.4) and labour market relations (see Sect. 6.7).

Nevertheless, it is important not to overstate the significance of automation processes in the Bangalore export-garment cluster. Up to day, in the sewing process comprising the highest share of labour in the production process, automation remains marginal and limited to the earlier-mentioned semi-automatic sewing machines for complex stitches. The gross of stitches continues to be carried out by workers on industrial sewing machines. Against this background, Bangalore garment manufacturers implement a set of practices summarised here under the notion of 'production targeting' to tightly monitor and standardise worker performance as a second strategy for hedging the 'indeterminacy of labour'.

6.1.2   Controlling Worker Performance Through 'Production Targeting'

'Production targeting' can be regarded as the central exploiting mechanism—i.e. the central mechanism for ensuring the extraction of surplus value from living labour—in Bangalore export-garment factories. Production targeting comprises a set of practices that structure the labour process, including production target calculating, monitoring and enforcing. *Production target calculating* firstly involves defining standardised performance times for each task involved in the garment assembly process. On the one hand, these standardised performance times align the various tasks involved in the assembly of a specific garment to guarantee a continuous workflow and minimise idle time for each worker (INT27). On the other hand, calculating standardised performance times for each task serves as the basis for defining hourly and daily production targets for workers to ensure that a previously fixed daily production output is reached. The specific practices through which production targets are calculated contribute to maximising surplus extraction from workers: Production targets are calculated based on the technical capacities of machines and standardised motion times for workers when doing a stitch, while additional working steps, such as placing the fabric aside, are not considered. As a result, production targets in most factories are impossible to achieve, as the quote of this garment worker illustrates:

> The calculated production targets are kind of impossible to fulfil. For example, I stitch sleeves now. The production target is 60 per hour, so that is one per minute. But with putting aside the piece, I can actually only do 40 in an hour. But the supervisors won't see that and they will scold us and say: "The production target has been carefully calculated, so it is possible, why can't you do it? No other worker has problems, only you cannot do it." (INT5, translated from Kannada)

In addition, production targets are calculated based on the unrealistic assumption of steady productivity levels for workers. This assumption ignores performance fluctuations due to learning curves and fatigue levels as well as due to workers' varying experience and physical conditions. The following quotes from a group interview with Bangalore garment workers illustrate the adverse effects for workers resulting from the imposition of standardised production targets on workers:

> Worker 1: I cannot move my right leg so well, so I have to work the machine with my left leg. But since the machine is made to be used with the right leg, the pedal is on the wrong side, and it is hard for me to use my left leg. And so, my left leg hurts a lot after some time. But even under these conditions, I still have to fulfil the same production targets as the other workers. (INT5, translated from Kannada)

> Worker 2: Sometimes the supervisor will make us switch machines and then we have to do a new shape, so at the beginning, it takes longer. Also, often workers are only trained to do sleeves, for example. But then the supervisor will shift them to a machine where they have to do collars and, obviously, the workers don't know how to do that. But still they have to fulfil the production targets. So, it is very stressful. (INT5, translated from Kannada)

As a result of this specific practice of calculating production targets based on machine times as a primary indicator, paired with unrealistic assumptions of uniform worker performance, workers usually have to skip breaks to finish their daily production targets within regular working hours. In a survey with 126 garment workers from the Bangalore area conducted by the Centre for Workers' Management between December 2014 and March 2015, 60% of workers stated that they needed to work during their lunch break. Moreover, 97% of workers stated that they skip tea and toilet breaks to complete their production targets within normal working hours (CWM 2014: 23). If workers cannot finish production targets within regular hours, they are ordered to work unpaid overtime hours to complete their targets (INT34). The work pressure exerted on workers through the practice of 'production targeting' places workers under severe psychological stress and has detrimental effects on workers' physical health. As a result of skipping breaks, many workers suffer from repetitive strain injuries or kidney damage.

Against this background, to ensure that workers reach their daily targets, tight *production target monitoring and enforcing* are necessary. Production target monitoring is performed in most Bangalore factories manually by counting the output of finished garments per batch, that is per production line. At the end of every working hour, the number of finished garments for each batch is noted on a board and visualised for all workers (see Fig. 6.2).

To achieve the hourly production targets, there must be no bottlenecks in the assembly line to avoid a garment getting stuck due to individual workers not meeting the calculated processing time for this particular task. Hence, supervisors closely monitor workers' performance by selectively measuring the time a worker takes to complete a specific stitch with a stopwatch, for example. Whereas manual worker performance monitoring is still the prevalent practice in Bangalore garment export factories, some factories have recently introduced digital sewing machine networks. In these digital sewing machine networks, each machine is equipped with a sensor that automatically records workers' stitching and idle times and transmits this data to a central cloud via Wi-Fi. A software then automatically stores data from each machine and calculates dexterity and efficiency levels for each worker. By looking at this data, production managers can monitor production outputs and individual worker performance in real time, leading to enhanced control over the labour process (INT53).

However, production target monitoring is not enough. Given that production targets are usually impossible for workers during regular working hours, active practices of *production target enforcing* are needed. Production target enforcing is performed through various practices of exercising direct or indirect control over workers. Supervisors exercise *direct control* over workers through practices of penalising workers for not meeting production targets. A common practice by supervisors to penalise workers, known as 'public shaming', is to make workers who have not met production targets stand in front of the whole production line. At the same time, the supervisor makes a public announcement through the floor microphone that this worker has worked too slow. Another common practice of penalising repeated failure to meet production targets is demoting workers' skill level classification. Sewing

Fig. 6.2 White board for documenting production targets in Bangalore export-garment factory. *Source* Photo by Hubert Thiemeyer; stylised by author to ensure anonymity of depicted persons and company

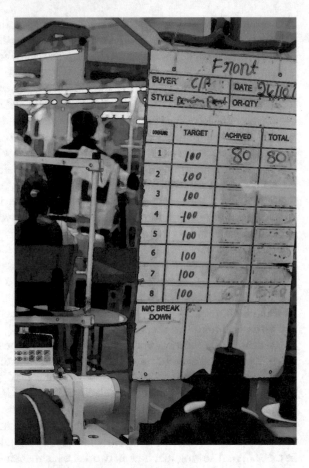

machine operators are assigned different skill levels based on their ability to operate only one or various different machines and to perform stitches of various levels of complexity. These skill levels are, in turn, linked to specific wage categories and bonus payments. Being demoted to a lower skill level classification hence results in monetary loss for workers (INT34).

Indirect control over workers is, in turn, performed through various practices of incentivising workers to meet their daily production targets. In many factories, workers receive incentives in the form of bonus payments: These bonus payments are granted to workers of a batch if the whole batch has achieved production targets on all days of the month. These bonus payments range from 200 to 400 Rupees and provide a vital wage component for workers, given the generally meagre wages in the Bangalore export-garment industry (see Sect. 6.3). In this light, giving the incentive not as an individual incentive but as a group incentive works as an indirect control mechanism in two ways: On the one hand, workers exercise social pressure on co-workers within their batch who are lagging behind to work faster. On the other, when workers in a batch are absent, the present workers usually try to distribute the workload of the absent worker among themselves to make up for the production loss (INT29).

6.1.3 Interim Conclusion

In summary, the labour process in the Bangalore export-garment cluster is linked to the material settings of large factory buildings and structured by a Taylorist workplace regime. This workplace regime is constructed through a tight network of interrelated labour process automating and production target calculating, monitoring and enforcing practices. These practices constitute and structure the labour process. Furthermore, these practices are directly shaped by retailers' sourcing practices. To respond to retailers' pressures for lower prices, shorter lead times and increased flexibility in styles, manufacturers need to hedge labour's indeterminacy through standardising tasks and tightly controlling workers' performance. Automating production and 'production targeting' are two central practices through which manufacturers in the Bangalore export-garment industry respond to these needs. Moreover, manufacturers' practices of labour process automating and production targeting allow garment manufacturers to increasingly substitute (semi-)skilled workers with unskilled workers, who receive lower wages. Thereby, garment manufacturers can offset statutory minimum wage increases and hire migrant workers to balance out the local shortage of semi- and unskilled labour. In this sense, the labour process also intersects with wage and labour market relations (for a more detailed discussion of these intersections, see Sects. 6.3 and 6.7).

For unions, manufacturers' practices of exercising tight control over the labour process create, on the one hand, several challenges that constrain unions' abilities to build associational power within the workplace. Since worker performance is closely monitored and workers are frequently forced to skip breaks, there are little opportunities for worker activists to talk to co-workers inside the factory. Moreover, manufacturers' practices of production targeting are closely intertwined with retailers' sourcing practices. Consequently, unions' attempts to negotiate wages or production target reductions are frequently fended off by managers, who argue that a reduction is impossible due to buyers' demands. As a result, to formulate informed demands and to develop strong arguments in negotiations with employers, unions need to fully understand the complex intersections between sourcing relations and the organisation of the labour process. This involves acquiring an enhanced understanding of the prices paid by retailers as well as of the requested lead times and potential penalties for production delays. However, such an understanding is difficult for unions because they usually do not have access to information from sourcing contracts.

However, the harsh labour process conditions in Bangalore export-garment factories also provide opportunities for 'hot-shop' organising. 'Production torture'—the term employed by workers to refer to abusive behaviour by supervisors—is one of the main issues motivating workers to organise collectively. Along this line, particularly harsh penalising practices by supervisors towards workers who have not met production targets have led to several worker sit-ins and day strikes in Bangalore garment factories over the past years, providing garment unions with opportunities for engagement. Building on such opportunities, Bangalore garment unions have managed to build a membership base in various factories and to establish an informal dialogue relationship with management to address workers' grievances. Through these kinds of interventions, GATWU, for example, has stopped particularly abusive practices of penalising workers for not achieving production targets, such as public shaming or verbal and physical abuse, in factories with a strong presence. GATWU's interventions hence provide an example of the 'small transformations' (Latham 2002) that unions can achieve. They can stop particular practices of labour control even in contexts where—due to constraints for building associational power—they do not have the strength to challenge broader processes and relations of labour control fundamentally.

6.2   Sourcing Relations

As mentioned in the previous chapter, practices of controlling the labour process in Bangalore garment factories are directly shaped by another set of networked practices that constitute and structure the relationship between lead firms and suppliers, which I call sourcing relations. Sourcing relations in the Bangalore export-garment cluster link various actors in multiple countries and places: retailers' sourcing officers in their US and EU headquarters, production managers in retailers' regional production offices in Bangalore, social auditors working in various countries and garment manufacturers in Bangalore and Karnataka. Two main sets of practices structure the relationships between US and EU garment retailers and garment manufacturers in the Bangalore export-garment cluster. These sets of practices are retailers' practices of managing supplier pools, on the one hand, and retailers' purchasing practices, on the other. As lead firms, retailers have the power to perform these practices in ways that allow them to produce and perpetuate relationships of unilateral dependence with suppliers. Suppliers' unilateral dependence on lead firms, in turn, enables retailers to define the terms and conditions of exchange with suppliers and thereby directly or indirectly influence labour processes, workplace relations and wage relations at specific nodes of a GPN. Retailers' sourcing practices hence need to be understood as fulfilling several exploiting and disciplining functions since their main purpose is to ensure that retailers appropriate a large share of the surplus value produced by workers. In the following, I will demonstrate in more detail the various sets of practices through which global garment retailers construct sourcing relations characterised by asymmetrical power relations with Bangalore export-garment manufacturers. Furthermore, I will demonstrate how these practices impact territorially embedded labour processes, wage relations and workplace relations in the Bangalore export-garment cluster.

6.2.1 Retailers' Practices of Managing Supplier Pools: Establishing Control Through Tight Auditing Regimes and Spatially Asymmetrical Buyer–Supplier Relations

Managing supplier pools is a set of practices through which retailers set up supplier networks and manage supplier relations. EU and US garment retailers, who are the main buyers of Bangalore export-garment manufacturers, usually construct their supplier pools by building complex, networked relations with garment suppliers in a large number of countries, predominantly in Asia, Latin America, North Africa, and Central and Eastern Europe. Relations between EU and US garment retailers and suppliers in these countries are constituted through various practices through which retailers exercise power over suppliers: First, retailers control the inclusion and exclusion of factories into their supplier networks through various practices of technical and social auditing. Auditing is performed either by auditors directly employed by retailers or by independent, globally acting auditing firms. The auditing process usually encompasses the following steps: After receiving a detailed briefing on the required technical and social standards, aspiring suppliers are granted a period of preparation, within which they need to ensure that the required technological and social standards are implemented in the factory. After this preparation period, an initial audit is conducted by a team of industrial engineers and social auditors. If a factory fails to meet one or more standards, a corrective action plan is developed. Thereafter, suppliers undergo regular follow-up audits, usually carried out by staff from retailers' regional production offices or—in some cases—by external auditing firms. When external firms carry out audits, usually suppliers have to bear the costs (FN5). In many cases, practices of conducting audits are also combined with practices of counselling and capacitating suppliers to improve their production capacities and efficiency through restructuring production processes (INT53). Together, practices of defining technical and social standards, auditing and counselling hence constitute tight auditing regimes that allow retailers as lead firms to exercise significant control over suppliers by influencing and intervening in the organisation of labour processes at suppliers' factories.

At the same time, technological and social standards prescribed by different brands and retailers often show great variations. Therefore, garment manufacturers in the Bangalore export-garment cluster usually seek to build stable and long-term relationships with a few core customers since buyer-switching entails significant transaction costs. These costs are associated with manufacturers needing to adapt their production system to the specific requirements of the new buyer. Examples of manufacturers having to tailor their production systems to buyer requirements include, for example, placing fire extinguishers at a specific height, abiding by overtime work restrictions or implementing specific workplace committees as part of buyers' social standards (INT9, 27; see also Locke et al. 2007). A social compliance manager at a major garment export company in Bangalore explains that it is challenging for manufacturers to reconcile the varying technical and social standards set by different key buyers. He reports that the company's key buyers—H&M, Walt Disney and Mothercare—set different standards regarding overtime work and worker representation. As opposed to the two other brands, Mothercare does not allow suppliers to habitually rely on overtime work beyond 48 weekly hours, which is the regular working time in Indian law. Walt Disney, in turn, requires all suppliers to establish separate grievances and works committees, whereas all other brands are satisfied when manufacturers install the legally prescribed works committee. To cope with these differences in buyer requirements, the garment manufacturing company has split up production for buyers across their various factory units, with each factory unit producing exclusively for one buyer (INT9). In addition to these social standards, retailers usually impose specific technical standards on suppliers, such as the introduction of specific machines or the obligation for manufacturers to source inputs such as fabric or buttons from specific producers (INT53). Manufacturers risk being excluded from retailers' supplier networks when manufacturers do not fulfil these technological and social standards controlled by buyers in regular audits.

Asymmetrical power relations between retailers and Bangalore garment manufacturers due to tight auditing regimes and linked transaction costs for buyer switching have been further accentuated in recent years as retailers are making further efforts to enhance the speed, flexibility and responsiveness of their supply chains by digitising the sourcing process. In this light, over the past years, retailers have encouraged manufacturers to adopt digital technologies to enhance digital end-to-end supply chain integration. Among these technologies, 3D sampling and digital production systems using Radio Frequency Identification (RFID) technology are the most prominent. According to an industry expert, in the face of pressures from retailers, Bangalore garment manufacturers are increasingly investing in these kinds of digital technologies, with about 10% of manufacturers having introduced these technologies currently (INT53). 3D design software allows manufacturers to create a digital sample of a specific garment and transmit it digitally to the retailer. By transmitting samples electronically, manufacturers and retailers avoid shipping a physical sample via courier, which may take up to two weeks. Hence, the sample approval process is significantly sped up, enabling manufacturers and retailers to reduce lead times for production orders and, thereby, time-to-market for new products. However, various types of 3D design software exist on the market that are incompatible. As a result, to share a 3D sample with buyers, manufacturers need to acquire the same software used by retailers. The same applies to RFID-based digital production systems that allow manufacturers and retailers to monitor the state of production orders in real time and enable automated replenishment orders. Against this backdrop, retailers' practices of digitalising the sourcing process are further accentuating asymmetrical power relationships between garment manufacturers and retailers since the introduction of buyer-specific digital production technologies transaction further increases transaction costs for buyers switching (see also López et al. 2021).

In contrast, retailers keep transaction costs for supplier switching comparatively low through practices of maintaining large, geographically dispersed supplier pools. As mentioned earlier, notwithstanding US and EU garment retailers' tendency to establish longer-term relationships with strategic suppliers (see Sect. 5.1.2), retailers usually still maintain large supplier pools. Retailers' supplier pools typically comprise several hundred or more than a thousand suppliers in various countries. H&M and Inditex, for example, two leading global fashion retailers sourcing from Bangalore, maintain geographically dispersed supplier pools that comprise more than 700 tier one supplier companies and over 1,600 factories located in Asia, Europe, Africa and Latin America (H&M 2021; Inditex 2021). Consequently, whereas Bangalore export-garment manufacturers usually concentrate their business on a few core customers, US and EU retailers can flexibly switch orders between suppliers in various countries.

As a result of these spatially asymmetrical buyer–supplier relationships, Bangalore export-garment manufacturers are facing increasing competition from garment manufacturers both from within and outside of India. Compared to other garment clusters in South India and many newly emerging garment production locations in Asia and Africa, wages in Karnataka and specifically in the Bangalore urban areas are relatively high. Bangalore export-garment manufacturers are hence under increasing competitive pressures from emerging garment production hubs in Asian and African low-wage countries, such as Bangladesh, Myanmar or Ethiopia. On the other hand, Bangalore garment manufacturers are also facing increasing competition from other production hubs within India that are located in states with lower statutory minimum wages, particularly from the nascent garment industry in the neighbouring state of Andhra Pradesh. Competitive pressure from lower-wage countries is also higher for manufacturers in the Bangalore garment cluster compared to other Indian garment clusters due to the cluster's specialisation in men's wear. Men's garments are characterised by relatively low complexity and low value added, and therefore represent entry-level products frequently offered by new, emerging garment production hubs in lower-wage countries.

Competitive pressure on Bangalore export-garment manufacturers is further exacerbated by the fact that—opposed to many other garment producing countries—India is not covered by the EU's 'Generalised Scheme of Preferences' providing import tariff waivers for 'vulnerable countries'. According to industry representatives, it is a common practice by retailers to exploit the competitive pressure resulting from the spatial asymmetries in buyer–supplier relations by using wage and tariff differentials as an argument for pressuring manufacturers from Bangalore into agreeing to lower prices and shorter lead times. This practice is illustrated in the following quote by the director of a Bangalore-based garment sourcing consultancy:

> There is a lot of price pressure! One, because labour costs are higher than in the other countries, and two, because there is also the customs issue. So, the brands will use these arguments also to get lower prices from the manufacturers. They will be like "I am going to buy from you, but I have to pay these extra duty costs, so you have to help me with the piece price". (INT44)

Along the same lines, a manager of a Bangalore export-garment factory states that:

> [If we don't agree to their conditions, buyers] will say: "Fine, Bangladesh will do it, we will give orders to Bangladesh'. So, buyers will blackmail us. Then we have to compromise and say 'Okay, we will take the order". (INT29)

The two quotes illustrate how the power imbalance created by retailers through constructing spatially asymmetrical buyer–supplier relations allows retailers to unilaterally define the terms and conditions of exchange and perform 'predatory purchasing practices'.

6.2.2 Predatory Purchasing Practices

Drawing on Anner (2019), I use the term 'predatory purchasing practices' to refer to a set of practices that structure the sourcing process and through which retailers maximise their capture of surplus value generated by workers in the labour process. These practices include squeezing prices, shortening lead times, penalising late deliveries and irregular order placing. Interviewed managers and industry representatives report that over the past years, buyers have not increased prices paid for ready-made garments while at the same time demanding shorter lead times and increased flexibility in styles (INT 9, 27, 28, 29). In the same period, production costs for Bangalore export-garment retailers have significantly risen due to increasing costs for production inputs statutory minimum wages. Against this backdrop, retailers' 'price squeeze' places significant economic pressure on Bangalore garment manufacturers. An industry representative summarises this dilemma in the following words:

> See, the cost of living and everything is going up in fact. Unions as well as the government desire [manufacturers] to pay the minimum wages. The minimum wages have actually been rising with the cost of living index. But what is happening is, since the minimum wages are going up […] the cost of production is continuously going up right from the day […] the industry was started in India, till date. And it is only progressing. On the contrary, if you see the prices fixed by the buyers, it is under progressive. It is not at all going up. So how do you think the manufacturers do this? They cannot do any magic. See, when everything is going up, the buyers should also realise and support them with an increased price, but unfortunately, it never happened. (INT27)

Economic pressures on garment manufacturers have become further accentuated due to two further 'predatory purchasing practices' performed by retailers: penalising delayed deliveries and placing orders in irregular rhythms. With retailers shifting towards lean retailing business models that work with minimised inventory and stock levels, the punctual delivery of orders has gained importance since delays cause retailers opportunity costs from not selling out of stock garments. Against this background, it is a common practice by retailers to include penalties for late delivery of products into contracts with manufacturers. If production is delayed, for example due to production input delays, poor planning or overbooking of production capacities, manufacturers usually only have two options. Either manufacturers must take on the costs for sending garments via air freight—a very costly mode of transport— or they must accept price cuts defined in the contract (FN5). Meeting ever-shorter lead times is hence crucial for garment manufacturers to avoid further reductions of already low prices paid by retailers.

While demanding shorter lead times and punctual delivery by manufacturers, retailers do not assure manufacturers of a certain number of orders per year. This assurance would give manufacturers more economic stability and allow for enhanced production planning. Additional economic pressures on manufacturers result from retailers' practices of placing orders irregularly, leading to periods with an accumulation of orders and others with order slumps. The fact that retailers do not place orders with Bangalore manufacturers throughout the whole year is also linked to the specialisation of production in the cluster on men's wear. Whereas most retailers now produce new women's collections throughout the whole year, men's collections still tend to be lesser in number and more seasonal. As a result, interviewed Bangalore export-garment manufacturers report that they frequently face order slumps May till September when production for summer collections has already been completed, and production for the winter collections and Christmas sales has not started yet. During this time, many manufacturers report that they take on smaller production orders from non-key customer brands or even from domestic brands just to cover running costs, as exemplified by this quote from a Bangalore factory manager:

> For example, the orders from H&M we do only during peak season. From May to September, we get only small orders. That means small quantities, maybe an order of 10.000 pieces. Then we work on that four weeks and then, after that, we have no more work. During that time, we also sometimes take on orders from local brands to cover the running costs for the factories and to pay workers' wages, but we do not make any gains. (INT9)

As a result, to cope with retailers' practices of squeezing prices, shortening lead times, penalising delayed deliveries and placing orders at irregular intervals, Bangalore garment manufacturers resort various practices to maintain tight control over production and labour processes (see Sect. 6.1) and keep wages low (see Sect. 6.3). As illustrated in the previous chapter, Bangalore export-garment manufacturers employ diverse practices of closely monitoring worker performance and enforcing daily production targets to avoid delays. When delays are caused by external factors, such as delays in the delivery of fabrics or other production inputs, workers are frequently forced to work overtime hours or Sunday shifts without receiving the legally prescribed double wage rate. Instead, workers receive a day off as compensation during periods with order slumps—a practice commonly known as 'comp-off' in Bangalore (for more details, see Sect. 6.3). Garment manufacturers hence transfer the economic pressure and risks from retailers' predatory purchasing practices directly to workers. The intersection between sourcing relations and exploitation in the labour process is exemplified in the following statement by the HR manager at a major Bangalore garment export company producing for US and EU buyers such as H&M, Inditex, the G.A.P. and American Eagle:

> When I am doing 60 per cent business with G.A.P […] the advantage is, I am assured of this business. But the disadvantage is that, every year, I have to compromise on price, you know. They reduce the price. Because they know [company name] can do it. [Company name] will not refuse. [Every year], we need to increase our volume, we need to increase our productivity. […] That is the challenge. That is the pressure we put on our people. (INT29)

The quote illustrates how power asymmetries between buyers and suppliers allow retailers to squeeze prices and thereby put manufacturers under economic pressure which is then transferred to workers.

6.2.3 Interim Conclusion

In summary, sourcing relations in the Bangalore export-garment cluster are characterised by network embeddedness since they link Bangalore export-garment manufacturers to sourcing and auditing officers of US and EU retailers located usually at retailers' headquarters. The relations linking Bangalore garment manufacturers with these actors are constituted and structured predominantly through two sets of practices: retailers' practices of managing supplier pools and retailers' purchasing practices. In this section, I have shown that retailers' value chain power stems from spatial asymmetries in retailer-supplier relations. These spatial asymmetries are actively constructed by retailers through practices of maintaining large, geographically dispersed supplier pools and tight auditing regimes. The resulting competitive pressures within supplier networks and high transaction costs for buyer switching lead Bangalore export-garment manufacturers to concentrate their business on a few core buyers. Consequently, sourcing relations in the Bangalore export-garment cluster are characterised by the largely unilateral dependence of manufacturers on retailers. Suppliers' dependence enables retailers to secure a large share of the surplus value produced in the labour process through predatory purchasing practices. Manufacturers transfer economic pressures and risks to workers in the labour process and in wage relations through various practices directed at maximising worker productivity while keeping wages low. Consequently, territorially dis-embedded sourcing relations intersect directly with localised labour processes and territorially embedded wage relations.

This strong intersection of labour processes with wage and sourcing relations has constraining effects for Bangalore garment unions. The economic pressure exerted by retailers constrains the 'wiggle room' (Castree et al. 2004: xvii) for workers and unions for negotiating production target reductions or wage increases. Manufacturers depend on maximising labour productivity and minimising labour costs to respond to retailers' 'predatory purchasing practices'. Consequently, most manufacturers regard collective worker representation as a threat to their business model and refuse to engage in collective bargaining with unions. Instead, managers frequently actively undermine any collective bargaining attempts through various 'union-busting' practices (see Sect. 6.5).

On the other hand, retailers' influence over manufacturers also opens up opportunities for unions to contest particularly cruel or illegal exploiting and disciplining practices in garment factories by framing them as violations of buyers' social standards. In these cases, unions can use retailers' leverage over manufacturers to achieve corrections of basic labour rights violations. However, as literature on private social regulation in GVCs has pointed out, retailers also have a limited interest per se in detecting and penalising violations of social standards at their suppliers since these violations ultimately enable manufacturers to comply with retailers' demands for quick and flexible yet cheap production (see, e.g., Bartley and Egels-Zandén 2015; Locke et al. 2007). Therefore, to seize retailers' leverage over manufacturers, local unions need to develop strategies of networked agency that activate the moral power of consumer campaigning networks in retailers' most important consumer markets. Following such a networked agency approach, the local union GATWU has been able to harness retailers' influence over Bangalore-based garment manufacturers in various labour struggles. By doing so, GATWU has, for example, achieved to stop large-scale minimum wage violations (for a more detailed discussion, see Sect. 7.1.1). Nevertheless, wages in the cluster currently remain below subsistence levels due to continued exploiting practices by Bangalore garment manufacturers directed at keeping wages low, which will be explored in more detail in the next chapter.

6.3 Wage Relations

Wage relations in the Bangalore export-garment cluster are constituted through a complex mesh of practices and relationships, including workers, employers, unions and state actors within the territory of the State of Karnataka. To ensure that the labour process is continuously reproduced in its profit-maximising form, employers perform two sets of practices to keep wages in the cluster low. On the one hand, employers perform various practices to undermine or circumvent the legal-institutional process of setting the statutory minimum wage for the garment sector in the State of Karnataka. On the other hand, employers perform various 'wage theft' practices at the workplace level to reduce labour costs. In the remainder of this section, I will introduce these two sets of practices and illustrate how they are enabled by state actors' 'pro-business' practices. Moreover, I will illustrate how wage relations intersect with sourcing relations and industrial relations.

6.3.1 Undermining the Legal-Institutional Statutory Minimum Wage Setting Process

In the absence of collective bargaining agreements, wages in the Bangalore export-garment cluster are primarily defined by the statutory minimum wage for the garment sector. For each sector, statutory minimum wages in India are fixed at the state level. According to the Indian Minimum Wage Act of 1948,[1] minimum wages for a specific

---

[1] The most recent Indian labour law reform has subsumed the regulations from the Indian Minimum Wages Act of 1948, together with various other laws related to wages, under the so-called Labour

sector may be fixed through two types of practices. As a first option, the state government can unilaterally fix the minimum wage for a specific sector based on the recommendations for a general wage span made by a tripartite minimum advisory board constituted by employer, worker and state representatives. As a second option, the state government may convene a minimum wage board for a specific sector constituted of employer, worker and state representatives, who shall negotiate the minimum wages for the respective sector. In both options, board members and the government shall consider various factors when fixing the statutory minimum wage for a specific industry, such as the prevailing conditions of the economy, the skill requirements and type of work, and basic living costs (INT52). To reflect differences in skill levels and living costs, different minimum wages are fixed for skilled, semi-skilled and unskilled workers, and for workers in urban, semi-urban and rural areas. Overall, the 15th Indian Labour Conference in 1957 and subsequent court rulings have defined and defended that statutory minimum wages should be calculated following a needs-based approach. According to this approach, statutory minimum wages should at least ensure workers' reproduction costs by covering the average expenditure of a worker and his family consisting of one spouse and two children below the age of fourteen (Mani et al. 2018). These guidelines exemplify that legislators originally conceived the Indian legal statutory minimum wage framework as a protection mechanism for workers to avoid wages below a subsistence level. In this line, the Minimum Wage Act also defines that intervals between statutory minimum wages should not exceed five years to ensure that minimum wages increase proportionally to the cost of living.[2]

However, the specific practices through which Bangalore-based manufacturers and state actors have enacted the legal-institutional minimum framework over the past decades undermine the historical purpose of the Minimum Wage Act. Manufacturers have, over the past decades, avoided or circumvented minimum wage increases through practices of delaying and challenging minimum wage revisions. These practices have, in turn, been enabled by state actors' pro-business practices. In the following, I illustrate the interplay of various employer and state practices that, together, construct exploitative wage relations in the Bangalore export-garment cluster, characterised by below-subsistence wage levels.

The first practice through which Bangalore garment manufacturers and Karnataka labour department officials have traditionally undermined the legal-institutional statutory minimum wage process is by deliberately delaying minimum revisions. As opposed to the legal provisions, minimum wages for the garment industry in

Code on Wages Act, enacted in August 2019. The provisions for the institutional statutory minimum wage setting process have, however, largely remained the same.

[2] Reflecting the Indian governments' attempts to further deregulate wage relations in India, the new Indian Labour Code on Wages of 2019 provides significantly less details on the procedure through which the statutory minimum wages should be fixed and the criteria according to which the minimum wage should be calculated (for more details Jayaram [2019]). Thereby, the new Labour Code on Wages dismantles the original purpose of the Minimum Wage Act of ensuring that wages cover workers' basic living costs and provides an official institutional frame for the 'pro-business' wage setting practices by state actors illustrated in this chapter.

the State of Karnataka have only been revised at intervals of 7 to 9 years from 1986 to 2010 (CWM 2014). Only since the emergence of independent unions in the garment sector—specifically since GATWU's minimum wage campaign in 2009 and 2010 (see Sect. 7.1.3)—minimum wage revisions have been implemented within the legally fixed three- to five-year periods. Nevertheless, statutory minimum wages for the garment industry in the State of Karnataka still remain significantly below subsistence levels: Currently, the statutory monthly minimum wage for garment workers ranges from about 9,000 Rupees (approx. 123 US$) for an unskilled worker in a rural area to slightly above 10,000 Rupees (approx. 136 US$) for a skilled worker in the Bangalore urban area (Labour Commissioner Office, Government of Karnataka 2021). According to a study by labour researchers and unionists, to guarantee a decent standard of living for a worker and his or her family, wages, however, need to amount to 18,000 Rupees (approx. 238 US$) at minimum (Mani et al. 2018).

To ensure that minimum wage increases in each round of minimum wage revisions remained as low as possible, Bangalore export-garment manufacturers have employed a second practice of lobbying the state government to take back and reduce statutory minimum wage increases: In 2001, 2009 and 2018, following the declaration of a new, increased statutory minimum wage for the Karnataka garment industry by the state government, Bangalore manufacturers refused to pay the new, increased minimum wage rate. Instead, manufacturers lobbied the state government to withdraw the new minimum wage notice arguing that the increased wage rates will force employers to relocate factories to neighbouring states with lower wage levels. Garment manufacturers' lobbying practices are exemplified in the following quote from a letter by the owner of one of Karnataka's major garment companies to the Labour Department after announcing new, increased minimum wage rates in 2018. In this letter published in a local online newspaper, the industry representative wrote in the name of all Bangalore garment manufacturers:

> We are adversely affected by the abnormal increase in the minimum wages to an extent of 50% increase in wage bill in the case of highly skilled workers and 18% in the case of semi-skilled workers. Minimum wages in Karnataka were comparable with neighbouring states before the proposed increase. Hence, there was no need for the hike in minimum wages for this industry. (Letter to the Karnataka Labour Department by a representative of Himatsingka Limited, quoted in Bath 2019)

In response to manufacturers' lobbying practices, it has been a common practice of the Labour Department to withdraw the original Minimum Wage Revision notification under the pretext of a 'clerical error' and to subsequently issue a new notification with a reduced minimum wage increase. Following legal complaints by local trade unions, the government's practice of withdrawing the original minimum wage notification was declared illicit by the High Court of Karnataka in two rulings from 2013 and 2019. In both instances, the High Court, however, did not order the implementation of the originally notified minimum wage. Instead, it ordered the formation of a tripartite minimum wage committee for the garment industry to fix a minimum wage that should be acceptable to both workers and employers in the next revision. Given the low unionisation levels in the Bangalore export-garment cluster, ordering for a tripartite committee needs to be regarded as a practice that serves employers' interests rather than workers' interests. This fact is exemplified in the outcome of the last tripartite minimum wage negotiations in 2019, which fixed statutory minimum wages at around 9,000–10,000 Rupees. Thereby minimum wages, however, remain significantly below the original (subsequently withdrawn) minimum wage notification issued by the Karnataka State Government in 2018, which had fixed monthly wages at 14,000 Rupees (approx. 185 US$) for skilled workers and at 11,500 Rupees (approx. 152 US$) per month for unskilled workers.

In this light, the state government's practice of withdrawing already issued minimum wage notifications and the high court's practice of ordering tripartite wage negotiations need to be interpreted as 'pro-business practices' that support manufacturers' interests. This alignment of state practices with the interests of manufacturers is indicative of a broader shift in the Indian state's strategic orientation since the 1990s. Historically, the Indian state's strategic orientation had been shaped by Fabian socialist ideals of strong state control over production and wages. However, with economic liberalisation, the role of the state as an active regulator of the economy shifted towards a focus on creating a business enabling environment for private capital actors. In this line, Indian state governments have been increasingly competing for national and foreign private investments understood as central conditions for fostering industrialisation and economic development. Due to its capacity to provide large-scale employment for low-skilled groups of the population and for attracting foreign currency, the export-garment industry has, in this context, been regarded as a sector of particular strategic importance by the Indian national and state governments. As a result, Indian government actors have sought to minimise state interventions in wage relations in the garment industry, as this labour researcher explains:

> Regarding [the] garment [sector], because of its labour intensity and because it's a huge potential of earning foreign currency through exports, the welfare of minimum standards of working or living conditions are getting compromised. […] [The] government basically now plays a passive role and withdraws itself from labour market interventions. Even with regard to the minimum wage, the government is not very clear about regulating the minimum wage, because […] foreign investments come when you have cheap labour. (INT15)

The quote illustrates how employer and state practices together construct exploitative wage relations in the Bangalore export-garment cluster characterised by below-subsistence wages. Manufacturers' practices of refusing to implement statutory minimum wages covering workers' basic living costs are enabled by state actors' practices of not enforcing subsistence level minimum wages. Instead, government actors leave it to unions alone to fight for wages that cover workers' basic needs while knowing that given low unionisation levels in the Bangalore export-garment cluster, unions' bargaining power is relatively limited.

In this sense, wage relations in the Bangalore export-garment cluster also intersect with industrial relations (Sect. 6.5) and with workplace relations (Sect. 6.4). On the one hand, unions' overall low membership compared to the total number of workers in the cluster constrains unions' capacities to push for a living wage in tripartite minimum wage negotiations. Unions' bargaining power in minimum wage negotiations is further constrained by the fact that they are independent unions. Bangalore-based unions organising garment workers have developed out of NGO-led community organising projects and therefore have no ties to any political parties. Hence, the three Bangalore garment unions also have little leverage vis-à-vis state actors in tripartite minimum wage negotiations. Moreover, the several thousand members that all three garment unions have together are distributed across a large number of factories, constraining unions' capacities to negotiate bilateral collective wage agreements with individual employers that exceed minimum wage rates. Unions' low membership levels inside specific factories can, in turn, be regarded as resulting from management practices of constructing tightly controlled and segmented workplace relations hampering unions' abilities to build a strong membership inside the workplace. As a result of unions' limited power to negotiate collective bargaining agreements at the workplace level, the legally prescribed statutory *minimum* wage at the industry level represents the de facto *maximum* wage paid by employers in Bangalore export-garment factories.

6.3.2 Wage Theft Practices at the Workplace

In addition to performing several practices directed at keeping statutory minimum wages at the state level low, Bangalore export-garment manufacturers employ a second set of practices of withholding a part of the wages that workers are rightfully entitled to. These practices are commonly subsumed under the notion of 'wage theft' practices. Common wage theft practices by Bangalore export-garment manufacturers include linking minimum wage increases to production target increases, stealing overtime wages and giving 'comp-offs'. The first practice of linking minimum wage increases to production target increases is commonly employed by Bangalore garment manufacturers to compensate for higher labour costs due to statutory minimum wage increases: To offset higher costs, manufacturers push workers to deliver higher productivity. In this context, workers and unionists report that after each increase of the statutory minimum wage, managers also increased workers' production targets. Similarly, when annually a legally prescribed wage component called 'Dearness Allowance' is increased to compensate for inflation, this increase also leads to a raise of production targets, as described exemplarily by this garment worker:

> And then, once a year our wages increase, but then also our production targets increase and then we cannot complain because the supervisor will just say: "You get more wage, so you have to produce more. We must make more money to pay you a higher wage". (INT5, translated from Kannada)

Since production targets in most factories are barely achievable during regular working hours, further production target increases result in de facto unpaid overtime work since extra time spent by workers to complete production targets is generally not paid. As a result, workers usually stay 15 or 30 min longer daily, amounting to almost two full days of extra unpaid work per month (INT47).

Besides linking wage raises to increases in production targets, employers use two other 'wage theft' practices directed at maximising productivity while minimising labour costs: stealing overtime wages and 'giving comp-offs'. According to the Minimum Wage Act, all work hours exceeding the regular hours of 48 h per week must be paid with the double regular wage rate. Hence, overtime work is a significant cost factor for management and therefore avoided, if possible. However, particularly during the peak production season, managers frequently resort to ordering additional overtime work. To avoid paying the double wage rate, workers report that managers make workers check out with their time stamp card at the end of the regular shift to then continue working informally for one or two hours. In some cases, workers are compensated for these informal extra working hours in the form of a 'productivity bonus', which is significantly less than the applicable double wage rate. In turn, giving *'comp-offs'* is used locally in Bangalore to refer to the practice of giving workers paid leave days during periods with little or no production orders, which must then be recovered through unpaid Sunday work during peak season (INT4,13; see also Jenkins and Blyton 2017). In doing so, workers are cheated out of the double overtime wage for Sunday work that they are legally entitled to, and of the half wages that workers are entitled to receive during lay off periods, i.e. periods when a factory does not have work. In some cases, managements go even one step further and deduct workers' comp-off days from their regular contingent of paid leave days, thereby de facto making the performed Sunday work completely unpaid work (INT33).

As previously mentioned, Bangalore export-garment manufacturers' wage theft practices are directly interlinked with retailers' predatory purchasing practices in two ways. On the one hand, retailers' practices of neutralising higher costs from wage increases through extracting higher productivity from workers enable manufacturers to ensure value capture despite retailers' 'price squeeze'. Accordingly, interviewed garment managers state that buyers do not increase their prices when the minimum wage is raised but rather expect manufacturers to make up for increased labour costs through heightened productivity (9, 27, 28). Only one manager states that their key buyer, with whom they have been doing business for many years, substantiates at least a part of minimum wage raises by paying higher prices (INT29). On the other hand, stealing overtime wages and 'comp-offs' is a strategic practice through which Bangalore garment manufacturers externalise the negative economic effects of unstable orders to workers. Through the practice of 'comp-offs', managers construct wage relationships as debt relations, in which workers owe employers working time for wages that have already been paid. In periods with low orders, workers consequently accrue a significant debt of hours to a managerially instituted 'time bank'—as Jenkins and Blyton (2017) put it—allowing managers to flexibly dispose over workers' reproductive time on Sundays and to transform it into productive time when orders are available.

6.3.3 Interim Conclusion

In summary, wage relations in the Bangalore export-garment cluster need to be understood as constituted through power-laden networked relationships between retailers, state actors, employers and workers. These relationships intersect with network sourcing relations (Sect. 6.2) at the global level and with territorially embedded industrial relations (Sect. 6.5) at the state level. Whereas retailers exercise pressure on manufacturers for lower prices by stressing competitive pressures from lower-wage locations, manufacturers pass this pressure on to the State Government of Karnataka by threatening to relocate production to neighbouring states if statutory minimum wages are raised beyond a certain threshold. The Government of Karnataka, in turn, responds to pressure from manufacturers by shifting the authority for fixing statutory minimum wages to tripartite minimum wage committees. Given the asymmetrical power balance characterising capital labour in the Bangalore export-garment cluster due to low unionisation rates, the last rounds of tripartite minimum wage negotiations fixed wages that remained significantly below a subsistence level. Economic pressures on workers are further increased by employers' 'wage theft' practices at the workplace level. These practices neutralise minimum wage increases through increases in production targets and construct wage relations at the workplace level as time debt relations between individual workers and management through 'giving comp-offs'.

The fact that exploitative wage relations in the Bangalore export-garment industry are constructed through networked state and employer practices poses significant challenges for Bangalore garment unions. 'Pro-business' state practices constrain the ability of unions to draw on the legal-institutional minimum wage framework as a source of institutional power. When the state acts as a regulator that actively ensures adequate wage levels, the presence of state actors in tripartite minimum wage negotiations can provide a counterweight to dominant employers and balance off capital-labour power asymmetries. When state actors, however, act primarily as business enabling agents—as is the case in the Bangalore export-garment industry—workers and unions need to activate associational power resources. To achieve significant wage increases, unions need to develop networked agency strategies that combine public campaigns pressuring the state to implement adequate minimum wages with collective action at the workplace to ensure that minimum wage increases neutralised through a raise of underpaid overtime work.

Following such a networked agency strategy, especially a large-scale minimum wage campaign conducted by GATWU from 2009 to 2010, has contributed to transforming some of the state and employer practices that have traditionally contributed to constructing exploitative wage relations in the Bangalore export-garment cluster. For example, GATWU has achieved that minimum wage revisions have since 2010 been undertaken within the legally prescribed periods of maximum five years. Moreover, GATWU has stopped the practice of giving 'comp-offs' in a selected number of factories where the union has a strong membership. In most factories without union presence, giving 'comp-offs', however, remains a common practice. The continued prevalence of 'comp-off' practices in Bangalore export-garment factories needs to be understood as also enabled by the fact that there is still a large number of factories in the cluster without union presence. Unions' capacity for establishing strong membership bases inside factories is, in turn, constrained by the various disciplining practices through which Bangalore garment manufacturers construct tightly controlled workplace relations. These practices will be illustrated in more detail in the next chapter.

6.4   Workplace Relations

I designate as workplace relations the sum of relationships between workers and supervisors or management and between workers in a specific workplace. Similar to the labour process, workplace relations in the Bangalore export-garment cluster are tied to the material settings of large factories with several hundred or even thousands of workers and are, therefore, highly localised. Workplace relations represent a traditional domain of labour control. To ensure workers' subordination under the labour process, managements need to construct workplace relations in a way that hedges or prevents potential labour resistance. In the following, I will demonstrate how managers in Bangalore garment factories construct workplace relations through various sets of disciplining practices that constrain opportunities for collective worker organising and (re-)produce asymmetrical power relations between supervisors and workers. These sets of practices include constructing the workplace as a tightly controlled space, segregating shop floors along gender lines and undermining workplace committees as spaces for collective dialogue.

6.4.1   Constructing the Workplace as a Tightly Controlled Space

The first set of practices through which manufacturers in the Bangalore export-garment cluster construct workplace relations as de facto disciplining relations is directed at constructing workplaces as tightly controlled spaces. Management and supervisors closely monitor workers' movements and interactions within these spaces in two ways. First, Bangalore garment manufacturers already design the physical-spatial layout of factory buildings in a way that allows control over incoming and outgoing persons while at the same time shielding any interactions happening inside the factory premises from the outside world. To this end, each factory building is usually surrounded by high walls, and the gate is secured by guards and security cameras (see Fig. 6.3).

To enter the factory premises, workers need to present their ID cards, and visitors need to sign in stating their organisation and reason for the visit. In this context,

Fig. 6.3 Export-garment factory compound on Mysore Road, Bangalore. *Source* Photography by Cosimo-Damiano Quinto; stylised by the author to ensure the anonymity of depicted persons

unionists report that guards usually deny them access to the factory, making it impossible for unionists to meet and organise workers inside the factory. Managements' efforts to control and restrict access of unionists to the workplace were also evident during my fieldwork, when visiting an export-garment factory located about 80 km outside of Bangalore with a group of German unionists and two union leaders of GATWU. Whereas the factory manager warmly welcomed our group of German visitors, the two local GATWU union leaders were denied access to the factory. Only after a lengthy discussion and the repeated assurance that they would refrain from any interactions with workers, GATWU leaders were finally allowed to enter the factory. Nevertheless, a manager followed the two unionists closely throughout the three hours factory tour and even waited in front of the door when they went to the washroom.

Besides controlling workers' interactions *inside* the workplace, managers also restrict and prevent interactions between unionists and workers *outside* the workplace. In this context, unionists report that when approaching workers or holding meetings in front of the factory gate after the end of workers' shifts, security guards frequently dissolve these interactions. Moreover, it is a common practice for managers to call workers who have been recorded interacting with unionists to the management office the next day. There, workers are advised not to engage with the union, as this union leader reports:

> They [the management] keep on 'advising' the workers. They will call the workers, saying: "Any problem you tell us, you don't go to unions". [Company name] constantly does that. And they have cameras fitted at the gate. So, if workers are talking to anyone outside the factory, they will call them the next day and advise them on how they should behave and take care of them so that they don't get 'misled'. (INT 30)

Management practices of controlling worker interactions inside and outside the workplace that prevent worker engagement with unions can be classified as union-busting practices since they actively seek to prevent collective worker organisation. In this sense, workplace relations need to be understood as directly intersecting with industrial relations (Sect. 6.5) in that managements' practices of controlling and preventing interactions between workers and unionists in the workplace significantly constrain unions' abilities to build bargaining power vis-à-vis employers.

Control over workers' interactions is, however, not limited to monitoring workers' interactions with external actors, such as unionists. Also, inside the factory, supervisors tightly monitor and restrict workers' movements and interactions. In this line, workers report that they are ordered to remain at their specific assigned workstations throughout the shift and are not allowed to leave their batch to talk to co-workers from another batch, even during breaks (INT36). Control over workers' interaction inside the factory is also supported by the specific spatial arrangements of workstations in assembly lines, which provides little possibility for interactions between workers. The assembly line's spatial arrangement moreover allows to construct spatial asymmetries between supervisors and workers. Whereas workers sit and have to remain in their designated places, supervisors walk around between the various lines and thereby oversee all activities on the shop floor. The spatial asymmetry between supervisors and workers hence helps to reproduce and reinforce hierarchical relationships between supervisors and workers.

Besides being supported by the factory floor's spatial layout, workplace hierarchies are also further stabilised and reinforced by management's practices of reproducing gendered power asymmetries through segregating shop floors along gender lines.

6.4.2 Segregating Shop Floors Along Gender Lines

The second set of disciplining practices through which Bangalore export-garment manufacturers construct workplace relations that ensure the subordination of workers under the labour process is hiring women, in particular for the large share of unskilled or semi-skilled tasks in the sewing process. As mentioned in Sect. 5.2.2, about 85% of workers in Bangalore export-garment factories are women. Hiring women can be regarded as a disciplining practice directed at preventing labour unrest and at ensuring the smooth subordination of workers to the labour process in two ways: First, by hiring women predominantly for lower-skilled tasks, power asymmetries enshrined in broader gender relations are reproduced in the workplace relations through intersecting lines of worker segregation according to gender and position. In most Bangalore garment factories, the unskilled or semi-skilled positions of sewing machine operator or helper are performed by women. In contrast, higher skilled positions such as operating digital printing or cutting machines and especially supervisor and manager positions are predominantly performed by men, as this NGO representative explains:

[…] we see that a large proportion of workers are women, at least, definitely in South India, whereas all the supervisors and managers and beyond are men. And even if a worker, a woman worker, sticks to the job, to the same factory, for five, ten years, whatever, there is no career mobility given to them by the industry, by the factory. It is also not in the thought because the industry has been there since so long. And we hardly see any women workers being promoted to supervisors. Or even hiring an outsider as a supervisor, a female supervisor, that's also not done. (INT11)

By segregating shop floors along intersecting lines of gender and skill levels, managements replicate broader patriarchal structures inside the workplace to reinforce power asymmetries between supervisors and workers. These power asymmetries, in turn, enable managers and supervisors to suppress worker complaints about high work pressure and abusive behaviour by supervisors, as this worker describes:

We have high production targets and all workers must fulfil them, even if a person doesn't feel well. Whenever we complain, the supervisor will tell us to pack our stuff and leave. […] Our […] HR manager is also not supportive at all to the workers. When we point these things out he will yell at us and say stuff like: "So you know what to do? You want to be HR? You know my job better than me? I am the manager here, and I know what to do, so you don't worry. You go back and do your work, and you let me do my work". (INT5, translated from Kannada)

Besides reinforcing power asymmetries between supervisors and workers, managers' practices of creating largely feminised shop floors also serve as an indirect disciplining mechanism by preventing collective worker organising and unionisation in the workplace. Given the large factory set-ups in the Bangalore export-garment industry, which could potentially provide a breeding ground for unionisation, managers are particularly interested in hampering attempts at collective worker organising in the workplace. In this line, Bangalore garment managers frequently mention in interviews that they prefer to hire women over men since women are more 'docile' in nature and therefore less likely to join a union or to create other sorts of 'trouble' (INT9). This narrative of women as naturally more submissive than men is not exclusive to the Bangalore garment export industry but underpins the feminisation of the garment industry across Asia (Chakravarty 2007). Not at last, the spontaneous mass strike of about 450,000 Bangalore garment workers to protest against a new law restricting workers' provident fund access in April 2016 has, however, debunked this narrative as socially constructed rather than based on biological facts.

Nevertheless, it remains a fact that the feminised nature of the workforce poses several challenges for unions with regard to organising workers. These reasons are, however, not linked to women's supposedly 'docile' nature but rather to women's embeddedness into broader patriarchal social relations that establish women as sole caretakers in the household. As a result, many female garment workers face the double burden of wage work and care work. Therefore, women workers are often unable to engage in union activities after work or on Sundays, as this unionist explains:

They [garment workers] are ladies who have little awareness [of their rights]. That is the main problem. After work, they have to go back home and do chores at home as well, so they won't pay much attention to all these [gate meetings]. Men get easily attracted to the union, but women, even if they come to a meeting, they'll be in a rush as they have chores to do at home […]. Management uses this as an advantage and hires only such women because they won't have time to take additional responsibility within the union. (INT36)

In particular, married women workers report that they need to seek permission from their husbands or parents-in-law to leave the house for meetings. Given that many workers are first-generation industrial workers who have migrated to the city with their families from rural villages, husbands are often sceptical of unions. They do not allow their wives to join union meetings because they either regard it as a waste of time or fear that it might cause the wife to lose her job. Therefore, it is due to these kinds of asymmetrical power relations enshrined in gender relations that feminising shop floors in the Bangalore export-garment industry de facto serves as a disciplining practice.

6.4.3   Undermining Collective Dialogue in Workplace Committees

The last set of practices through which managers construct workplace relations as de facto disciplining relations is linked to various national and state legal provisions that prescribe the implementation of four types of workplace committees in industrial establishments.[3] These committees are: (1) a works committee for resolving grievances that may arise in the daily work between management and workers; (2) a safety committee responsible for carrying out health and safety surveys and raising awareness among workers for health and safety provisions; (3) a canteen committee that shall be consulted inter alia on the quality and quantity of food; and (4) an Internal Complaints Committee for processing any worker complaints related to sexual harassment (FWF 2018). The legal rationale for these committees is to provide an institutionalised space for worker-management dialogue in which workers' grievances can be addressed in a structured manner. As such, workplace committees can potentially provide a source of institutional power for workers and promote collective worker organisation in the social dialogue process, e.g. when workers jointly identify collective issues to raise with the factory management. However, managers enact the legal provisions for workplace committees through practices that undermine the original purpose of workplace committees. Rather than constructing workplace committees as spaces for institutionalised social dialogue, managers have instead constructed dis-functional workplace committees that exist merely on paper. Dis-functional workplace committees are constructed by management through practices of holding committee meetings as short gatherings during lunch breaks reduced to making workers sign the attendance list. As a result, most interviewed workers show little awareness of the existence of any workplace committees in their respective factories. Workers' little awareness of workplace committees also stems from the fact that usually there is no democratic election process for

---

[3] For an overview of the respective laws, see FWF (2018, p. 5).

workplace committees. Instead, selected workers are called in an ad hoc manner by managers for committee meetings, as described by this unionist:

> Committees should meet once every two months, usually. But those meetings are just for like 10 minutes. The management makes the workers sign the attendance list, and that's it. We from GLU, we try to give the workers awareness about the committees and what they are for and that they can raise issues in the committees. But that is also not so easy. When workers raise questions or issues during a committee meeting, the management will not call them for the next meeting. (INT4, translated from Kannada)

Where workers are aware that workplace committees exist, they are usually unaware of the purpose of these committees, since appointed committee members receive no training or introduction regarding their role and the purpose of committee meetings. When actual committee meetings are held, managers construct these meetings as spaces of unilateral communication by addressing workers with 'motivational speeches' rather than engaging in dialogue, as exemplarily described in this statement by a union representative:

> If the committee meetings really take place, workers often have no space or time to raise issues. It is just the management who gives a motivational speech. Like, they will tell the workers how important it is that they work fast and produce good quality to satisfy the buyers because workers' jobs depend on buyers' orders. Also, they will tell the workers not to speak about conflicts or negative things with anybody outside the factory, and especially with unions because that might put the factory in a bad light, and, then, they might not receive orders from buyers any more. [...] So, what they do is really emotional blackmailing. They will say stuff like: "This factory is your house, this is your family. Your family is giving food for you and for your children, so you must not speak against the family". (INT4, translated from Kannada)

The second quote illustrates once more the intersection of workplace relations with industrial relations. By undermining any worker-management interaction that might encourage collective organisation and using committee meetings to actively discourage workers from joining the union, manufacturers actively constrain unions' possibilities of building bargaining power in the workplace and therefore, to engage in collective bargaining.

On the other hand, workplace relations intersect with sourcing relations and, more specifically, with retailers' CSR practices: Many retailers have, over the last decade, set up so-called Social Dialogue Programs in response to criticisms from consumer organisations regarding abusive behaviour by supervisors and managers. Retailers' Social Dialogue Programs prescribe detailed practices through which managers at suppliers shall establish or implement workplace committees as spaces for an institutionalised worker-management dialogue. These practices include training workers regarding the purpose and functioning of workplace committees, conducting democratic and secret elections of worker representatives, and liberating workers to participate in committee meetings during working hours. Retailers provide financial means for training that are usually conducted by external NGOs.

Workers, unionists and managers report that in Bangalore garment factories where Social Dialogue Programs have been implemented, these programs have transformed managers' practices of conducting committee meetings. As an outcome of the training conducted in the Social Dialogue Program, workers report that workplace committee meetings are not held during lunch breaks anymore but that workers are liberated during their regular working hours to participate in committee meetings. Moreover, they report that managements now allocate sufficient time, i.e. 30–40 min, for committee meetings. Workers and unionists also find that overall worker awareness about committees as institutionalised spaces for addressing workers' individual or collective grievances has increased (INT9).

At the same time, workers' and unionists' reports, however, illustrate that the success of retailers' Social Dialogue Programs ultimately depends on the power relations between management and workers in a specific factory. Workers report that in factories with low unionisation levels, only non-controversial issues can be addressed during committee meetings, such as the provision of clean drinking water or decisions over which holidays should be leave days for all workers. More controversial issues—such as sexual harassment by supervisors, excessive production targets or unpaid overtime work—are fended off immediately by management as not falling under the scope of workplace committee discussions. Workers report further that where individual workers have tried to raise such controversial issues in committee meetings, these workers were simply not called for the next committee meeting (INT4, 5). In one extreme case, several GLU worker activists who had been voted into the works committee at their factory were even dismissed after repeatedly addressing labour rights violations in the works committee (INT36). Barriers to the effectiveness of Social Dialogue Programs resulting from power asymmetries and hierarchies in the workplace are illustrated in the following observations of an NGO representative who had participated in a newly set-up Internal Complaints Committee[4]:

> When I was there, no complaint came. Also, because I was in one of the first few meetings, in fact in the first, or the second meeting of the ICC [Internal Complaints Committee]. And the welfare officer was the senior woman from the factory. And she would just make some gestures, no, [just] give some signs from the eyes. And then no one would… I mean, they would start and then, they would look at the welfare manager and they would keep quiet and not say [anything]. […] I mean for everything that they had to say, they were first looking at the welfare manager and then saying it. And she would nod or not nod. And then they would get the message of whether they should go on or be shut. (INT11)

Therefore, where asymmetrical power relations between managers and workers exist, these power asymmetries significantly limit the potential of workplace committees to function as spaces for management-worker dialogue even after the implementation of Social Dialogue Programs.

---

[4] As per the Indian Sexual Harassment of Women at Workplace (Prevention, Prohibition and Redressal) Act of 2013, the Internal Complaints Committee should be constituted by: (1) a chairperson and presiding officer who should be women "employed at a senior level" in the respective factory; (2) at least two employees "preferably committed to the cause of women"; (3) one member from an NGO or association "committed to the cause of women" (Indian Ministry of Law and Justice [2013]).

6.4.4 Interim Conclusion

In summary, workplace relations in Bangalore export-garment factories are constructed as de facto disciplining practices through various management practices of constructing workplaces as tightly controlled spaces, reproducing patriarchal power asymmetries in the workplace and undermining workplace committees as spaces for social dialogue. As such, workplace relations directly intersect with broader industrial relations (Sect. 6.5) and with sourcing relations (Sect. 6.2). Intersections of workplace relations with industrial relations result from the fact that power asymmetries between management and workers in the workplace hamper unions' capacities to build a strong membership and engage in collective bargaining with employers at the workplace and industry level. Intersections with sourcing relations, in turn, result from the fact that practices of enacting legal provisions for workplace committees are, in some factories, shaped by the guidelines of buyers' Social Dialogue Programs. Whereas in these cases, opportunities for social dialogue at the workplace are improved, managers nevertheless make sure to limit the scope of manager-worker dialogue to non-controversial issues that do not imply significant costs for employers.

For unions, the tight control exercised by management over workplace relations presents a great challenge since it constrains opportunities for union organisers to engage with workers and thereby build associational power at the workplace level. As a result of the significant barriers for organising workers inside the workplace, Bangalore garment unions have for a long time relied on alternative organising strategies that focus on organising garment workers in their communities (see Chapter 7). This strategy has, however, been increasingly challenged through the increasing geographical fragmentation of the workforce resulting from employers' practices of expanding labour market frontiers by recruiting workers increasingly also from villages in the rural areas surrounding Bangalore (see Sect. 6.7).

Nevertheless, in some cases where unions have—through strategic organising practices—been able to form strong worker leaders and build a membership inside the factory, unions have shifted the capital-labour power balance in the workplace and constructed more collaborative worker-management relations. This is highlighted for example in the following experience of GATWU worker leaders at an export-garment factory located in the rural town of Srirangapatna in about 150 km distance from Bangalore. At this factory producing exclusively for H&M, GATWU had organised the majority of the 1100 workers and pushed for democratic elections to the workplace committee. In these elections, GATWU worker leaders were elected committee members with 800 out of 900 total votes. According to GATWU, this overarching victory led the management to recognise GATWU worker leaders as workforce representatives and to engage in regular dialogue with them about worker grievances.

It is important to note, however, that while the management recognised GATWU worker leaders as elected worker representatives in the works committee, they still refused to recognise GATWU as a collective bargaining partner. Consequently, discussions with GATWU worker representatives in the works committee strictly excluded any issues referring to a change in service conditions, such as bonuses or wages. According to Indian labour law, works committees do not have the competence to negotiate on these issues since these issues belong to the sphere of industrial relations—that is to relations between employers and unions. Whereas in works committees, workers—unionised or not—can address problems at the factory, improvements for workers that go beyond the legally prescribed labour standards can only be negotiated by unions and management as part of industrial relations. Against this background, managers in the garment industry in Bangalore have a great interest in preventing collective bargaining and seek to maintain capital-labour power asymmetries in industrial relations through various union-busting practices. These practices will be illustrated in more detail in the next chapter.

6.5 Industrial Relations

I use the term industrial relations here to refer to relationships between employers, unions and state actors in the Bangalore export-garment cluster. Whereas industrial relations *include* relationships between unions and employers at the workplace level, industrial relations stretch *beyond* the workplace since they also involve practices of industrial dispute settlement and collective bargaining at the industry and state levels. These practices are constructed around various legal frameworks that have traditionally instituted workers' rights to Freedom of Association and Collective Bargaining and laid out several rules that facilitate interactions between unions, employers and state actors in industrial disputes. These legal frameworks are: (1) the Indian Constitution, 1949, granting all workers the right to Freedom of Association and Collective Bargaining; (2) the Indian Trade Union Act, 1926, specifying the criteria and process for union formation and registration; and (3) the Indian Industrial Disputes Act, 1947,[5] specifying the procedure and practices for settling industrial disputes between employers and workers or unions. According to the Trade Union Act, workers have traditionally been able to register a union at the factory or industry level when their membership comprises either 100 workers or at least 10% of the workforce in a specific factory or industry.[6] Registered trade unions may then act as official representatives of the workforce and negotiate with employers regarding any issues related to working conditions and economic benefits for workers, including inter alia wages, bonus payments, leave days, lay-offs, production norms and the terms and conditions of service. To enter into bilateral negotiations with an employer or group of employers, the respective management needs to recognise the union as

---

[5] As part of India's latest labour law reform, provisions from the Industrial Disputes Act and the Trade Union Act have been merged under the Industrial Relations Code, implemented in 2020 (see Indian Ministry of Law and Justice [2020]).

[6] The Indian Industrial Relations Code of 2020 has raised this threshold to 20% of the workforce in a respective industry or factory.

a bargaining partner. Indian labour laws, however, do not legally oblige employers to recognise a trade union as bargaining partner. Hence, recognition as collective bargaining partners is usually a contested issue, which requires unions to exercise associational power through industrial action to receive the management's recognition. Suppose a union has no membership base in the workplace and is, therefore, unable to exercise associational power in the workplace: In that case, it is virtually impossible for a union to realise the management's recognition and negotiate a collective bargaining agreement at the company or factory level.

Manufacturers employ two sets of practices to avoid collective bargaining processes in the Bangalore export-garment industry: discursively constructing the garment sector as exempt from industrial relations and union-busting practices. Manufacturers'—mostly illegal—union-busting practices are in turn enabled by a set of 'pro-business' state practices that undermine unions' opportunities for leveraging institutional power in the industrial dispute settlement process. These three sets of practices together lead to the construction of industrial relations characterised by employer dominance and a passive state. In the following section, I will illustrate these three sets of practices in more detail.

6.5.1 Discursively Constructing the Garment Sector as Exempt from Industrial Relations

In interviews, Bangalore garment factory managers give various reasons why the concept of collective bargaining is not feasible or applicable to the garment industry. The most frequent argument is that collective bargaining that leads to wage increases would ruin the industry in light of buyers' price squeeze and competition from other lower-wage garment production clusters in India or other Asian countries (INT1, 9, 27). This discursive construction of the garment industry as a unique sector to which industrial relations are not applicable is exemplified in the following quote from a Bangalore garment factory manager:

> I think unions are only mandatory for other industries. There are a lot of unions in, for example, the automotive sector and in the engineering industry. But these are capital-intensive industries. So, they have more machines and less people. In the garment industry, for 200 machines, you employ 500 people. So, if you provide unions for the textile industry, all the factories will probably have to close because buyers will go to other places. You see, the mindset of the workers is different. They don't understand this. They will ask for too much. (INT9)

As the quote illustrates, industrial relations in the Bangalore export-garment clusters intersect with wage and sourcing relations. The economic pressures for lower wages resulting from buyers' predatory purchasing practices provide additional incentives for Bangalore garment manufacturers to undermine unionisation and collective bargaining processes present in many other industrial sectors in India.

Managers' narrative of industrial relations and collective bargaining not being applicable to the garment industry is further underpinned by the discursive framing of unions as 'troublemakers', who seek to 'create problems' and disturb the industrial peace. As part of this narrative, Bangalore garment managers frequently refer to buyers' social standards and regular audits as substitutes for industrial relations since these mechanisms supposedly ensure acceptable working conditions and therefore make collective bargaining obsolete. This discursive framing is highlighted in the following statement of an HR manager at one of Bangalore's leading garment export manufacturers:

> You know, in theory, unions and collective bargaining are good concepts. But bargaining is supposed to happen when there is a problem, when minimum standards for working conditions are denied. But if nothing is denied, what should we bargain about? With the brands coming in, we got that whole regime of codes of conduct and audits and now the situation is much better [than in the beginning of the export-garment industry], we have really achieved a lot. I mean nobody is perfect, but we have achieved 90%. So why would you turn the whole thing upside down just for 10%, which has not been achieved? (INT1)

The discursive construction of unions as 'troublemakers' whose demands for better working conditions are harmful to the industry is manifested in the following quote by this regional representative of a national garment industry association as well:

> Unions are required for the industries, no doubt about it. For developing, for the sake of staff welfare. No doubt. But what happens is, everything is not taken in the right sense by the union, also. People should be more... they should also understand the ground realities. Unfortunately, [...] [in the export-garment sector] the role of the union is misunderstood. The union does not look into the [...] requirements of the industry. [...] it is misunderstood by the unions that it is only for the sake of workers. Unions should also go one step further then [and] make the employees understand the need of the day, that is, higher productivity and higher quality. [...] See, when the union is able to create an impact with the employees, that means they are powerful. So, with the same power, they should also start spreading some positive attitudes among the employees. (INT27)

The two manager quotes exemplify the overall 'anti-union' stance common among Bangalore export-garment managers and industry representatives. This stance also informs other sets of practices by managers directed at avoiding unionisation and collective bargaining at the workplace level.

6.5.2   Union-Busting Practices at the Workplace Level

As mentioned in the previous chapter, industrial relations directly intersect with workplace relations through 'union-busting' practices that are usually performed in the workplace. I use the term 'union-busting' to refer to a set of practices directed at repressing, mitigating or preventing collective worker organisation and collective bargaining. Union-busting practices performed by Bangalore garment manufacturers include leading an anti-union discourse, closing down factories with high unionisation levels and various practices of victimising union activists.

Leading an anti-union discourse refers to managers' widespread practice to discourage workers from relating to union members or activists by discursively framing unions as an external threat in interactions with workers. As the following quote by a Bangalore garment union leader illustrates, when unions intensify their organising activities in a specific factory, managements seek to intimidate workers and make them refrain from interacting with unions:

> They [the management] tell them [workers] that if you keep doing this union thing, the factory will close down. [...] see, [in that factory] there's a considerable number of workers who are not unionised yet. So, although we are saying we have the majority, the management doesn't accept that [...]. But the other workers who are not part of the union have all been kind buying the management narrative that union is bad for the industry. So, that also works as a pressure tactic. (INT50)

This practice of framing unions as a threat that will cause the factory to close and workers to lose their jobs is particularly powerful in the Bangalore export-garment industry due to the specific local historical context. The image of the garment industry and unions prevalent among workers is still shaped by the historical practice of Bangalore garment factories under the quota regime just to close down after production quotas had been met (see Sect. 5.2.1). Moreover, many workers still remember an incident from the 1990s when a central trade union conducted a strike at a garment export factory for several months forcing the factory to close down (see Sect. 5.2.3). As mentioned in Sect. 5.2.2, today, tier one supplier factories in the Bangalore export-garment industry are predominantly owned by larger export-garment companies with several production units. These large export-garment companies have a higher economic capacity to withstand strikes by unions or economic slumps. Nevertheless, the historical fear among garment workers that union activity may lead to factory closures persists and is further reinforced through continued management practices of discursively framing unions as external threats to the factory. As a result, unionists report that many workers are afraid to engage with unions, which presents unions with severe challenges for organising and constrains their ability to build associational power resources.

Workers' fear of engaging with unions is further reinforced by management's practices of closing down factories with high unionisation levels. In this context, GATWU leaders report several incidents where, when their membership had reached a significant level inside a factory, the management closed the factory and reopened it in rural areas, sometimes up to 100 km outside of Bangalore (INT46). According to legal provisions, management is obliged to offer laid-off workers employment in the new factory in case of a factory relocation. In practice, long commuting times, however, often make it infeasible for most women workers to work in the new factory, given their double burden of wage work and care work. Hence, workers usually prefer to seek employment in another factory closer to their living areas or even in the growing service industry, offering growing job opportunities for unskilled workers. As a result, the union membership base often built in year-long organising work can be effectively destroyed by management through factory relocations. That managers deliberately select factories with high unionisation levels for relocation has become particularly obvious during the COVID-19 pandemic. During the pandemic, several large 'mega-suppliers' owning between 10 and 50 factory units in and around Karnataka seized the slack in orders to restructure their operations by closing down unionised factories in and around Bangalore and reopening these factories in remote, closed-off apparel parks with restricted access for unions (ExChains 2020).

In addition to reproducing an anti-union discourse and closing down factories with high unionisation levels, managers in Bangalore garment factories seek to prevent the unionisation of workers through various practices of victimising active union members or union worker leaders inside the factory. A common practice of victimising active union members or worker leaders is locking them out of the factory or not giving them any work for several days. This often happens under the pretext of some excuse, as exemplified in this quote from a unionist and former garment worker:

> In our factory, we have founded a factory-level union committee about two years ago. [...] But it was not an easy process. There was a lot of harassment from the factory management. They tried to intimidate workers with different means. For example, there were 34 colleagues who are also union members who had been in the cleaning team. But when we founded the factory union, those workers were "promoted" to tailors, but they did not receive any training. So, they were not prepared for the job and made mistakes. Because of that, they were left out of the factory for 27 days and not allowed to work. (FN4, translated from Kannada)

Locking workers out of the factory is an effective disciplining practice for two reasons: First, it takes an emotional toll on workers due to the insecurity of whether they will be able to work in the factory again and due to the embarrassment of having to wait in front of the factory. Second, it puts workers under economic pressure since being unable to work for several days causes significant wage losses for workers, given the already meagre wages in the garment industry.

The most extreme practice of victimising union members is dismissing union members permanently. According to the Indian Industrial Disputes Act, any worker who has served in a factory for longer than one year cannot be dismissed without reasonable justification and prior government permission. When firing union activists, managers construct false allegations against the worker to circumvent this legal provision. This management practice is illustrated in the following report by a worker and active union member:

> I am a union member, but the management does not know. If they find out, I will suffer. [...] The management already has an eye on me because I participated in the strike on 2 September. I am a tailor, and I make collars for shirts. [...] At the end of the day, I store the collars at my workplace. From there, they are collected every morning and brought to a different department where the collars are sewn to the shirts. One morning, when I came to my workplace, the collars were not there anymore. So, the management said it is my fault, and I must bring them back or they will fire me. (FN4, translated from Kannada)

Hence, victimising active union members and union leaders needs to be understood not primarily as a practice directed at preventing these workers from doing union work but rather to make them leave the job. In many cases, union leaders inside the factory succumb to managers' harassment and resign due to the emotional stress, as this unionist reports:

> In [factory name], they have also targeted our worker leaders. In the cutting section, they made five of our workers who are very active in the union stand for three days. They were so humiliated [...] they were made to stand, and they were not even given chairs. They are senior workers in the cutting section, and they are men. So, they felt very humiliated [to stand] in

front of thousands of people. When such things happen and when they feel humiliated, they
leave the job. Such cases are increasing. (INT36, translated from Kannada)

To further intimidate fired unionists and prevent them from appealing the dismissal,
workers and unionists report that managers frequently hire groups of 'rowdies'.
These rowdies wait for the dismissed worker in front of the factory or even visit
the worker's home telling him or her to stay away from the factory. 'Union-busting'
practices by managers are hence not limited to the workplace but also stretch into
workers' reproductive spaces.

Since freedom of association and collective bargaining are legal worker rights in
India, managers' union-busting practices represent de facto labour law violations.
Hence, unions can file a complaint with the labour department and seek rectification
of these violations through the legal-institutional dispute settlement framework laid
out in the Industrial Disputes Act. However, pro-business state practices constrain
unions' possibilities for activating institutional power resources, as illustrated in the
next section.

6.5.3 Undermining the Industrial Dispute Settlement Process

In case of conflicts between workers and employers, the Indian Industrial Disputes
Act of 1948 has introduced an institutionalised tripartite settlement process. The
legal framework for industrial dispute settlement foresees three practices: (1) filing
a complaint at the labour department to settle the dispute, (2) conducting a tripartite
conciliation process involving a labour department officer, management and union
representatives, and (3) adjudicating the dispute in the labour court. As such, the
legal framework for settling industrial disputes can represent a potential source of
institutional power for unions. However, in the Bangalore export-garment cluster,
unions' opportunities to use this legal framework as a source of power vis-à-vis
employers—especially to contest illegal dismissals—are constrained by the specific
practices through which labour department officers and judges enact the framework.

Unions' possibilities for filing a complaint at the labour department against the
illegal dismissal of a union activist are constrained by labour department officers'
practice of insisting on the provision of proof that the dismissal was indeed unjus-
tified. For unions, it is, however, difficult to provide such proof since managers
frequently bribe or intimidate co-workers into giving false statements supporting
management's version of events. At the same time, labour officers seldom make
use of their right to conduct an independent inspection of the case that could refute
management's false allegations. Hence, when filing a complaint, usually the union's
version of events stands against the management's word (INT15).

Furthermore, when conciliating a dispute, union representatives report that state
officers tend to take on a passive position. Instead of demanding evidence from
management representatives for their allegations against a dismissed worker, labour
officers usually limit their intervention to facilitating dialogue between the parties to

arrive at a compromise. As a result, power asymmetries between management and workers are reinforced through the intervention of labour officers. The leader of a Bangalore garment union attributes this relatively passive role of labour officers in industrial disputes to a very peculiar interpretation of their official mandate to resolve the conflict peacefully:

> The management will present their version of events, and the conciliation officer will say "Yes, yes". Because their mindset is also to promote the ease of doing business and, you know, not to cause unrest so that the industry doesn't get affected. So, they'll try to pacify the workers […] It's very rare that an officer has the guts to say "I'm sorry, I'll take more action. I'll come and take all your records and verify it". Most of the industrialists already have some political cloud and the labour commissioners know that if they wag their tail too much, then managers will go beyond them and get the issue resolved somewhere else. So, they [labour commissioners] are very limited in their abilities to actually push them [management] in the conciliation. The mandate for the conciliation is to resolve it [the conflict] peacefully. And peacefully would mean that you have to somehow compromise the interest of the workers. (INT48)

As a result of this particular interpretation of resolving a conflict peacefully, the conciliation officer in charge seldom declares the dismissal of a union activist as illegal nor orders management to reinstate the fired worker. Instead, labour officers seek to promote a settlement between management and the worker—a settlement that, however, usually compromises the workers' and unions' interests, given the asymmetrical capital-labour power relations in the Bangalore export-garment cluster. Union leaders report that, in this vein, labour officers often propose that management merely pay the dismissed worker a compensation or reinstate the worker in a different factory unit. In the latter case, managers often deliberately select a factory unit far away from the original, knowing that the worker won't be able to commute there. This practice of labour department officers pushing for a compromise between unions and workers, therefore, de facto, enables management to effectively deploy illegal dismissal of union activists as a union-busting practice. By removing the worker from the factory, management effectively breaks the unionisation process in the factory. Moreover, the fact that the union activist could not be reinstated even after the union filed a legal complaint enables management to use the dismissal as a showcase to discourage other workers from joining the union. When the worker and the union consequently refuse management's offers for a compensation payment, the individual case is referred to the labour court for adjudication.

The practice of *adjudicating* a labour dispute involves convening court meetings, conducting hearings for evidence-taking and issuing a ruling. Due to severe under-staffing with vacancy rates of about 50% in Indian labour departments and labour courts, the rhythm with which hearings for evidence-taking are held is, however, very slow, with several weeks or months passing between two hearings. As a result, it takes, on average, seven years until a ruling is made. This delay often results in fired union activists taking on a job in a different factory or industry, given that unions usually do not have the financial means to support fired activists for such a long time. Accordingly, it is a common practice of employers to drag on or block the conciliation process so that the case is transferred to the court for adjudication, as this labour researcher explains:

> Normally, the management knows that the judiciary is overburdened. So, they know, they
> keep on lingering the cases so much that it automatically ends at the court [...] For the
> court, labour laws or the workmen are not a priority. So, they are almost at the bottom of
> the pyramid at the court. So that is the problem that workers face, that trade unions face.
> (INT15)

Labour researchers and unionists argue that state actors' pro-business practices in
the industrial dispute settlement process need to be understood as shaped by the
general neoliberal policy turn in India since the 1990s. The industrial dispute-settling
process and Indian labour laws, more generally, were designed post-independence
as frameworks for worker protection and ensuring workers' well-being. Whereas
the legal-industrial frameworks have not significantly changed since the Indian post-
independence period, the atmosphere for implementation, however, has, as this labour
rights researcher explains:

> In India's industrialisation, collective bargaining has played a huge role. And I think that has
> helped both industrials as well as workers. It is only post-liberalisation, that the atmosphere
> for labour rights has changed. And that is largely through this idea of flexible labour which is
> propagated in India through the Washington Consensus. It is exactly part of the augmented
> Washington Consensus, which some people called 'Post-Washington'. And even though
> laws were not amended, the atmosphere for implementation completely changed. (INT34)

With the shift towards neoliberal policies, the role of the Indian state has shifted from
an enforcer of law to a facilitator of the ease of doing business. As a consequence, legal
mechanisms and institutionalised processes for settling industrial disputes have lost
their function as institutional power resources for unions, as this unionist explains:

> Actually, there was a time when the labour commissioner had a lot of...quite a bit of the
> industries used to be frightened to come to the Labour Department. There was at least
> the security that violations will need to be rectified. But ultimately, even the judiciary has
> changed. Earlier [...] lots of judicial decisions went in favour of the workers. So, the atmo-
> sphere was in a sense that there was almost neutrality. But now it is definitely biased against
> the workers' interest and pro-management and corporate interests. And that actually also
> affects unionisation, because workers don't want now a long-drawn process which ultimately
> ends up in some defeat. (INT48)

Most recently, the neoliberal shift in India's state apparatus' practices has also been
complemented by several pro-business amendments to the long-standing labour laws
as part of India's latest labour law reform. In 2019 and 2020, the Indian government
passed four new Labour Codes[7] that subsume existing labour laws intending to make
labour legislation more comprehensive and easier to apply. In this context, the new
Industrial Relations Code has introduced several provisions that constrain unions'
institutional power resources. As such, whereas unions could formerly, under the
Trade Union Act, form a factory union and engage in collective bargaining with
employers when they had organised 10% of the employees in a specific factory, the
new Industrial Relations Code foresees a threshold of 20% of worker organisation in a

---

[7] These Labour Codes are the Code on Wages, 2019; the Code on Social Security, 2020; the Occu-
pational, Safety, Health and Working Conditions Code, 2020; and the Industrial Relations Code,
2020.

specific factory. In addition, the new Industrial Relations Code curtails workers' right to strike by introducing a fourteen-day notice period for any strike. Moreover, the new Industrial Relations Code prohibits strikes while a conciliation or adjudication of a matter is in place. This restriction was only applicable to workers in public utility services before the reform. As a result, workers and unions in the Bangalore garment export sector have been confronted with increasing constraints on their legal-institutional power resources over the past years due to pro-business state practices on the one hand and recent pro-business reforms of labour laws on the other hand.

6.5.4   Interim Conclusion

In summary, industrial relations in the export-garment industry in Bangalore link unions, employers and state officials within the State of Karnataka. In particular, industrial relations are shaped and constructed through various employer union-busting practices in the workplace, which are, in turn, enabled by pro-business state practices in the legal-institutional industrial dispute settlement process. Employers' union-busting practices constrain unions' capacities to build a solid membership base in the workplace. State actors' pro-business practices in turn constrain unions' capacities to contest employers' union-busting practices through complaints at the labour department or legal appeals. While India has traditionally had strong labour laws, conceived to offset employer dominance through strong state engagement, over the past two decades, pro-business state practices are increasingly undermining labour law frameworks as sources of unions' institutional power. In the Bangalore export-garment cluster, employer union-busting practices and pro-business state practices construct industrial relations as conflictive and antagonistic relations characterised by strong employer dominance.

To understand state actors' and employers' interest in creating antagonistic industrial relations and suppressing collective bargaining, it is important to understand the interrelations of industrial relations in the Bangalore export-garment cluster with two other sets of relations: sourcing relations (Sect. 6.2) and wage relations (Sect. 6.3). In light of retailers' practices of 'squeezing prices' combined with the highly labour-intensive nature of the production process, employers and state actors seek to construct the garment industry as a union-free space to prevent potential wage increases resulting from collective bargaining.

Against this background, engaging employers in collective bargaining represents a major challenge for Bangalore garment unions. In this context, all three local garment unions have sought to develop networked agency strategies over the past five to seven years, focussing specifically on building associational power and advancing collective bargaining processes in selected target factories (see Chapter 7). A significant challenge for unions to build a stable membership base in selected factories, however, results from the highly volatile employment relations in the Bangalore export-garment cluster and resulting high turnover rates in garment factories. The following section sheds light on the various employer practices creating volatile employment relations in the Bangalore export-garment cluster.

6.6 Employment Relations

Employment relations are an essential element of the labour control regime in the Bangalore export-garment cluster. They are constituted through various networked practices performed by employers to keep labour costs down and outsource economic risks to workers. Employment relations in the Bangalore garment cluster link workers, garment manufacturers and temporary employment agencies in Bangalore and across the State of Karnataka. Employment relations are primarily constructed around specific territorially embedded workplaces. At the same time, employment relations in the Bangalore export-garment cluster are also shaped by sourcing relations at the vertical dimension of the GPN. Since retailers do not guarantee a minimum of orders per year even to their core suppliers and place orders at irregular intervals, Bangalore garment manufacturers employ various practices to flexibilise employment relations to avoid excessive labour force during periods with low or no orders. These practices usually circumvent India's relatively strict legal frameworks regulating employment relations: Traditionally, Indian labour laws have not provided employers with the opportunity to hire workers with fixed-term contracts. Moreover, as illustrated in Sect. 6.5, employers traditionally needed to seek prior permission from the labour department to dismiss workers. Against this backdrop, Bangalore garment manufacturers have established two major practices to circumvent these legal restrictions, leading to a de facto flexibilisation and informalisation of employment relations. These practices are: (1) 'hiring and firing' and (2) using contract labour.

6.6.1 Flexibilising and Informalising Employment Relations Through 'Hiring and Firing'

'Hiring and firing' refers to a common practice performed by employers in the Bangalore export-garment cluster of hiring and firing workers without following legally prescribed procedures. Unions report, for example, that—despite legal provisions and buyer requirements—workers traditionally did not receive a formal appointment letter or contract. This practice of informally *hiring* workers was widespread in the cluster until some years ago. However, due to various legal complaints filed by unions, this practice has become less prevalent. Nevertheless, informally *firing* workers is still a prevalent practice by Bangalore garment manufacturers. As mentioned in Sect. 6.5, managers in Bangalore garment factories frequently dismiss or retrench workers without seeking the approval of the labour department and without giving workers a formal letter of termination. Instead, workers are just denied access to the factory one day. Managers use this practice of informally dismissing workers for several purposes, from undermining unionisation processes (see Sect. 6.5) to avoiding legally prescribed benefits and gratuities for workers. Managers informally dismiss workers to avoid paying maternity leave benefits or gratuities accruing to workers for continued service (INT21). In the latter case, workers and unionists report that workers are often locked out of the factory for one or several days shortly before completing five years of service. Once workers are allowed back into the factory, their social security ID number has changed, meaning they have been registered as newly employed. Thereby, employers circumvent the provisions installed by the Indian Gratuity Act of 1972: According to this law, employees are entitled to a gratuity of 50% of the monthly wage for each year of completed service after five years of continued service with a company when leaving the job.

Illegal practices of 'hiring and firing' have led hence not only to the flexibilisation but also to the de facto informalisation of employment relations in the Bangalore export-garment cluster. For workers, the informalisation of employment relations means having to live with high levels of insecurity regarding social benefits and regarding employment, as the experiences described by this worker illustrate:

> There was a case of a worker who had worked for five years at the factory. Then she resigned because when you quit after five years, you are entitled to a gratuity. And she is from a poor family, so she could really use the money. But then she rejoined the factory. So, until today she has never received her gratuity. […] Also, there have been several cases where workers have been fired all of a sudden. So, we are all afraid of this. Because today we work at this factory, but we never know if we will still have our job the next day. (INT5)

This de facto informalisation of employment relations also constrains unions' potential for worker organising and building associational power resources in two ways. First, workers are generally afraid of being fired for becoming a union member and, therefore, harder to organise, as illustrated in the previous section. Second, since workers generally perceive their employment as insecure and unstable and expect no benefits from working at a factory for a prolonged time, they frequently quit their jobs for various reasons. These reasons include taking a prolonged leave to return to their native villages during harvest, accessing provident fund contributions, taking on a new job with slightly higher wages in another factory or avoiding particularly abusive supervisors or managers. As a result, attrition rates in Bangalore export-garment factories range around 10% per month, meaning that, on average, the whole workforce of a factory changes within a year. High attrition rates and an unstable workforce in Bangalore garment factories consequently constrain unions' capacities for building a stable membership and associational power at the workplace level, as this garment union leader explains:

> We have been organising garment workers in Bangalore city since 2006. With the growth of the industry, organising in the city has however become more difficult. There are so many garment factories now in the city area, and there is a high turnover rate of workers at the factories. When we start organising at a factory, and we have been organising there for one year, then in the second year most of the workers with whom we started working in the beginning won't be there anymore. If workers in the city are facing problems within their

factory, instead of struggling, they rather leave the factory and search for a job somewhere else since it is easy to find work in another factory. (FN1)

Whereas employers have traditionally used 'hiring and firing' practices to circumvent legal restrictions for fixed-term employment and worker lay-offs, several pro-business labour law reforms have recently provided legal ground for ending workers' services after a certain period of time without justification. In 2016, the Indian government passed a reform of the Indian Industrial Employment Act of 1946, introducing the exclusive option for employers in the garment sector to hire workers based on fixed-term employment. The government justified the reform with the need for garment manufacturers to cope with order fluctuations and retailers' seasonal sourcing practices, requiring higher employment flexibility than other sectors[8] (Business Standard 2018). Since the reform, unionists and labour researchers report that manufacturers have further reduced the number of permanent workers while hiring additional workers with fixed-term contracts, specifically during peak season. More recently, the Indian government passed a new Industrial Relations Code that merged several older acts in labour legislation, including the Industrial Disputes Act and the Industrial Employment Act. As part of this reform, the government raised the threshold of employees above which employers need to seek approval from the labour department in case of factory closure, lay-offs or retrenchments from 100 to 300 workers. A unionist comments that it is an increasingly common practice for export-garment companies to register the different departments within the factory, such as the cutting department, the sewing department and the washing department, under different companies belonging to the same company group (INT13). In doing so, manufacturers can avoid seeking government approval when closing down or moving factories—a fact that further increases employment insecurity for workers.

6.6.2 Using Contract Labour to Reduce Permanent Labour Costs and to Undermine Unionisation Processes

Besides 'hiring and firing', employers in the Bangalore export-garment manufacturing cluster flexibilise employment relations through a second practice: using contract labour. The Indian Contract Workers Act (1977) has traditionally restricted employers' use of contract labour in two regards. First, employers may only employ contract labour in non-core activities. Second, contract workers can work only up to 240 days per year at the same factory. If a contract worker completes more than 240 days of work at a factory, he or she should automatically become a permanent employee. De facto practices of using contract workers in Bangalore export-garment factories, however, circumvent these legal provisions: Garment manufacturers tend

---

[8] Nevertheless, the option to employ workers on a fixed-term contract was subsequently extended to all other industrial sectors by the government in 2018 as part of its general policy focus on improving the 'ease of doing business' (*The Economic Times* [2018]).

to rely on using male contract workers, in particular for the lower-skilled segments of the finishing process such as ironing or packing. Unlike permanent employees, contract workers tend to be paid by a piece-rate system and can be laid off quickly so that employers avoid having to pay these workers during periods with few orders.

In addition, employers are also increasingly using contract workers as an additional labour force for night shifts during peak order periods. According to the Indian Factory Act of 1948, it was traditionally prohibited for women to work shifts between 6 pm and 7 am. The Government of Karnataka has introduced the possibility for women to work nights in 2020. However, barriers to women working night shifts persist due to social norms and women's care responsibilities. Therefore, employers rely on male contract workers to maximise productivity during peak order periods. By relying on contract workers for specific tasks such as ironing and packing, employers can reduce the permanent workforce and thereby avoid paying workers during unproductive times when no orders are available or during times of sickness or annual leave.

Besides using contract labour to flexibilise employment, using contract labour is also performed by employers in the Bangalore export-garment cluster as a deliberate disciplining practice to gain concessions from workers and to undermine worker organising. In this manner, workers at the warehouse of a major export-garment company report that the management strategically increased the share of contract labour following the unionisation of the warehouse to divide the workforce and weaken worker organising. Since contract workers are formally employed by a contract labour agency (and not by the garment manufacturer), they cannot join the same factory union committee representing workers directly employed by the company. Moreover, since contract workers in this warehouse are employed on a piece-rate system, they can earn higher wages than regularly employed workers. Hence, the use of contract labour creates tensions and new lines of segmentation along the lines of employment type and payment among the workforce. During an international union meeting, a union activist describes management's use of contract labour as a disciplining practice in the following words:

> Since workers have started to organise, the management has been hiring more and more contract workers to split up the workforce. Before we founded the union, all workers were paid per hour. Now the unionised workers are being paid per hour, while new contract workers are being paid per unit, and they can make a lot more money. While the workers who are paid per hour make around 7,000 Rupees per month, the workers paid per unit can make up to 20,000 Rupees per month. This splits the group of workers and makes joining the union unattractive. But the group of contract workers is also divided. At our warehouse, we have around 200 workers, but they are employed by three different firms. (FN3, translated from Kannada)

The union activist's statement illustrates how Bangalore garment manufacturers employ contract workers to undermine unionisation processes at the workplace, thereby constraining unions' capacities for building bargaining power vis-à-vis employers. In this sense, employment relations also intersect with industrial relations (Sect. 6.5). The specific practices through which employment relations are constructed contribute to reproducing employer dominance and hamper unionisation and collective bargaining processes.

6.6.3 Interim Conclusion

In short, manufacturers' practices of hiring and firing and using contract labour contribute to the construction of employment relations in the Bangalore export-garment cluster as de facto informalised and flexible relationships between employers and workers. These relations are frequently mediated by contract labour agencies as third parties. Employment relations are constructed first and foremost around national legislation, with manufacturers taking advantage of provisions allowing for labour flexibilisation while circumventing restricting provisions. Therefore, employment relations in the Bangalore export-garment cluster are territorially embedded. However, employment relations also intersect with sourcing relations characterised by network embeddedness (see Sect. 6.2). The main arguments with which employers and legislators justify informal practices and formal labour law reforms to flexibilise employment relations are retailers' seasonal sourcing practices and the resulting order fluctuations, requiring employers to adapt the size of the workforce flexibly. However, coping with order fluctuations is not the only motive for Bangalore garment manufacturers. As illustrated, manufacturers also employ practices of 'hiring and firing' as and of using contract labour to reduce labour costs by avoiding paid maternity leave or seniority benefits for workers. Lastly, manufacturers also employ these practices as deliberate disciplining practices to instil fear among workers of being fired for joining the union or to divide the workforce. Therefore, employment relations also intersect with industrial relations (see Sect. 6.5) because they reproduce employer dominance and hamper unionisation and collective bargaining. As illustrated, the specific practices through which employers construct flexibilised and informalised employment relations lead to high attrition rates and workforce segmentation along lines of contract. These conditions, in turn, constrain unions' capacities for building a stable membership inside specific factories.

6.7 Labour Market Relations

Labour markets are an essential part of any local labour control regime. Ideally, labour markets fulfil the crucial function of ensuring adequate labour supply—an essential precondition to guarantee the continuous reproduction of the labour process. To fulfil this function, labour markets, however, need to be actively constructed by capital and state actors through training and skilling practices as well as through recruiting and placement practices. In the Bangalore export-garment cluster, labour market relations link garment manufacturers located in the State of Karnataka with workers, training centres and recruitment agencies located in Bangalore and in the broader State of Karnataka and other Indian states. Labour market relations in the Bangalore export-garment cluster have traditionally been highly localised, linking workers from Bangalore and surroundings with factories in urban Bangalore (see Sect. 5.2.1). However, over the past decade, employers have implemented various practices to expand labour market relations. These practices need to be understood as responding to a growing shortage of unskilled or semi-skilled labour in the Bangalore urban area. At the beginning of the 2000s, the labour market in the Bangalore metropolitan area still provided an abundant supply of unskilled workers due to the incorporation of women into the labour market. However, with the rapid growth of the IT industry and other international industries, such as the pharmaceutical industry, and the rapid development of the service and transport sector, many alternative jobs have emerged for unskilled and semi-skilled workers in Bangalore.

Given the low wages and high work intensity in the garment sector, many former garment workers from Bangalore now prefer to seek work in the growing service sector, offering more attractive conditions. The minimum wage for messengers—an unskilled position—in the Bangalore urban areas, for example, is around 13,300 Rupees (approx. 176 US$). However, a semi-skilled shop assistant in Bangalore already earns roughly 14,500 Rupees (approx. 192 US$). In contrast, the minimum wage for semi-skilled garment machine operators in Bangalore is only roughly 10,000 Rupees (approx. 132 US$) (Labour Commissioner Office, Government of Karnataka 2021). In addition to offering better wages, work in the service sector is characterised by less rigid performance control and less physical strain on workers compared to the garment industry, where work is characterised by rigid production targets and monotonous tasks (see also Sect. 6.1). As a result, Bangalore-based garment manufacturers report a labour shortage of around 10 to 20%, meaning that 10 to 20% of installed sewing machines are not operated due to a lack of operators (INT9, 28). The factory manager of an export-garment factory located in the industrial area Yelahanka near the airport even estimates a labour shortage of around 50% if the factory had to recruit workers exclusively from Bangalore:

> In Bangalore, we can't get manpower anymore. See, here, a helper gets 299 Rupees a day. The shopping malls give a better wage, and there, the worker can even sit in the AC. Even here in our factory in Yelahanka, we have almost no workers from the city. Most of them come from outside. In Bangalore city, it's like this: We maybe have capacity for 1000 workers, but we can only get 500. (INT 9)

In the face of these increasing challenges for recruiting workers from the Bangalore urban labour market, Bangalore garment manufacturers have, over the past years, developed various practices directed at territorially expanding the labour market frontier. These practices have, in turn, been enabled by the various intertwined state policies and practices by employer associations and training and recruiting agencies that, together, have constructed a complex training and migration regime for the

Indian garment industry. In the following, I will first illustrate the policies and practices underpinning this regime and afterwards lay out the various practices through which Bangalore garment manufacturers draw on the rural and migrant workforce trained under the regime to territorially expand labour market relations beyond the Bangalore urban area.

6.7.1 Constructing a Complex, Multi-Level Training and Migration Regime for the Indian Garment Industry

To ensure continued labour supply in the Bangalore export-garment industry (and in other major garment clusters), the Indian government and industry associations have established a complex vocational training regime to train the rural population—especially women—to become machine operators in garment factories. In this context, the Government of Karnataka and the Apparel Training and Design Centre (ATDC), the training arm of the Indian Apparel Export Promotion Council, have since the late 2000s set up a large number of apparel training centres in rural parts of Karnataka as part of the state government's Textile and Garment Policy. These training centres are part of a broader landscape of vocational training centres in India that are run by public and private agencies as well as through public–private partnerships and that fulfil the function of providing basic skills for India's vast rural population (Ramasamy and Pilz 2020; Wessels and Pilz 2018). In this context, apparel training centres offer various courses to train high school dropouts or graduates for employment in the garment industry. The most widely offered course is a course for sewing machine operators that has a duration of six to eight weeks and only requires the completion of the 5th grade. The majority of trainees are women between 18 and 35 years. In apparel training courses, trainees learn how to operate industrial high-speed sewing machines, including various machines for specialised operations such as making button holes. In addition, workers are prepared for the work in a factory environment through lessons in social skills and professional ethics, as this Karnataka state government representative from the Department of Handlooms and Textiles explains:

> So, [trainings are] regarding punctuality, then safety measures. How to treat the industry as their own industry. Like, if the workers treat the industry as their own, they will work more. The efficiency will be more. They [the trainers] will make them understand the industry better. The atmosphere, friendly atmosphere. So, they have to be cordial with the hierarchy, cordial with the co-workers, like that. (INT 38)

As this quote shows, a vital part of the training is to hedge the 'indeterminacy of labour' by preparing women from rural areas, who are usually first-generation workers, for subordinating themselves under the labour process in a factory environment.

In addition to short-term courses for semi-killed sewing machine operators, selected training centres also offer courses of longer duration (six months or one year), enabling trainees to work in skilled or high-skilled positions related to programming, operating and maintaining computer-assisted design and manufacturing machines. Currently, the Government of Karnataka and ATDC maintain roughly 300 garment training centres all over Karnataka. Government-sponsored centres are run either directly by government agencies or by private training and placement agencies acting as contractors. The number of training centres has been significantly boosted between 2010 and 2017 with the introduction of the Integrated Skill Development Scheme (ISDS) by the Indian National Ministry of Textiles. The ISDS represents one of several schemes that the Indian government has introduced over the past 15 years to provide basic skills to (primarily rural) population segments with lacking or low formal education. Thereby, the government aims to boost employment while at the same time producing labour supply for the country's growing urban industrial sectors (Pilz and Regel 2021).

In this context, the Indian Ministry of Textiles has introduced a subsidy under the ISDS that covers 75% of the costs per trainee undergoing short-term vocational training as a sewing machine operator. The Government of Karnataka funds the rest of the training costs plus an additional transport stipend for workers coming to the training centres from other villages. Agencies carrying out training, in addition, receive a financial bonus if 75% of candidates from a training batch are employed within three months upon completion of the course. Whereas training centres provide trainees with the basic technical and social skills required to work in a factory, workers usually undergo an additional in-company training of four weeks after being placed in a factory. In this in-company training, workers' motor skills are assessed and workers receive further, specialised training for the machines used in the factory (FN5, INT32).

In addition to establishing regional training regimes in states with major garment clusters, under its PAN India component, the ISDS has also incentivised the construction of an inter-state training and migration regime. This regime links manufacturers in major garment clusters with rural youth in the Northern and North-Eastern states of India. For-profit agencies play a central role in training and recruiting young, unmarried women between 18 and 23 years from rural villages in Northern and North-Eastern, predominantly agricultural states such as Odisha, Assam and Bihar. In addition, these agencies facilitate the trainees' placement in a garment factory in one of India's major garment industry clusters, organise their migration process and, in some cases, maintain hostels to accommodate migrant workers when arriving in the garment cluster. To receive training under the PAN-India component of the ISDS, young women need to give their consent to migrate to a different city and work there for at least six months. The training consists of a two-month course in a training centre in the trainees' home state and one additional month of on-the-job training at the factory (Gram Tarang 2020). The emergence of a complex training and migration regime at the state and national level has hence laid the relational base for Bangalore garment manufacturers' practices of expanding the labour market frontier, which will be described in the next section.

6.7.2 Securing Adequate Labour Supply Through Expanding the Labour Market Frontier

The complex vocational training regime has enabled Bangalore export-garment manufacturers to compensate for the labour shortage in the Bangalore cluster by expanding the labour market frontier through three sets of practices: (1) recruiting workers from rural villages within Karnataka, (2) moving production facilities for garment assembly to rural villages and textile parks and (3) hiring inter-state migrant workers originating from Northern and Eastern, economically weaker states of India such as Odisha, Assam and Bihar.

First, Bangalore export-garment manufacturers have started *recruiting workers from rural villages up* to 80 km away from Bangalore, who have undergone training in one of the numerous garment training centres. To ensure that workers can commute daily to the factory, garment companies provide company transport for workers from these areas. To recruit workers from villages, Bangalore garment manufacturers actively carry out recruitment campaigns in rural areas, as this HR manager of a major Bangalore export-garment company explains:

> So, there is a lot we do in terms of providing transportation. We carry people from distant locations, sometimes up to 80 km away. That is because locally, in the city, people are not available. So, we have to get them from distant places. We do job mailers, we have to work on rest days, go search for people, do road shows, get people to sell the company to them, so that they say 'Okay, this is a company I will work for'. We have to woo them. If we don't poach them, we woo them. (INT29)

As a result, a significant part of the workforce in Bangalore export-garment factories today comes from villages around Bangalore and is transported to the factories and back home in company-provided buses. This spatial division between workers' living areas and workplaces poses new challenges for unions' organising work: since most workers now have a long commute ahead of them, especially women workers have little time to engage with union organisers for a chat or meetings after work, given their care work responsibilities at home. As a result, organising workers has become more difficult for unions, as the following statement by the leader of a Bangalore garment union illustrates:

> Earlier we would go and talk to workers while they were walking back after work, but nowadays, the management has put them in buses, so we can't do that. [...] Even if we try to meet them at the gate, workers cannot talk as they are in a rush to board the bus, because there are never enough seats for everyone [...]. Also, workers don't stay close, they come from different communities, and the van will go drop them off one after another. The minimum journey in the van is 30 minutes, but it might also be an hour or more. So, it is also difficult for our worker leaders to search the communities where workers from a particular factory live to go meet them there. Because all these workers live in different locations, in different areas. The workers' leaders won't be able to gather them all in the same place. (INT36, translated from Kannada)

Export-garment companies' practices of recruiting and transporting workers from rural villages around Bangalore hence fulfil the strategic function of ensuring labour market supply. In addition, these practices have a de facto disciplining effect since they hamper union organising and thereby mitigate the conflict inherent to production.

As a second practice of expanding the labour market frontier, Bangalore garment manufacturers are increasingly *moving production facilities for labour-intensive steps from the Bangalore urban area to rural areas.* Over the past decade, Bangalore export-garment companies have opened up many new factories in villages located in rural areas, towns and industrial parks up to 150 km away from Bangalore. These new factories are connected to the Bangalore urban areas—where most garment companies in the cluster have their national or regional headquarters—through central highways, ensuring good connectivity for buyers, auditors and central management. By setting up these new factories, in many cases, Bangalore export-garment manufacturers have established a regional division of labour that coincides with urban–rural geographical divisions. In this division of labour, more capital and skill-intensive steps of the production process, such as sourcing, designing, sample making, dyeing and washing are carried out in a centralised manner for all factories in Bangalore. In contrast, the labour-intensive and rather low-skilled cut-make-trim process is relocated to new factory units in rural villages. This practice of setting up a new regional division of labour has two important benefits for garment manufacturers: On the one hand, manufacturers can tap into rural labour pools located too distant for workers to commute. On the other hand, employers can save labour costs since minimum wages in rural areas are lower than in urban or semi-urban areas.

Managers' practice of setting up new factories in rural villages is further incentivised by the Government of Karnataka through various types of subsidies. These subsidies are granted to new factories in so-called industrially backward areas. Subsidies were first introduced in the state's Textile Policy for 2008–2013. They include inter alia a subsidy of up to 50% of the employer's contribution to the Employee State Insurance and Provident Fund and an investment subsidy covering up to 25% of all investments made into a factory set-up, including costs for construction or machinery. In addition to the various financial subsidies, the Government of Karnataka actively develops land slots for industrial use through the Karnataka Industrial Areas Development Board, which buys plots of land, develops them and then leases them to companies for industrial use. Given these incentives combined with pressures from the increasing labour shortage, Bangalore-based garment companies have increasingly closed down factory units in the Bangalore urban area over the past ten years and opened new factory units in rural villages. As this officer of the Apparel Export Promotion Council explains, this spatial expansion strategy benefits Bangalore garment manufacturers not only in terms of availability of labour supply but also in the quality of labour attained:

> Various companies [...], are now taking the production centres to the villages. So, it is like a 'walk-to-factory'. But they have not taken the whole production there. They have only taken cut, make and trim there. The other issues like logistics and fabric issues are still maintained from their central offices, but the more labour-intensive parts they have taken to the villages. So, they say 'we find workers coming with fresh faces. A worker travelling

for one or 1.5 hours and a worker that has just walked to the factory for 10 minutes are two different workers'. (INT42)

For unions, the spatial proximity between workers' workplaces and living areas characterising new factory set-ups in rural villages also brings benefits since it allows unions to gather workers from the same factory more easily in one place. In addition, workers in rural villages are more likely to see their work in this specific factory from a long-term perspective, given the lack of alternative employment opportunities. Employers' practices of setting up new large-scale factories in remote rural villages can open up new opportunities for unions to build a stable membership in these factories and thereby enhance their associational and bargaining power in the workplace. Besides moving factories to rural villages where the local population provides the workforce, Bangalore export-garment companies have also been setting up new large, technologically upgraded factory units in the growing number of Textile and Apparel Parks in rural areas of Karnataka. These so-called mega factory units house all production steps under one roof, from fabric dyeing, over cutting to cut-make-trim, washing and packaging. In these textile parks, access for unionists is highly restricted, and workers are transported to the factories in company-provided buses from villages in up to 100 km distance, limiting opportunities for unions to engage with workers outside the factory. Hence, garment manufacturers' practice of expanding the labour market frontier by setting up new factories in rural areas within the State of Karnataka has both enabling and constraining implications for union organising.

Third and last, Bangalore export-garment manufacturing companies are increasingly seizing the mobile workforce created through the national training and migration regime under the ISDS PAN India program through practices of *hiring inter-state migrant workers*. Unionists estimate that currently, about 20% of workers in garment factories in the Bangalore urban area are inter-state migrant workers. In the Bangalore export-garment industry context, hiring inter-state migrant workers needs to be understood as a practice that combines exploiting and disciplining elements. On the one hand, unionists report that inter-state migrant workers are particularly vulnerable to various exploiting practices performed by managers. Being predominantly young women with no prior work experience, no social networks in the city and no knowledge of the local language Kannada, inter-state migrant workers usually have little capacities to interact with local co-workers or labour rights organisations. Moreover, inter-state migrant workers tend to have more limited capacities for protesting or resisting illegal exploitation practices by managers, since they cannot complain or look for a job in another factory due to their limited local language skills. As a result, unionists report that inter-state migrant workers are frequently made to stay about 60 minutes longer than local workers without receiving extra payment for that time. In this light, hiring inter-state migrant workers represents an exploiting practice since it serves not only to ensure labour supply but also to maximise surplus value extraction from labour power. Moreover, manufacturers' practices of hiring inter-state migrants also fulfil strategic disciplining functions since these migrant workers usually live in hostel accommodations representing tightly controlled spaces where outsiders are not allowed access and workers' interactions with externals are strictly monitored.

6.7.3   Interim Conclusion

In summary, labour market relations in the Bangalore export-garment cluster link employers, workers and state-led and private training and recruiting agencies within the State of Karnataka and across India. In the face of the increasing shortage of unskilled labour in the Bangalore urban area, garment manufacturers have over the last five to seven years implemented various practices to ensure adequate labour supply. Through these practices, manufacturers aim to territorially expand the labour market frontier and tap into rural pools of unskilled workers outside of Bangalore. In this context, Bangalore garment manufacturers have, firstly, started to hire workers from rural areas within Karnataka in up to 80 km distance and to transport these workers to Bangalore factories daily with company-owned buses. Second, Bangalore garment manufacturers are increasingly relocating the labour-intensive cut-make-trim part of the labour process to rural villages within Karnataka. There, manufacturers are opening new, large factories employing workers from the village and surrounding villages. Lastly, Bangalore garment factory managers have been increasingly hiring migrant workers from the 'poorer' Northern and North-Eastern Indian states. Employers' practices of expanding labour market relations have, in turn, been enabled and supported by various policies and initiatives under the national Integrated Skill Development Scheme and the state-level Karnataka Textile and Garment Policy. These policies have supported and incentivised the construction of a complex vocational training and migration regime through subsidising practices of setting up specialised garment training centres in rural areas within Karnataka and across India.

For unions, employers' practices of expanding labour market relations have had ambiguous consequences: On the one hand, practices of recruiting and daily transportation of workers from and to rural villages have created a division between workers' working and reproductive spaces. This separation creates significant barriers for garment unions' traditional community organising strategies. Moreover, employers' practices of hiring inter-state migrants who do not speak the local language, Kannada, and stay in tightly secured hostels have created further organising challenges for unions. However, on the other hand, employers' practices of opening up new factories in rural villages have also opened up new opportunities for garment unions, who have expanded the territorial reach of their organising activities. In rural factories where the workforce comes from the local area, employment relations tend to be more stable than in the Bangalore urban area. Workers have greater social capital due to their stronger community embeddedness. Therefore, garment unions have been able to build associational power resources in the form of strong membership bases in various rural factories. As a result, garment unions have established working relationships with the management stopped various exploitation practices such as 'production torture' (see Sect. 6.1) or 'comp-offs' (see Sect. 6.3) in these factories.

The following section summarises the exploiting and disciplining practices and labour control relations that constitute the labour control regime in the Bangalore cluster. Moreover, it highlights how the labour control regime constrains unions' capacities for building and activating power resources in various ways.

6.8 Interim Conclusion: Networked Labour Control and Resulting Constraints for Local Union Agency in the Bangalore Export-garment Cluster

In this chapter, I have illustrated how the labour control regime in the Bangalore export-garment industry emerges from the intersections of various networked labour control practices and processual relations that ensure the reproduction of the labour process in its profit-maximising form. I have shown how the relations and practices that are intertwined in the labour control regime link actors across various distances. Processual relations of labour control encompass the highly localised labour process and workplace relations within garment factories, labour market relations linking workers, managers and recruiting and training agencies across Karnataka and India, and sourcing relations linking Bangalore garment manufacturers with retailers' sourcing offices in Europe and the US. In this light, the analysis of the labour control regime in the Bangalore export-garment cluster has highlighted that the place specificity of the labour control regime results not primarily from the localised nature of the practices and relations that constitute it. Instead, the place specificity of the labour control regime results from the place-specific articulations of a multitude of practices and relations of varying territorial extension.

Moreover, the empirical analysis has demonstrated that the labour control regime as a structural framework for surplus extraction and capital accumulation is relatively stable due to the routinised nature and complex interrelations of the practices and relations that constitute it. These practices are, in many cases, directed at circumventing or undermining specific legal-institutional frameworks originally conceived to protect workers' well-being, such as the Indian Minimum Wage Act or the Industrial Disputes Act. Hence, many practices that constitute the labour control regime in the Bangalore export-garment cluster are rather informal. Nevertheless, the complex interrelations between various sets of labour control practices and relations constituted through them make it difficult for workers and unions to transform single sets of practices.

Two types of interrelations have figured particularly important in the empirical analysis. These are, on the one hand, interrelations where one set of practices *shapes* the nature of another set of practices. On the other hand, practice interrelations were salient in which one set of practices *enables* another set of practices. As illustrated, retailers' predatory purchasing practices at the vertical dimension of the

GPN directly *shape* the practices through which manufacturers construct territorially embedded labour processes, workplace relations, wage relations, employment relations and industrial relations at the horizontal dimension. Retailers implement various 'predatory purchasing practices' (Anner 2019) directed at maximising their value capture, including price squeezing, placing irregular orders and demanding shorter lead times and increased flexibility from manufacturers. These practices in turn shape the practices through which Bangalore garment manufacturers construct the labour process as well as wage, workplace, employment and industrial relations. To comply with retailers' demands while at the same time ensuring surplus production and capture for themselves, Bangalore export-garment manufacturers employ various exploiting practices directed at maximising surplus value. In addition, manufacturers employ various disciplining practices directed at mitigating the conflict inherent to production.

Employers, for example, perform tight control over workers' performance in the labour process through practices of production targeting and digitising the labour process with smart sewing machine networks. Moreover, to keep labour costs down, workers employ various practices to keep wages down, such as lobbying governments for lower statutory minimum wages at the state level or giving 'comp-offs'. To respond to retailers' demands for flexibility and to cope with irregular orders from retailers, Bangalore garment manufacturers, in turn, construct flexibilised and informalised employment relations with workers through 'hiring and firing' practices. Lastly, to prevent collective bargaining and wage increases, employers construct workplaces as tightly controlled spaces and perform various sets of union-busting practices to undermine collective worker organising and unionisation. Through these practices, employers construct workplace and industrial relations characterised by high power asymmetries and employer dominance, allowing employers to fend-off unions' attempts to negotiate wages beyond the statutory minimum wage.

Bangalore garment manufacturers' exploiting and disciplining practices are in turn *enabled* by 'pro-business' state practices (Pattenden 2016) performed by labour department officers and judges at the state level as well as by national legislators. Karnataka labour department officials and judges have, for example, repeatedly enabled employers' practices directed at keeping minimum wages low by withdrawing notifications of statutory minimum wage increases and by making ineffective rulings in response to unions' legal complaints. Similarly, when unions file complaints against illegal union-busting practices by managers, such as firing union activists, labour department officials seldom order an independent inspection of the case during the conciliation process. Instead, labour officers tend to push workers and unions to accept settlement offers. By not enforcing legal-institutional frameworks conceived originally to protect workers' rights, Indian state actors' pro-business practices enable employers to perform de facto illegal exploiting and disciplining practices. Going even further, at the national level, legislators have passed various labour law reforms that formally legitimise various employer practices of disciplining or exploitation. Recent national labour law reforms have, for example, introduced fixed-term contracts specifically for the garment industry and raised the threshold for mandatory prior government approval for worker lay-offs from factories with more than 100 workers to factories with more than 300 workers. These labour law reforms have provided legal ground for Bangalore garment manufacturers' long-standing practices of 'hiring and firing'.

Difficulties for unions to challenge the exploiting and disciplining practices that constitute the labour control regime in the Bangalore export-garment industry do not exclusively result from the intertwining of the various practices and relations of labour control but also from the intersections between employers' labour control practices with broader social power asymmetries along the lines of age, gender and migrant status. Employers deliberately reproduce wider social power asymmetries linked to these categories within workplace, employment and labour market relations to secure employer dominance and to prevent collective worker organising. In this vein, employers deliberately hire women for the lower-skilled tasks in the labour process since women are less likely to speak up to supervisors or to engage in collective organisation—albeit not due to their inherent 'docile' nature but rather due to broader patriarchal structures and the double burden of care and wage work. Moreover, employers have started to recruit young, female inter-state migrant workers through regional and national vocational training and migration regimes constructed by state and private actors under the national Integrated Skill Development Scheme. Whereas hiring inter-state migrant workers is in the first place a practice through which employers seek to cope with an increasing local labour shortage, intersections with relations of age, gender and migrant status also make hiring inter-state migrant workers a de facto exploiting and disciplining practice. Since inter-state migrant workers are predominantly young women who do not speak the local language and stay in tightly controlled hostels, inter-state migrant workers are harder to approach by unions. As a result, migrant workers are also more vulnerable to employers' exploiting practices such as forced unpaid overtime work.

In summary, the complex intersections and interdependencies between the various relations and practices of labour control that constitute the labour control regime in the Bangalore export-garment cluster constrain the terrain and capacities for the agency of local garment unions in three critical ways:

First, the specific ways in which retailers construct spatial power asymmetries in supplier relations and the relatively low-skilled nature of the majority of jobs in Bangalore garment factories constrain workers' *structural power* resources. In light of retailers' practices of maintaining large supplier pools, employers use competitive wage pressure as an overarching argument for fending off any collective wage bargaining attempts by unions. Competitive wage pressures are also higher in the Bangalore export-garment cluster compared to other Indian clusters. Production in the cluster specialises in men's casual wear, characterised by less time-sensitive, lower value-added products that less experienced suppliers can produce in lower-wage locations. At the same time, producing men's wear requires mainly unskilled and semi-skilled labour due to the lower complexity of products. Lower-skilled workers are, in turn, more easily replaceable through intra- and inter-state migrant workers who receive a short training of only three months duration. Hence, both dimensions of workers' structural power—labour market power and workplace power—are constrained due to workers being relatively easily replaceable and to workers' limited capacities to cause broader disruptions of the production network. Workers' limited capacity to cause disruptions to the production process is a result of workers' limited capacity for constructing stable membership bases inside factories that could be mobilised for industrial action.

Second, the various direct and indirect disciplining practices performed by employers pose significant constraints on unions' capacities for building *associational power* at the workplace or industry level: At the workplace level, employers' practices of constructing garment factories as tightly controlled spaces, for victimising union supporters, and of flexibilising and informalising employment relations pose significant constraints for unions' capacities to build strong and stable membership bases in specific workplaces. At the same time, employers' practices of expanding labour market frontiers and the resulting increasing geographical separation of workers' working and living spaces make it harder for unions to organise workers in their communities and thus to build associational power at the industry level.

Third, state actors' pro-business practices hamper unions' capacities to activate *institutional power* resources. Traditionally, labour inspectorates and institutionalised industrial dispute settlements have represented a source of institutional power for worker and unions in India. However, these sources have been unravelling over the past three decades in the context of the economic liberalisation and shift towards neoliberal policies. In the post-liberalisation era, the state's role has shifted from regulating the economy to securing business enabling conditions for national and foreign private investors. Within this neoliberal framework, the garment industry has traditionally enjoyed a special protective status due to its capacity to provide mass employment for India's unskilled or low-skilled rural 'reserve army of labour' (Breman 1996; see also Ramasamy and Pilz 2020). Against this backdrop, labour department officials rarely take a proactive worker stance in institutionalised tripartite industrial dispute settlements or collective bargaining processes. Moreover, due to the chronic understaffing of the Indian labour judiciary, court cases take, on average, seven years until a ruling is made. As a result, workers' and unions' capacities to exercise institutional power through invoking legal frameworks and institutionalised procedures for settling disputes or negotiating statutory minimum wages are significantly constrained. Unions' constrained institutional power, in turn, also further limits unions' capacities to challenge employers' union-busting practices that constrain unions' capacities for building associational power resources.

In summary, the various interrelations and interdependencies between employers' exploitation practices and state actors' pro-business practices make it harder for unions to challenge, stop or transform labour control practices and achieve lasting improvements for workers. Due to the intersections of the labour process and wage relations with sourcing relations, for example, to stop or transform 'production targeting' practices or achieve significant wage raises, unions need to tackle employer and retailer practices simultaneously. While constraining the 'wiggle room' (Castree et al. 2004: xvii) for workers and unions, retailers' power over manufacturers can serve as an enabling factor for unions as well. Unions may seize retailers' leverage over manufacturers, for example, in cases where manufacturers' exploiting and disciplining practices also provide violations of local labour laws and of retailers' codes of conduct. In these cases, unions may be able to push retailers to enforce manufacturers' compliance with the code of conduct. Since retailers' central managements are usually located in geographically distant places in the Global North, it is difficult for unions to push retailers through direct interactions. Consequently, as literature on networks of labour activism (see Sect. 2.2.4) has pointed out, local unions need to develop networked agency strategies that target multiple actors in various places simultaneously and that make use of coalitional power resources through building alliances across borders. Through such networked agency approaches, Bangalore garment unions have achieved numerous 'small' and not so small transformations (Latham 2002) of exploiting and disciplining practices in the Bangalore export-garment cluster. For example, garment unions have achieved to stop employer and state practices of delaying minimum wage revisions as well as employers' practices of giving 'comp-offs' in factories where unions have established strong membership bases through networked agency strategies.

The next chapter examines the networked agency strategies that Bangalore export-garment unions have developed over the past 15 years and assessed the potential of different strategies for building sustained local union power. To this end, I draw on the relational framework for analysis developed in Chapter 3, which proposes to analyse the agency strategies of local unions through the lens of intersecting spaces of organising, collaboration and contestation.

References

Anner M (2019) Predatory purchasing practices in global apparel supply chains and the employment relations squeeze in the Indian garment export industry. Int Labour Rev 158:705–727. https://doi.org/10.1111/ilr.12149

Bartley T, Egels-Zandén N (2015) Responsibility and neglect in global production networks: the uneven significance of codes of conduct in Indonesian factories. Global Netw 15:S21–S44. https://doi.org/10.1111/glob.12086

Bath P (2019) A fight for minimum wages: K'taka garment workers rekindle 18-month protest: The News Minute, Friday 13 Sept 2019. https://www.thenewsminute.com/article/fight-minimum-wages-ktaka-garment-workers-rekindle-18-month-protest-108820. Accessed 31 Dec 2021

Breman J (1996) Footloose labour: working in India's informal economy. In: Contemporary South Asia, vol 2. Cambridge University Press, Cambridge

Castree N, Coe NM, Ward K, Samers M (2004) Spaces of work: global capitalism and the geographies of labour. Sage, London

Chakravarty D (2007) 'Docile oriental women' and organised labour. Indian J Gend Stud 14:439–460. https://doi.org/10.1177/097152150701400304

CWM, Centre for Workers' Management (2014) State of garment workers in Bangalore. CWM, New Delhi

ExChains (2020) „Die Arbeiter*innen sind entschlossen, ihre Arbeitsplätze zurückzubekommen und bleiben standhaft in ihren Forderungen": interview mit Prathibha R., Vorsitzende der

Gewerkschaft GATWU. ExChains Newsletter 13/2020. http://www.exchains.org/exchains_new
    sletters/2020/exchains_NL_13_2020_screen_dt.pdf. Accessed 31 Dec 2021
Fair Wear Foundation (2018) Worker-management dialogue in Indian legislation—a guidance
    document. https://api.fairwear.org/wp-content/uploads/2020/09/Worker-management-dialogue-
    in-Indian-legislation.pdf. Accessed 5 Apr 2022
Gram Tarang (2020) Gram Tarang at a glance. https://gramtarang.org.in/index.php/at-a-glance/.
    Accessed 31 Dec 2021
H&M (2021) H&M group supplier list. https://hmgroup.com/sustainability/leading-the-change/sup
    plier-list.html. Accessed 23 May 2021
Indian Ministry of Law and Justice (2013) The sexual harassment of women at workplace (Preven-
    tion, Prohibition and Redressal) Act, 2013. https://dst.gov.in/sites/default/files/1.%20shc_act
    s2013.pdf. Accessed 24 Sept 2021
Indian Ministry of Law and Justice (2020) The industrial relations code, 2020. The
    Gazette of India. https://prsindia.org/files/bills_acts/bills_parliament/2020/INDUSTRIAL%20R
    ELATIONS%20CODE,%202020.pdf. Accessed 14 Sept 2021
Inditex (2021) Supplier map. https://www.inditex.com/about-us/inditex-around-the-world#contin
    ent/000. Accessed 23 May 2021
Jayaram N (2019) Protection of workers' wages in India: an analysis of the labour code on wages,
    2019. Econ Polit Wkly 54(49)
Jenkins J, Blyton P (2017) In debt to the time-bank: the manipulation of working time in Indian
    garment factories and 'working dead horse.' Work Employ Soc 31:90–105. https://doi.org/10.
    1177/0950017016664679
Labour Commissioner Office, Government of Karnataka (2021) Minimum rates of wages for
    the year 2021–2022. https://karmikaspandana.karnataka.gov.in/info-4/Minimum+Wages+Notifi
    cation/Minimum+Rates+of+wages+for+the+year+2021-2022/en. Accessed 31 Dec 2021
Latham A (2002) Retheorizing the scale of globalization: topologies, actor-networks, and
    cosmopolitanism. In: Herod A, Wright MW (eds) Geographies of power: placing scale. Blackwell,
    Malden, MA, pp 115–144
Locke RM, Kochan T, Romis M, Qin F (2007) Beyond corporate codes of conduct: work organi-
    zation and labour standards at Nike's suppliers. Int Labour Rev 146:21–40. https://doi.org/10.
    1111/j.1564-913X.2007.00003.x
López T, Riedler T, Köhnen H, Fütterer M (2021) Digital value chain restructuring and labour
    process transformations in the fast-fashion sector: evidence from the value chains of Zara &
    H&M: Online First. Glob Netw:1–17. https://doi.org/10.1111/glob.12353
Mani M, Mathew B, Bhattacharya D (2018) Critiquing the statutory minimum wage: a case of the
    export garment sector in India. National Law School of India University, Bangalore
Pattenden J (2016) Working at the margins of global production networks: local labour control
    regimes and rural-based labourers in South India. Third World Q 37:1809–1833. https://doi.org/
    10.1080/01436597.2016.1191939
Pilz M, Regel J (2021) Vocational education and training in india: prospects and challenges from
    an outside perspective. Margin J Appl Econ Res 15:101–121. https://doi.org/10.1177/097380102
    0976606
Ramasamy M, Pilz M (2020) Vocational training for rural populations: a demand-driven approach
    and its implications in india. IJRVET 7:256–277. https://doi.org/10.13152/IJRVET.7.3.1
The Economic Times (2018) Fixed-term employment extended to all sectors to boost ease of doing
    business. 21 March 2018. https://economictimes.indiatimes.com/news/economy/policy/govern
    ment-extends-facility-of-fixed-term-employment-for-all-sectors/articleshow/63382807.cms?fro
    m=mdr. Accessed 31 Dec 2021
Thompson P (2010) The capitalist labour process: concepts and connections. Cap Class 34:7–14.
    https://doi.org/10.1177/0309816809353475
Wessels A, Pilz M (2018) International handbook of vocational education and training: India. Verlag
    Barbara Budrich, Leverkusen

Chapter 7
Union Agency in the Bangalore Export-garment Cluster: Linking Spaces of Organising, Spaces of Collaboration and Spaces of Contestation

Abstract This chapter analyses the networked agency strategies of three local garment unions in the Bangalore export-garment cluster. Drawing on the heuristic of three interrelated spaces of labour agency constructed by unions—spaces of organising, spaces of collaboration and spaces of contestation—the chapter highlights the various challenges for building sustained union bargaining power vis-à-vis employers. These challenges result on the one hand from the tight labour control regime and on the other hand from unions' engagement with consumer organisations and donor NGOs from the Global North: When unions rely on financial support from NGOs instead of members' contributions to fund their operations, and on moral power exercised by consumer organisations instead of associational power exercised by workers, unions risk constructing spaces of organising, collaboration and contestation that provide limited opportunities for workers and organisers to develop strategic capacities. Consequently, unions' associational and organisational power remains limited. In contrast, when unions strategically use moral power resources from consumers to open up spaces for workplace organising and collective bargaining, this can enable unions to enhance their bargaining position vis-à-vis employers and thereby bring about sustained improvements for workers.

Keywords Bangalore · Garment industry · Union agency · Spaces of organising · Spaces of collaboration · Spaces of contestation · Collective bargaining

This chapter analyses the agency strategies of three local garment unions that are active in the Bangalore export-garment cluster: the Garment and Textile Workers Union (GATWU), the Garment Labour Union (GLU) and the Karnataka Garment Workers Union (KGWU). As stated in Sect. 5.2.3, the roots of all three unions lie in an NGO-led community organising project with garment workers financed through Oxfam International and carried out by the Bangalore-based labour rights NGOs, FEDINA and Cividep. All three unions qualify as local unions since the geographical distribution of their members is limited to the Bangalore garment cluster.

All three garment unions can, furthermore, be classified as independent unions, since—as opposed to India's twelve central trade union federations—they do not maintain close ties to any specific political party. Local independent unions face particular challenges in building bargaining power vis-à-vis employers. These challenges result from the complex, networked structure of the labour control regime (see Chap. 6), which is constituted through practices and relations that stretch beyond the territory of the Bangalore export-garment cluster and, therefore, beyond the local unions' direct sphere of action. On the other hand, local garment unions in the Bangalore export-garment cluster face challenges in building bargaining power vis-à-vis employers due to their limited political and institutional power and financial resources. Whereas India's central trade unions can use political leverage and ties as well as financial and associational resources from historically grown, large political membership bases to exercise power over employers (and state actors), the three Bangalore garment unions had to build their membership bases from scratch over the past 10 to 15 years. Since their foundation, all three unions have engaged in building alliances at various levels—i.e. over various distances—to secure financial resources and leverage coalitional power resources in labour disputes.

In the following, I lay out the different agency strategies through which GATWU, GLU and KGWU have sought to build bargaining power vis-à-vis employers and the state actors to improve conditions for workers in the Bangalore export-garment cluster. To this end, I draw on the relational framework for analysing the agency of local unions at specific nodes of a GPN developed in Sect. 3.3. Following this framework for analysis, I understand the agency strategy of each union as emerging from the intersection of the various networked sets of practices and relations through which unions construct three spaces of labour agency: (1) *spaces of organising* linking union organisers, workers and union members; (2) *spaces of collaboration* linking local unions to other external labour and non-labour actors in solidary ways; and (3) *spaces of contestation* constructed around specific labour struggles, linking unions and their allies with employers or state actors as 'targets' of unions' actions and demands. As laid out in Sect. 3.3, I argue that within these spaces, unions and workers can develop strategic capacities and power resources which, in turn, enable unions to build bargaining power vis-à-vis employers and thus to win concessions for workers. At the same time, the networked relationships constituting these spaces are themselves structured by power relations. As a result, resources and strategic decision-making competencies may be distributed asymmetrically within these relationships. When resources and strategic decision-making competencies are centralised and controlled by few actors within the union or by actors external to the union, unions' capacities to build bargaining power and strategic capacities remain limited.

In the remainder of this chapter, I will analyse the specific practices and relations through which each union constructs spaces of organising, collaboration and contestation and analyse to which extent these practices and relations enable workers and unionists to develop strategic capacities. Developing workers' and unionists' strategic capacities is vital for building associational and organisational power resources—the two power resources understood as central for building bargaining power vis-à-vis employers and the state (see Sects. 2.4.3 and 2.4.4). The analysis is structured as follows: Sect. 7.1 introduces the agency strategies of GATWU at two different points in time—first, in GATWU's early years as a union project led by the NGO Cividep (Sect. 7.1.1), and second, in the period since GATWU's strategic break with Cividep

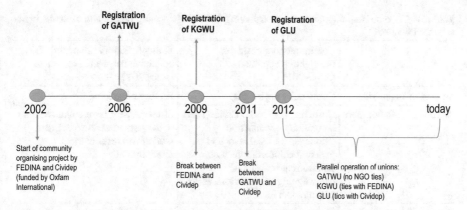

Fig. 7.1 Timeline of the evolution of the three Bangalore-based local garment unions. *Source* Author based on interview data

in 2011 (Sect. 7.1.2). Section 7.2 then turns to the second local garment union, GLU, which emerged as a spin-off from GATWU in 2012. Section 7.3 finally analyses the agency strategy of KGWU, the third local garment union in the cluster, founded as a spin-off from GATWU in 2009. While GLU, until date, maintains close ties with Cividep, KGWU maintains close ties with FEDINA. In contrast, GATWU has sought to remain financially and strategically independent from NGOs after breaking with Cividep in 2011. Figure 7.1 visualises the evolution of the three local unions in the Bangalore export-garment cluster.

In the next section, I focus on GATWU as the oldest of the three Bangalore garment unions and lay out the evolution of the union's strategic agency approach over the past 15 years since its foundation.

7.1 Garment and Textile Workers Union (GATWU)

This section analyses the strategic agency approach by GATWU, the oldest of the three Bangalore garment unions. At the time research was conducted, GATWU had about 10,000 members, according to its own reports. Starting as the union arm of the local NGO Cividep, GATWU took a strategic turn in 2011 with the decision to become strategically and financially independent from Cividep. The decision to become independent from NGO project funding led to a wide-ranging transformation of the practices and relations through which GATWU constructed spaces of organising, spaces of collaboration and spaces of contestation. Therefore, in this section, I distinguish between two different strategic approaches underpinning GATWU's agency in two different periods: (1) a strategic approach based on networked practices of community-based organising, transnational campaigning and 'fire-fighting' (i.e. tackling basic labour rights violations), which characterised GATWU's early years under the lead of Cividep; and (2) a strategic approach that combines a focus

Table 7.1 Two strategic approaches underpinning GATWU's agency before and after the break with Cividep in 2011

	Community organising & fire-fighting approach (2006–2011)	Strategic factory organising & collective bargaining approach (since 2012)
Spaces of organising	Community organising	Factory organising
Spaces of collaboration	Close ties with NGO Cividep and community organisation 'Munnade' at local level and punctual collaborations with transnational consumer campaigning organisations at the international level	Strategic long-term collaborations with international union and worker networks at the international level
Spaces of contestation	Seeking redressal of basic minimum labour rights violations	Seeking collective bargaining of improvements beyond minimum labour standards

on factory organising with the strategic goal to engage employers in collective bargaining and thereby to win concessions for workers beyond the mere implementation of basic labour rights. This second strategic approach has guided GATWU's strategic reorientation process after its break with Cividep in 2011. Table 7.1 sums up the main characteristics of each strategic approach.

In the following, I will illustrate how GATWU has constructed spaces of organising, collaboration and contestation through very different practices and relations under both strategic approaches. When doing so, I will also assess the potentials and limits of each strategic approach for enabling GATWU to build lasting bargaining power and bring about sustained improvements for workers.

7.1.1   The Origins: Community-Based Organising, Transnational Campaigning and Fire-Fighting

This section lays out GATWU's strategic approach from 2006 to 2011, i.e. the years right after GATWU's foundation, when the union was still closely tied to Cividep. In interviews, GATWU activists and leaders referred to the strategic approach characterising GATWU's agency during these years as an 'NGO-led' union model due to the strong influence of NGO funding on GATWU's practices and internal relationships. Another common term used by GATWU organisers to describe their agency during this period was 'fire-fighting' due to the focus on remediating basic labour rights violations rather than pushing for collective bargaining. Therefore, in this chapter, I use the term fire-fighting approach to refer to GATWU's agency strategy during their time of close collaboration with Cividep. The remainder of this section analyses the practices and relationships through which GATWU constructed spaces of organising, spaces of collaboration and spaces of contestation under the community organising, fire-fighting approach. I will show the various improvements in the working conditions that GATWU achieved through this approach. At the same time, I will discuss the limits of this approach for building sustained bargaining power vis-à-vis employers.

7.1.1.1   Spaces of Organising

To understand how GATWU constructed spaces of organising in its early years, it is essential to look back at the historical origins and evolution of GATWU. As previously mentioned, GATWU's origins lie in an NGO project focussing on community work with garment workers, which started in 2002 and continued until 2011. The project was formally carried out by the local NGOs Cividep and FEDINA and financed by the international NGO Oxfam (INT8). The project aimed to start organising garment workers in Bangalore through a community-focussed 'pre-union' concept. This concept was directed at building awareness among women workers for their rights and at introducing the concept of collective action. The project was formally institutionalised in 2004 with the founding of the community-based women's organisation Garment Mahila Karmikara Munnade (Engl.: Garment Women Workers' Progress)—or short Munnade. However, organisers soon noted the limitations of a community organisation for tackling labour rights violations in the workplace: Since Munnade was registered as an NGO and not as a union, organisers could not formally intervene in workplace conflicts on behalf of workers. Against this backdrop, in 2005, Munnade and Cividep organisers founded the Garment and Textile Workers Union (GATWU), officially registered in 2006. In its early years, from 2006 to 2011, GATWU continued to work with the community-based organising approach introduced by Munnade, and there was no clear personal separation between Munnade organisers and GATWU organisers.

Following the pre-union concept, GATWU constructed relationships with workers in its earlier years in close collaboration with Munnade through two main sets of practices: forming saving groups and setting up area committees. Savings groups were formed by GATWU and Munnade organisers by gathering women garment workers from the same neighbourhood around the goal of collectively saving money: women paid a monthly amount and took turns in receiving the groups' collected money (INT8, 39). The rationale underpinning these practices was twofold. On the one hand, saving groups were conceptualised as a tool for women workers' economic empowerment, since many female garment workers are the primary breadwinners in their families—either because their husbands cannot find work or because they are separated. On the other hand, saving groups were conceptualised as a safe space for women to discuss shared experiences related to struggling economically and to carrying the double burden of wage and care work. Discussing these shared experiences served union organisers as an entry point for raising women garment workers' awareness of the structural conditions of capitalism and patriarchy and develop a collective mindset—two vital pre-conditions for the unionisation process.

Building on relations with workers constructed in the saving groups, GATWU and Munnade organisers then engaged in setting up area committees to deepen relationships with those workers who appeared ready to participate in the unionisation process. As the name suggests, area committees were set up in specific areas where many garment workers lived. These areas were primarily located in the adjacent neighbourhoods to the (back then) still-growing geographical concentrations of garment factories along Mysore Road and Hosur Road (see Sect. 5.2.2). In area committee meetings, which usually brought together about 15 women, GATWU and Munnade organisers combined discussions of workers' collective experiences in the community and workplace with more structured practices of educating women workers about their civic and labour rights. When workers reported any workplace issues in saving group or area meetings, they were offered to join GATWU. Then GATWU organisers would intervene with the respective management on behalf of the worker. To become a member of GATWU, workers had to pay an initial administrative fee of 20 Rupees and an annual payment of 60 Rupees (approx. 0.80 US$).

In addition to forming saving groups and setting up area committee meetings, GATWU organisers also sought to build relationships with workers through more proactive and direct organising practices such as distributing leaflets in front of factory gates and engaging with individual workers in conversations after their shifts in front of the factory. Given the overall community organising approach, GATWU's organising practices in these first years were not primarily directed at gaining a strong membership base in specific factories but rather at increasing GATWU's overall membership in specific areas. The focus on organising workers in specific areas rather than in specific factories needs to be understood as the result of three intertwined conditions shaping GATWU's organising strategy during its first years: First, as mentioned before, GATWU's organising practices during the years following its foundation showed a strong path-dependency to the 'pre-union' community organising project, out of which GATWU had emerged. GATWU continued to collaborate closely with Munnade to the point where there was no clear personal separation between GATWU and Munnade organisers. Hence, during these early years, GATWU's organising practices constructed the community rather than the factory as the primary space for organising.

Second, GATWU's initial community-focussed organising strategy was favoured by the specific geographical and labour market conditions of the Bangalore export-garment industry during the 2000s: As mentioned earlier, during these years, the industry was still concentrated in two main areas along two major traffic outlets, Mysore Road and Hosur Road, and workers lived in these same areas, often in walking distance to the factories. Therefore, at the end of the workday, workers were not in a rush to catch a bus or Rickshaw already waiting for them after their shift (see Sect. 6.7) and hence were able to engage in conversations with organisers. In this line, GATWU leaders report that it was a frequent organising practice under the community organising approach for union organisers to accompany groups of workers on their walk home and use this time to discuss problems in the factory and the benefits of unionisation.

Third and last, the community-based organising model that aimed at increasing membership at the community rather than at the factory level also needs to be understood as shaped by the specific relationships that GATWU had with international NGOs. These NGOs were financing GATWU's activities and staff. For donor NGOs, GATWU's capacity to engage with as many workers as possible, e.g. through training sessions or family counselling, was more important than GATWU's capacity to build strong ties with a few workers in selected factories. As donor organisations, NGOs aimed to maximise the outreach of their funded projects (see Sect. 7.1.1.2). In response to donors' focus on increasing membership quantity rather than quality, GATWU organisers' interactions with many union members were limited to collecting complaints from workers, which GATWU organisers would then discuss with management on behalf of workers.

Against this backdrop, in interviews, GATWU leaders today spoke of having constructed the union as a 'service-providing' entity for their members during these years. This union model stands in contrast to the concept of the union as a membership-based organisation, in which members are actively involved in the union's strategic decisions. Consequently, under the strategic community organising approach, GATWU's intra-union relations were characterised by the centralisation of strategic activities and capacities on full-time GATWU organisers. Strategic capacities included for example the capacity to organise workers or to discuss with management.

As a result, GATWU's community-based organising strategy had mixed outcomes in relation to building the unions' associational and organisational power resources. On the one hand, increasing overall membership numbers helped GATWU organisers to gain legitimacy as representing garment workers and to engage in dialogue with management about labour rights violations. Moreover, through the community organising approach, GATWU mobilised several thousand workers for punctual public protests. Through these mobilisations, GATWU achieved inter alia substantial raises in the statutory minimum wage (see also Sect. 7.1.1.3). On the other hand, the area-based organising approach had its limits with regard to translating associational power resources into workplace bargaining power. Despite having achieved a membership of about 4000 members by 2011, GATWU did not have the associational power resources to engage in prolonged industrial action at the workplace for two reasons. First, as a result of GATWU's focus on organising workers in specific communities rather than in specific factories, GATWU's members were distributed across a large number of factories with relatively low membership in individual factories. Second, where GATWU had a significant number of members in a specific factory, these members did not have the strategic capabilities to plan, execute and sustain prolonged industrial action. These capabilities were centralised with GATWU's full-time organisers, who however, did not have access to factories. The role of workers inside the factories was, in turn, mainly limited to informing full-time organisers about problems or labour rights violations in their factories.

As stated earlier, GATWU's community organising practices were closely intertwined with the practices through which GATWU built and maintained relationships with Munnade at the local level and international donor NGOs at the international level. The next section will analyse these practices and relations in more detail.

7.1.1.2 Spaces of Collaboration

Two types of collaborative relationships constructed by GATWU with external actors played a significant role in GATWU's initial strategic community organising and fire-fighting approach: (1) relations with the community-organisation Munnade and the NGO Cividep at the local level and (2) relations with donor NGOs and consumer campaigning networks from the Global North at the international level. First, GATWU's close collaboration with Munnade for organising workers in their communities through saving groups and area committees represented a central element of GATWU's strategic approach. The collaboration with Munnade represented a source of coalitional power for GATWU since closely working with Munnade allowed GATWU access to a workforce that would otherwise not have been accessible to GATWU for two reasons. First, given factories as tightly controlled spaces (see Sect. 6.4), GATWU organisers could not organise workers inside the workplace and hence had to seek other ways to get in touch with workers. Second, many workers were first-time industrial workers from a rural background who perceived unions mainly as political organisations or as 'troublemakers' according to the dominant management discourse. Therefore, approaching workers through Munnade was an important practice for building workers' trust and familiarising them with the idea and concept of unionisation. Accordingly, approaching workers through a community organisation rather than directly through the union also allowed GATWU to build workers' collective mindset starting from women workers' collective experiences in the household and the community before transferring this mindset to the work-place. Underlying this community-based organising approach was a deeply intersec-tional understanding of workers' identity as not only shaped by capitalist relations of production but also by relations of gender, caste or geographical provenience. This understanding provided the base for the close collaboration with Munnade.

Closely intertwined with GATWU's collaboration with Munnade was also GATWU's collaboration with the local labour rights NGO Cividep, through which both GATWU and Munnade full-time activists were formally employed. Cividep acted as an intermediary organisation between Oxfam International, who funded Munnade's and GATWU's organising work. This intermediation by Cividep was necessary since, according to Indian law, unions are not allowed to receive funds from international organisations. In this sense, GATWU's collaboration with Cividep also represented a source of coalitional power since it allowed GATWU to access financial resources from Oxfam. GATWU's dependence on financing from Oxfam, however, had mixed effects on the union's capacities to build associational power: On the one hand, GATWU's collaboration with Cividep and Oxfam enabled GATWU to fund full-time organiser positions as well as various training sessions and activities with workers.

On the other hand, however, GATWU's relationship with Oxfam and Cividep was characterised by the largely unilateral dependence of GATWU on these organisations. As a result, the administrative requirements of Oxfam and Cividep largely shaped GATWU organisers' work profiles. A significant part of full-time organisers' time was spent preparing research reports or documenting the activities conducted within the project context—time resources that could not be invested in actual organising work (FN10). Moreover, the collaboration with Cividep and Oxfam shaped internal union relations. Since funding for the unions' activities came from external project funding rather than members' financial contributions, relationships of accountability inside the union were oriented towards Cividep as the formal employer and Oxfam as the funding organisation. Rather than discussing strategic decisions with union members, GATWU organisers took decisions in coordination with Cividep. Consequently, the external accountability relations with Cividep and Oxfam created significant barriers to fostering internal union democracy and participation. Democratic and participatory internal union relations are, however, a pre-condition for building lasting associational and organisational power that can be transformed into workplace bargaining power (see Sect. 2.4.1).

Besides the collaborations with Munnade and Cividep, GATWU, secondly, also engaged in constructing collaborative relations with consumer campaigning networks from the Global North. As highlighted in Chap. 5, a significant challenge for local unions in the Bangalore garment cluster to improve working conditions lies in the fact that geographically distant retailers as lead firms significantly shape labour processes as well as wage and employment relations through their sourcing practices. Against this backdrop, from its early days on, GATWU organisers understood that besides targeting local manufacturers, they also had to tackle retailers from the Global North and strategically use their leverage over manufacturers. To this end, GATWU organisers engaged in building relationships with various consumer-led organisations and campaigning networks from the Global North. In particular, GATWU built relationships with the Clean Clothes Campaign (CCC), a European network of consumer and civil society organisations, and with the Worker Rights Consortium (WRC), a US-based labour rights monitoring organisation led by students and universities. The practices through which GATWU built relations with these organisations were mainly focussed on using the urgent appeal mechanisms provided by the CCC and the WRC (see also Merk 2009). Urgent appeal mechanisms allow workers and unions to lodge a complaint about labour rights violations in a specific factory, which the CCC and the WRC then bring to the attention of the brands sourcing from that factory. If brands are unwilling to ensure that labour rights violations in their supplier factories are corrected, the CCC and the WRC conduct public consumer campaigns appealing to brands' responsibility for ensuring workers' rights in their supply chains. Actions by the CCC or WRC usually draw on leveraging moral power resources vis-à-vis brands through public 'naming and shaming' campaigns. The result of interactions between GATWU and the CCC or WRC around a specific urgent appeal was either determined by the fact that all sources for leveraging moral pressure had been seized without a result or—in successful cases—that a settlement between the union and the local management could be reached.

Given GATWU's lack of a strong membership inside factories, leveraging the coalitional power of the CCC and the WRC was an essential element of GATWU's community organising and 'fire-fighting' approach: Since GATWU could not exercise power over employers through engaging in collective industrial action at the workplace level, the union had to rely on the moral power exercised by consumer organisations to activate retailers' leverage over manufacturers. In GATWU's early days, large-scale labour rights violations were still present in the Bangalore export-garment cluster. The fact that GATWU pushed employers (and state actors) to implement basic legal labour standards through transnational campaigning brought about important improvements for workers. The following section illustrates how GATWU activated relationships with consumer campaigning organisations from the Global North when constructing spaces of contestation around specific labour struggles. It will further assess to what extent these practices of transnational campaigning enabled GATWU to build lasting bargaining power vis-à-vis employers.

7.1.1.3   Spaces of Contestation

In the years following its foundation, GATWU constructed antagonistic relations with employers and state actors mainly through practices of what their activists today refer to as 'fire-fighting'. The term 'fire-fighting' refers to the fact that their interventions with employers and state actors focussed primarily on rectifying labour law violations. GATWU organisers mainly reacted to the various illegal practices of labour control performed by factory managers: they sought to stop these rather than proactively formulating and negotiating demands beyond implementing basic labour rights. GATWU's focus on rectifying labour rights violations and ensuring minimum labour standards in its early years also needs to be understood in the context of the general state of the garment industry in the Bangalore export-garment cluster during the 2000s: During these years, large-scale, very severe violations of basic labour rights were still present in Bangalore export-garment factories, especially in the many smaller, informalised factories. It was common, for example, for workers to not receive employment contracts and to not receive wages for several months. Moreover, as illustrated already in Sect. 6.3, during these years, the legal-institutional minimum framework was not implemented, with periods of up to nine years passing between minimum wage revisions (as opposed to the legally prescribed revisions at intervals of three to five years). Against this background, during the first years of GATWU's existence, GATWU organisers constructed spaces of contestation mainly through two sets of practices: (1) intervening in cases of labour rights violations at individual factories and (2) campaigning for minimum wage increases at the state level.

Selectively Intervening in Labour Law Violations at the Factory Level

The first set of practices through which GATWU constructed spaces of contestation under the early fire-fighting approach was through selectively intervening in specific incidents of labour law violations, which were reported to GATWU by workers during saving groups or area committees or gate meetings. Common violations included, for example, non-payment or late payment of wages, lack of drinking water or illegal factory closures. GATWU's interventions in these cases consisted of three intertwined practices at various levels: (1) writing an official complaint letter to the respective factory management at the workplace level, (2) lodging a legal complaint with the competent government authorities at the state level and (3) involving consumer campaigning organisations at the international level, if the management remained unresponsive.

Given GATWU's lack of workplace bargaining power due to low membership levels in individual factories, GATWU usually combined actions at the factory and state levels as a first intervention. On the one hand, GATWU informed the management about workers' complaints and asked the management to engage in dialogue with GATWU to find a solution. On the other hand, GATWU also usually filed a complaint with the labour department or the Department of Factories, Boilers, Industrial Health and Safety, depending on the type of complaint. Issues representing industrial disputes such as non- or late payment of wages, illegal dismissals or factory closures are covered under the Indian Industrial Disputes Act. They, therefore, require a complaint at the Labour Department, where a tripartite conciliation process is initiated. If no settlement can be reached, the issue is referred to the labour court for adjudication.

As illustrated in Sect. 6.5, court processes, however, take several years and are therefore limited sources of institutional power for unions. Issues concerning factory infrastructure and health and safety provisions are, in turn, covered under the Indian Factory Act and, therefore, require a complaint at the Department of Factories, Boilers, Industrial Health and Safety. Complaints at this department are usually followed by an inspection carried out by labour inspectors from the department. As opposed to officers from the labour department, labour inspectors have the authority to give the factory management direct orders for correction if any violations of the Factory Act are detected in the inspection. Whereas institutional power resources accruing to unions from filing a complaint with the labour department is hence somewhat limited (see also Sect. 6.5), filing a complaint with the Department of Factories, Boilers, Industrial Health and Safety can give unions leverage over employers when a labour inspection is conducted. Consequently, GATWU's combined practices of writing complaint letters to factory managers and parallelly filing legal complaints with the competent state authorities represented an effective strategy, particularly in cases where labour law violations concerned basic factory infrastructure, e.g. when a factory lacked drinking water or ventilation.

In cases of more severe and less easily documentable labour rights violations, merely combining practices of writing complaint letters to management and filing legal complaints was, however, usually not enough to push employers for corrective action. In these cases, GATWU, therefore, relied on drawing international consumer campaigning organisations into spaces of contestation to leverage additional coalitional power resources. Cases of labour rights violations in which GATWU drew international consumer campaigning organisations into the space of contestation included, for example, physical abuse of workers in a factory producing inter alia for the brand G-Star. Another case was an incident in which a worker's baby died at a factory creche due to a lack of medical provisions at the factory. In both cases, when the management did not respond to GATWU's demands for corrective action and compensation payments, GATWU resorted to involving international campaigning and consumer organisations such as the CCC and the WRC. By writing letters to the brands sourcing from the respective factories and through public media campaigns, these consumer organisations pushed brands to intervene and to exercise pressure on local factory management for corrective action. As a result, the working environment in the respective factories could be improved, and compensation payments for individual workers achieved.

To what extent did GATWU's practices of 'fire-fighting' and activating alliances with international consumer campaigning networks enable GATWU to build sustained bargaining power vis-à-vis employers? Overall, their strategy of drawing international consumer organisations into spaces of contestation constructed around labour rights violations in individual factories had mixed effects with regard to strengthening GATWU's positions vis-à-vis employers. On the one hand, activating relationships with international consumer campaigning organisations strengthened GATWU's standing in relation to local factory managers. Thanks to the influence of consumer campaigning organisations, GATWU could exert pressure on management and achieve corrections of labour rights violations despite low membership numbers in these factories. As a result, GATWU leaders report that in both cases mentioned above, following interventions by the WRC and the CCC, managers established informal relationships with GATWU. While managers still did not officially recognise GATWU as a collective bargaining partner, they engaged in dialogue with GATWU organisers to solve everyday problems and grievances at the factory level.

On the other hand, GATWU's capacities and practices of involving international campaigning organisations into spaces of contestation constructed around individual labour rights violations, however, contributed little to building GATWU's associational and organisational power resources and thereby to shifting capital-labour power relations to the benefit of workers. Limitations for building lasting bargaining power through transnational campaigning resulted from four conditions. First, by relying on interventions and campaigns carried out by international consumer organisations as central leverage, the centre of gravity within spaces of contestation was moved from the factory space to the international network space. Many central practices constituting the space of contestation, such as consumer organisations' practices of writing emails to brands or conducting public campaigns targeting brands in consumer countries, were performed in geographically distant places and rather disconnected from workers' everyday agency spaces.

Second, interactions with international consumer campaigning organisations and local factory managements were limited to GATWU full-time organisers and did not include workers. The workers who had been the victims of the labour rights violations usually did not participate in any interactions with consumer organisations or in negotiations with the management. Therefore, only GATWU full-time organisers developed strategic capacities of planning, communicating and negotiating within interactions with managements and international consumer organisations.

Third, due to GATWU's predominant community organising approach, GATWU's full-time organisers did not systematically use victories achieved with the support of international consumer organisations to start organising campaigns in selected factories. This lack of strategic engagement by GATWU organisers to use victories as a tool for workplace organising can be explained through two main reasons. On the one hand, struggles involving consumer campaigns were centred on achieving compensation for labour rights violations concerning individual workers. As a result, in these cases, any compensations or corrections benefitted only individual workers and therefore did not serve as a base for large-scale workplace organising campaigns. On the other hand, the time frame of transnational consumer campaigns limited organisers' abilities to construct larger workplace organising campaigns around the victories achieved through these campaigns. Usually, GATWU's struggles involving transnational consumer campaigns took at least one or two years of sustained campaigning until a settlement with the management could be reached. Since most of the campaigning practices, however, took place in network spaces that were rather disconnected from workers' everyday spaces, attention for a specific case at the local and international levels was often quite disconnected from each other. Whereas incidents such as the child's death in the factory crèche caused outrage and spontaneous protests among workers in the respective factory right after the event happened (see also Text box 7.1), it took several months for the consumer campaign to take off due to lengthy administrative processes. Before consumer organisations such as the CCC or the WRC take action, a fact-finding mission has to take place, and an official report needs to be prepared. This process usually already takes several months or weeks. Consequently, by the time the international consumer campaign takes off, the incident is no longer present in workers' minds.

Last, limits for translating moral and coalitional power resources from involving consumer organisations into sustained associational and organisational power also resulted from the power asymmetries characterising GATWU's relations with international consumer organisations. These power asymmetries resulted from the fact that once international consumer organisations were drawn into spaces of contestation constructed by GATWU, they usually took on the lead regarding strategic planning and decision-making. In the case of the deceased baby at the factory crèche, GATWU had already settled with management for a compensation to be paid to the mother of the deceased child. The WRC, however, decided to conduct an international consumer campaign to push for a much larger compensation (for more details, see Text box 7.1).

International organisations taking the lead on these kinds of strategic decisions can have negative effects on local unions, since the institutional logic of international campaigning organisations and local unions are inherently different and not necessarily compatible. International campaigning organisations are interested in generating as much public attention as possible for a specific issue since public attention is the criterion against which their power is assessed by their stakeholders. For a union, however, the main criterion against which their power is assessed, is the size and strength of their membership. Hence, when drawing international consumer organisations into a struggle, interventions by these organisations followed an institutional logic of achieving maximum public attention for a case, independent of whether this public attention would benefit GATWU's union-building struggle.

In summary, GATWU's practices of constructing spaces of contestation around individual labour rights violations by intertwining practices of contacting the management, filing legal complaints and—in severe cases—involving international consumer organisations enabled GATWU to stop various particularly harsh practices of exploitation and disciplining performed by Bangalore export-garment manufacturers. However, at the same time, this strategy had three critical limitations regarding the scope of issues that could be addressed and regarding its potential for enhancing GATWU's bargaining power vis-à-vis management. First, the fire-fighting approach was merely reactive because it only targeted labour rights violations after they happened and sought to correct them. Second, the transformations that could be achieved through the fire-fighting approach were limited to ensuring that legally prescribed minimum labour standards were met. Third and last, the victories achieved through this fire-fighting strategy only helped GATWU build associational and organisational power to a very limited extent, since they neither contributed to increasing GATWU's membership nor to developing workers' strategic capacities.

In contrast, a second set of practices through which GATWU constructed spaces of contestation in its early years was more successful in building GATWU's membership base and associational power. This set of practices focussed on contesting for higher minimum wages and achieving regular revisions of the minimum wage for the garment industry in the State of Karnataka. In the next section, I will describe this set of practices in more detail.

> **Text box 7.1 Case in focus—Struggle for compensation for a worker's deceased child at Gokaldas Exports**
>
> In 2015, a two-year-old child passed away in the crèche of a factory owned by the Bangalore export-garment manufacturing company Gokaldas Exports. During feeding hour, the child allegedly got rice into his lungs and had trouble breathing. Against legal provisions, the factory did not have an ambulance, so the child was brought to the hospital in a manager's car but declared dead on arrival (INT4). The mother informed GATWU, and their organisers asked the management for financial compensation for the worker's loss given that the child might have survived if an ambulance and all required medical facilities

had been in place at the factory. GATWU finally agreed with the management on a compensation of 150,000 Rupees (approx. 2,300 US$ or two years of a garment worker's basic wage) to be paid to the mother of the deceased child. Simultaneously, GATWU notified the Indian representative of the WRC about the incident. GATWU has thought of this notice merely as an information to involve the WRC in case the management refused to engage in negotiations. The WRC, however, found the compensation too low and not in line with international standards. Thus, the WRC proposed to GATWU to renegotiate the case—this time with a public campaign targeting Gokaldas Exports' main buyers, asking them to ensure legal health and safety provisions in their supplier factories and adequate compensation for the mother of the deceased child. According to the WRC, the compensation should amount to 40,000 US$ equalling to an expected income of 25 years, with which her son could have supported her. This calculation was based on standards that had been defined by the ILO in the compensation process for the victims of the Rana Plaza factory collapse in Dhaka, Bangladesh in 2014 (WRC 2015b). After a year of sustained campaigning, brands finally arranged a tripartite negotiation involving brand representatives, the Gokaldas Exports management and GATWU. In this negotiation, GAWTU and Gokaldas Exports management agreed on an additional compensation payment of 800,000 Rupees (approx. 10,500 US$ or the equivalent of about ten years' basic wages of a garment worker) for the mother of the deceased child (WRC 2015a). Moreover, Gokaldas Exports ensured that all legally required medical and health facilities were put in place. The enforcement of such a high compensation payment can be considered a historical victory for GATWU in the Indian context, exceeding any compensation payments for injured or deceased workers fixed by Indian courts so far (INT17). However, for three reasons this historic victory did not lead to a significant increase in membership or bargaining power for GATWU vis-à-vis the Gokaldas Exports management. First, GATWU's victory did not imply any benefits for the factory workers except for the fact that medical facilities were put in place. Workers, however, seldom use these facilities because high production targets do not allow them to take breaks. Second, the WRC campaign relied on activating moral power resources by framing the health and safety violations in the Gokaldas factory as de facto human rights violations. This framing could, however, not easily be replicated in other struggles around less tragic incidents such as dismissals or factory closures. Third, most strategic interactions and practices that were decisive for leveraging moral power were carried out by WRC staff: WRC calculated the compensation payment, developed the strategy for a public campaign targeting brands, undertook a 'fact finding' investigation, published a report, and communicated with brands throughout the whole process. Therefore, practices of campaigning and negotiation took place primarily in transnational 'network spaces' and were only brought back to the local level in the end for the final tripartite negotiation between GATWU leaders, brands and the Gokaldas Export company management. As a result, GATWU leaders and workers had little involvement in the space of contestation and only limited chances to develop strategic capacities.

Campaigning for Minimum Wage Increases

As illustrated in Sect. 6.3, minimum wages in the garment industry in Karnataka have been traditionally low due to various employer practices of evading legally prescribed periodic minimum wage revisions. When GATWU was formally registered in 2006, the last minimum wage revision had taken place in 2001, and there had been no announcement by the state government about the date for the next round of minimum wage revisions. The average statutory minimum wage for a garment worker in Bangalore at that time was around 3000 Rupees per month (approx. 40 US$)—an amount that was not enough to cover the living expenses of a family. Consequently, low wages and resulting economic struggles were frequent topics of discussion in area committees and factory gate meetings. Against this background, GATWU's first major *collective* struggle was centred on achieving a revision and significant increase of the statutory minimum wage.

During this period, GATWU did not have a strong membership base inside individual factories. Therefore, the unions' capacities to push for collective bargaining at the factory level were limited. In the face of this constraint, public campaigning for a higher statutory minimum wage was a more viable strategy for achieving a wage increase. From 2007 to 2010, GATWU carried out a public campaign that combined two sets of interwoven practices: (1) organising and mobilising workers at the community level and (2) pressuring the labour department at the state level. At the community level, GATWU's and Munnade's primary organising practice involved building awareness among workers about wages as a collective, structural issue by discussing workers' daily expenses in area committee meetings. As a Munnade organiser explains, these group discussions served to de-individualise problems of struggling economically. Through sharing individual experiences of not being able to afford rent, food and children's education due to low wages, difficulties to make ends meet "became a collective experience" (INT39). Discussions in area committees were combined with a systematic expense survey among garment workers to calculate demands for a new minimum wage based on workers' real expenses. The survey showed that workers needed at least 200 Rupees per day to cover their monthly expenses—double the minimum wage at that time.

Based on this figure, in 2007, GATWU organised a public campaign targeting the State Government of Karnataka. In this campaign, GATWU demanded the legally due minimum wage revision to take place and for an increase of minimum wages to 200 Rupees per day, amounting to a wage raise of 100%. To spread awareness among workers, GATWU and Munnade activists distributed posters and stickers in workers' living areas stating "I am a garment worker and I need at least 200 Rupees per day to survive" in Kannada. As a Munnade activist explains, these stickers contributed significantly to building workers' collective consciousness since "at some point, almost every garment worker had this sticker at their front door" (INT39). Parallel to organising and mobilising workers in their communities, GATWU sent a memorandum to the labour department sketching the survey results and demanding a minimum wage of 200 Rupees per day (INT39). To put pressure on the state government, GATWU conducted several public rallies throughout 2007 and 2008, addressing the labour department and employers. GATWU's practices of conducting public rallies were, in turn, enabled by GATWU's organising practices at the community level, which allowed GATWU to gather and mobilise several thousand workers from different neighbourhoods for central public rallies.

It was hence the intertwining of practices of community organising around wages as a collective issue with practices of organising and conducting public rallies that allowed GATWU to exercise associational power in the streets and thereby influence the Karnataka state government. After two years of sustained campaigning, the labour department finally issued a new minimum wage notification that increased wages by 27 Rupees per day. Garment manufacturers, however, refused to pay this new minimum wage and continued to pay the old minimum wage for a whole year (see also Sect. 6.3). Simultaneously manufacturers started a lobbying campaign, asking the state government to withdraw the new minimum wage notification, arguing that it would ruin the industry. In reaction to manufacturers lobbying campaign, in March 2010, the Government of Karnataka formally withdrew the original minimum wage notification due to a 'clerical error', and issued a new notification which increased the minimum wage by only 22 Rupees per day (as opposed to 27 Rupees in the original notification). Nevertheless, garment manufacturers continued to ignore this new, reduced minimum wage notification.

GATWU reacted to the withdrawal of the original minimum wage notification and manufacturers' continued refusal to pay the new minimum wage rate by combining protest practices at the international, state and community levels and— for the first time—also at workplace level. At the international level, GATWU decided to activate support from international consumer organisations for additional leverage. Since brands' codes of conduct usually state that legal minimum wages must be paid, consumer organisations were able to put significant pressure on brands sourcing from Bangalore factories through public campaigns. Following a particularly powerful campaign by the WRC, various large US brands finally threatened to suspend sourcing from Bangalore suppliers until manufacturers would pay the statutory minimum wage. At the state level, GATWU filed a case in the Karnataka High Court against the withdrawal of the notification. At the community and workplace level, garment workers conducted a symbolic protest by only wearing black clothes for eight days. Whereas this symbolic protest did not mobilise workplace bargaining power by stopping or slowing down production, it nevertheless attracted significant attention in the local and national media and enabled GATWU to leverage moral power resources vis-à-vis employers. Parallel, GATWU held regular factory gate and area meetings to inform workers about the newly fixed minimum wages. Even though GATWU did not achieve the reinstatement of the original minimum wage notification, GATWU's protest practices at the community and factory level and the resulting public attention, however, ensured that employers paid the wage increment of 22 Rupees per day (INT39).

In the end, the wage increase achieved through GATWU's minimum wage campaign of 2007–2010 remained at 20%, significantly under the union's original demand of a 100% wage increase. Nevertheless, the campaign still had significant enabling effects for building GATWU's bargaining power vis-à-vis employers and the state in subsequent minimum wage revisions. Following GATWU's legal complaint about the withdrawal of the original minimum wage notification, the Karnataka High Court ruled in 2013 that for the next round of minimum wage revisions in 2014 a tripartite Minimum Wage Board for the garment sector should be constituted with GATWU as worker representative (INT39). As a result, since 2014, GATWU has been a member of the tripartite Minimum Wage Board for the garment sector and of the general Minimum Advisory Board for the State of Karnataka. Membership in these boards represents a source of institutional power: as a member of the Minimum Wage Advisory Board, GATWU has ensured that, since 2014, minimum wage revisions have been implemented in the legally prescribed intervals of three to five years. Moreover, as an official member of the Minimum Wage Board for the garment industry, GATWU was finally able to push for a minimum wage increase of about 100% and to raise monthly minimum wages to an average of 7,000 Rupees (approx. 100 US$) in the next round of minimum wage revision, which took place in 2014.

In summary, GATWU's initial minimum wage struggle in 2009/10 has brought about positive wage effects for workers in the mid- and long-term and contributed to enhancing GATWU's associational and institutional power resources. In terms of institutional power, GATWU established itself as the official representative of garment workers in the institutionalised minimum wage bargaining process. In terms of associational power, GATWU doubled its membership through the minimum wage campaign. Whereas in 2007, GATWU had around 3,000 members, by 2014, GATWU had managed to increase its membership to 6,500 members. This increase in membership was achieved through sustained community organising work and active involvement of workers in public rallies and collective, symbolic workplace action. Rallies and workplace action served for workers to experience collective organisation first-hand and to provide spaces for workers to develop an 'oppositional consciousness' (Katz 2004: 251), i.e. the capacity to understand and analyse one's individual situation as shaped by broader power structures. Interactions with international consumer organisations, in turn, helped to build GATWU's organisational power by allowing GATWU's leadership to develop strategising capacities. These capacities included understanding the structure of the value chain, communicating with international consumer organisations, and strategically employing brands' leverage to reinforce the exercise of associational and moral power through public campaigns and protests at the local level.

It is important to note that GATWU's minimum wage campaign of 2007 till 2010 was nevertheless still in line with their overall fire-fighting approach. The primary rationale for the campaign was to rectify prevalent violations of the Minimum

Wage Act by ensuring that the legally due regular minimum wage revisions were implemented and that employers paid the statutory minimum wage. By leveraging associational power through public protests and moral power through transnational consumer campaigns, GATWU stopped large-scale minimum wage violations and ensured regular minimum wage revisions. However, GATWU's bargaining power was still insufficient to push for a living wage in tripartite negotiations with employers and the state. Limitations to GATWU's bargaining power resulted from GATWU's still relatively low membership of 6,000 workers in 2014 compared to a total of about 450,000 garment workers in the cluster. Moreover, given that with the community organising approach, GATWU's members were distributed over a large number of factories, GATWU was unable to put pressure on employers through industrial action at the workplace.

In the face of these limitations, GATWU leaders and organisers have, over the past decade, shifted their organising activities from the community to the workplace. In the same line, they started to construct spaces of contestation around struggles for collective bargaining rather than around individual labour rights violations. The following section lays out GATWU's strategic reorientation process in more detail. It describes how GATWU has constructed spaces of organising, collaboration and contestation in the context of its strategic turn towards a new factory organising, collective bargaining approach.

7.1.2 Towards a Strategic Factory Organising and Collective Bargaining Approach

GATWU's strategic reorientation process was initiated by GATWU's split from Cividep in 2010, which also ended funding for their activities from Oxfam International. After the split from Cividep, GATWU aimed to restructure internal union relations to build accountability relations primarily between workers and union leaders— as opposed to external accountability relations with international funders. In this line, GATWU also aimed to strengthen the role of worker activists in workplace organising and building collective bargaining processes with employers. It is important to note that GATWU's strategic reorientation process went on for several years and is still not concluded.

GATWU's first attempt to form a factory union and initiate collective bargaining at a factory called Arvind Ltd. in 2013 failed: After notifying the management about the newly founded factory union, the management immediately fired all factory union representatives. Despite filing legal complaints and urgent appeals with various international labour rights networks, GATWU could not reinstate the fired worker leaders, and the organisation at the factory was crushed. According to GATWU leaders, this first failed attempt to initiate a collective bargaining process at the factory level represented a decisive moment in GATWU's strategic reorientation process for two reasons. First, they learned that to engage in factory-level collective bargaining, the union needed to achieve a significant level of organisation of about 60% in the respective factory to push for the reinstatement of dismissed workers leaders through industrial action. In the case of the collective bargaining attempt at Arvind Ltd., GATWU had only around 30 members in the factory, and these workers had not been sufficiently prepared to stay organised in case of repression by the management. Second, GATWU realised that neither filing legal complaints with the Labour Department nor filing urgent appeals with international consumer organisations represented effective tools to counter employer union-busting due to the lengthy processes of conducting fact-finding missions required for both measures.

The experience with the failed collective bargaining attempt at Arvind Ltd. led GATWU organisers to restructure how they constructed spaces of organising, spaces of collaboration and spaces of contestation in a more systematic way. Whereas they had formerly constructed the community as the main space for organising, GATWU now sought to build spaces of organising in selected target factories. Moreover, GATWU began to construct new spaces of collaboration by building relations with international worker and union networks that could help them develop their strategic capacities. Lastly, while GATWU had constructed spaces of contestation to rectify labour rights violations under the fire-fighting approach, GATWU now aimed to construct spaces of contestation around *proactive* struggles for collective bargaining. In the following, I lay out the practices and relations through which GATWU has been constructing spaces of organising, collaboration and contestation since 2015, when they first implemented the new factory organising, collective bargaining approach systematically.

7.1.2.1   Spaces of Organising

After the failed collective bargaining attempt at Arvind Ltd., GATWU reoriented its organising practices from community organising to factory organising. This strategic reorientation also involved a shift in focus from quantity of member relations—that is, from signing up as many members as possible—towards quality of member relations, as explained by GATWU's president:

> […] from 2006 we were in many factories. […] We would go to anybody, take their member-ship, and we were also not so worried about consolidating our membership. Only members had to give their names, and that was it. So, then we saw that after six years, there was little outcome. [..] workers would come to the union whenever they had a problem. And they would not take the membership or continue with the membership. Only when they wanted something they would come to us. So, there was no real union perspective. Now we are working on that. We have trainings with workers on their role and responsibilities as union members. (INT26)

As the quote exemplifies, with the strategic reorientation, GATWU's focus has shifted from merely increasing membership numbers to building stable and active member-ship bases in selected factories. Moreover, GATWU now invests a lot of time and resources into forming strong worker leaders inside factories who can organise and mobilise workers and negotiate issues with the management. Hence, with the reorientation towards a factory organising approach, central strategic functions and competencies have been shifted from full-time, paid organisers external to the factory space to worker leaders in the workplace (INT46). In this vein, internal union responsibilities and decision-making competencies have also been de-centralised: Worker leaders in factories now act largely independently when solving day-to-day grievances at the factory level. Union leaders, in turn, maintain relationships with external collaborators at the local and international levels. They also intervene with management in strategic struggles requiring additional leverage by involving senior union leaders.

In order to build the strong membership base and worker leader capacities needed to push for collective bargaining at the factory level, GATWU has developed a strategic approach for organising in selected target factories since 2017. Under this new factory organising approach, GATWU has been focussing its organising activities on five target factories, which were selected according to three main criteria. First, GATWU selected factories where they already had a significant number of members or where there had already been incidents of collective worker action. Second, factories were selected where GATWU had some members with strong leadership capacities and motivation to organise inside the factory. Third and last, GATWU selected key tier one suppliers for EU and US brands in line with their new strategy of forging collaborations with worker and union networks from brands' retail sectors. These collaborations enabled GATWU to mobilise quick solidarity action when union activists at target factories were dismissed or victimised, as will be described in more detail in the next section.

In the five selected factories, GATWU has formed union committees and provides continuous training to the members of these committees. As opposed to training sessions with area committees under the community organising approach, training sessions with factory union committee members now go beyond just informing workers about their rights. Instead, these sessions focus on building workers' strategic leadership skills by capacitating them to use labour laws as strategic tools and building awareness about their role and responsibilities as union representatives. Moreover, factory union committee members learn how to develop demands for collective bargaining (FN6).

The strategic shift from community organising towards factory organising has also gone along with the geographical restructuring and expansion of GATWU's organising practices. Of the five target factories selected in 2017, three are located in semi-rural or rural zones in a distance of between 50 and 130 kms from Bangalore. Only two target factories are located on the outskirts of the Bangalore urban area along Mysore Road. In this area, GATWU has traditionally had the highest concentration of members. The expansion of GATWU's geographical reach to include factories in rural areas needs to be understood against the backdrop of the general geographical restructuring of the sector over the past seven to eight years (see Sect. 5.2). The high worker turnover and presence of alternative job opportunities make it more difficult to organise workers in the city since workers just tend to look for another job when confronting problems in the workplace (see also Sect. 6.7). These challenges for organising are exemplified in the following report by GATWU's president during an international union meeting:

> There are so many garment factories now in the city area, and there is a high turnover rate of workers at the factories. When we organise at a factory, in the second year most of the workers with whom we started working won't be there anymore. If workers in the city are facing problems within their factory, instead of struggling, they rather leave the factory and search for a job somewhere else. It is easy to find work in another factory, or even in a shop or a mall. In the rural areas it is different. There are less companies and therefore it is not that easy to change jobs, so workers are more willing to struggle back. (FN1)

In rural areas, workers, in turn, tend to regard work in garment factories as a favourable alternative to working in agriculture. Hence, workers are more prepared to organise and struggle for better conditions, as this Munnade organiser explains:

> [...] between agricultural work and work in the garment factory, the garment factory is already an improvement. In agriculture, workers are outside in the sun the whole day. And in the garment factory there is shade, there are fans. Also, working at the factory, workers will get regular wages and ESI [Employee State Insurance] and EPF [Employees' Provident Fund]. And then the whole family can be insured on the workers' ESI. Agriculture depends so much on the weather. If the weather is bad, there is no income. And in the factory, you also get the Sunday off. In agriculture, there is no such thing as a day off. (INT39, translated from Kannada)

Against this background, GATWU has built its strongest membership bases under the new factory organising approach in target factories in rural areas. In addition to providing a more stable workforce, factories in rural areas also provided enabling conditions for organising because workers in these factories often had previous experiences of collective action, e.g. through conducting spontaneous protests. In one of GATWU's rural target factories called European Clothing II, for example, workers had organised a sit-in protest that lasted several weeks after the management had terminated 43 helpers. This protest provided an entry point for GATWU, who took up the issue, filed an official complaint with the labour department and negotiated the workers' reinstatement with the management.

This initial success allowed GATWU to quickly expand their factory membership base and thereby to build workplace bargaining power. This new workplace bargaining power, in turn, enabled GATWU to establish the leaders of the union factory committee as dialogue partners of the management. Now worker leaders from this factory independently negotiate with management on everyday problems and grievances in the factory. However, it is important to note that building a strong membership base at the factory level and engaging in collective action alone was not enough to achieve official union recognition. So far, GATWU has won official recognition by the management and signed a collective bargaining agreement only in one factory belonging to the company Avery Dennison. To achieve the collective bargaining agreement, GATWU had to combine practices at various levels, including leveraging coalitional power resources from new collaborations with international union networks. Before I lay out in more detail how GATWU constructed the space of contestation around the collective bargaining campaign at Avery Dennison, in the next section, I will first lay out which new collaborations enabled GATWU to conduct this campaign.

7.1.2.2 Spaces of Collaboration

As mentioned above, GATWU's shift towards a factory organising, collective bargaining approach also involved reorganising the practices through which GATWU constructed relationships with external collaborators. After ending the collaboration with Cividep, GATWU decided to construct collaborative relationships with external organisations differently: As part of the strategic reorientation process, they now sought to construct relationships with external organisations not primarily as a space for acquiring financial resources but as a space for developing strategic capacities for union building and collective bargaining. According to GATWU's leaders, two international union networks played a crucial role in helping them to construct these types of spaces: the International Union League for Brand Responsibility and the TIE ExChains network. The International Union League for Brand Responsibility (short: the League) is a network of 13 unions from export industries in Asia and Latin America, many of which are active in the export-garment industry. Founded in 2013, the League aims to build transnational solidarity among workers to collectively pressure brands into ensuring workers' rights to Freedom of Association at their supplier factories (IULBR 2021).

The TIE ExChains network, in turn, brings together workers from the fashion retail and logistics sector in Europe and workers from the South Asian garment industry. The TIE ExChains network aims to strengthen local union building and bargaining power through transnational solidarity and to develop joint demands of workers along the value chain vis-à-vis lead firms (ExChains 2015). GATWU leaders report that participating in these two networks has helped them to deploy coalitional power resources in form of transnational labour solidarity in ways that also strengthened GATWU's associational and organisational power. One GATWU leader stresses that it was vital for GATWU to prioritise collaborations with international union and worker networks. As opposed to NGOs or consumer networks which have different institutional logics (see Sect. 7.1.1), worker and union networks can relate to GATWU's struggles from first-hand experience. A GATWU leader expresses this in the following words: "They are giving us the strategic and ethical strength. Because they also believe what we believe" (INT46).

As opposed to former project-based collaborations with international NGOs, collaborations within these transnational union networks focussed on sharing experiences from workplace struggles. GATWU leaders argue that learning from Central American garment unions' experiences with pushing for collective bargaining and withstanding employer repression was crucial for their process of developing a factory organising strategy (INT46). In particular, learning how to use transnational solidarity to strengthen local union bargaining power was a crucial takeaway for GATWU from discussions with other unions in the League and the TIE ExChains network. GATWU leaders have also built territorially embedded networks with other

Indian unions, inter alia through their affiliation to the New Trade Union Initiative. However, these unions usually have no experience with using transnational mechanisms. While from these territorially embedded union networks, GATWU can hence get advice, for example, on how to use Indian legal mechanisms most effectively, these networks are not spaces where GATWU can develop capacities of networked agency. Therefore, network spaces of collaboration in transnational union networks represent a critical complementary space for GATWU, in which union leaders can develop strategic capacities, which they cannot develop through interactions within territorially embedded union networks at the national level.

Planning joint networked action with the members of the TIE ExChains network played a crucial role in developing GATWU's strategic factory organising approach. As mentioned earlier, one important criterion for GATWU's selection of target factories was that these factories should produce for international brands where retail and logistics workers from the TIE ExChains network have strong representation. Linking factory organising practices with practices of constructing transnational labour solidarity relationships enabled GATWU to proactively plan and coordinate parallel interventions at the factory level as well as at the international level to support struggles for collective bargaining in target factories. To this end, GATWU regularly informs the European coordinators of the TIE ExChains network about the state of organising at target factories and plans for initiating the collective bargaining process. As a result, solidarity actions directed at mobilising brands' leverage can be activated immediately in case of employer repression.

Contrary to consumer campaigns, which usually rely on leveraging moral power resources, solidarity actions within the TIE ExChains network use workers' associational and institutional power to exercise leverage over brands and retailers. Works councils in Germany, for example, use weekly meetings with store managers to address GATWU's struggles for collective bargaining and ask management to ensure the rights to freedom of association and collective bargaining along their supply chains. Whereas these practices do not necessarily lead brands to push suppliers to enter into negotiations with GATWU, brands usually contact their suppliers to inquire about the situation. This inquiry, in turn, raises awareness with the local manufacturer that there is attention from brands on the case and can thereby prevent repression in reaction to GATWU's demand for collective bargaining.

Besides constructing collaborations with international union networks as spaces for strategic capacity development and leveraging coalitional power in struggles for collective bargaining, GATWU has constructed several collaborations with local organisations as spaces for securing financial resources. Even though membership fees are collected more strictly under the new factory organising model, membership fees are still not enough to cover costs for the union's office and activities. Therefore, to acquire additional financial resources, GATWU has built a network of local organisations and individuals who support the union's work through financial and material donations.

Lastly, GATWU has transformed their practices of constructing relationships with international consumer networks from the Global North under the new factory organising, collective bargaining approach. However, as opposed to GATWU's early years, under the new factory organising, collective bargaining approach GATWU has filed urgent appeal mechanisms with international consumer networks only very selectively. In an interview, the leaders stress that GATWU no longer uses urgent appeal mechanisms for individual workers' cases due to the high time and personnel resources involved in filing such an appeal compared to low benefits for the union building and organising process:

GATWU leader 1: You know it is a small case, but then you have to do so many things… The energy and time…And international campaigns, they want 100% details. (INT46)

GATWU leader 2: And after that, any positive thing that comes about is only for one worker. For only one worker, we cannot do the whole thing. […] It will take a lot of our time, effort, resources, everything. But only for one worker, not for every worker. And after that, the worker will leave the job and then everything was in vain, (INT46)

Therefore, currently, GATWU exclusively resorts to leveraging moral power through international campaigns to address collective issues linked to the union-building process in a specific factory. This is the case for example, when key worker leaders have been dismissed or when a collective issue affects a large group of workers in a target factory. In the following subsection, I will lay out in more detail how GATWU deployed transnational campaigns as one of several tactics in their struggle at an Avery Dennison factory in Bangalore. It was in this struggle that GATWU achieved the first collective bargaining agreement in the Bangalore export-garment industry.

7.1.2.3 Spaces of Contestation

As mentioned earlier, under the new factory organising approach, GATWU currently constructs spaces of contestation primarily around collective issues at target factories or even at non-target factories when these issues have the potential to serve as catalysts for building a strong membership at that factory. Only in these cases, GATWU deploys strategies of networked agency that intertwine actions at the factory, state and international levels to target multiple actors simultaneously. The remainder of this section analyses GATWU's networked agency strategy for constructing spaces of contestation under the factory organising, collective bargaining approach through the lens of a labour struggle led by GATWU from 2017 till 2020 at the Bangalore factory of the multinational company Avery Dennison. This struggle represents an important milestone in GATWU's new factory organising, collective-bargaining strategy since it was the first struggle in which GATWU signed a collective bargaining agreement at the factory level. In the following, I first set out the background of the struggle and a chronology of events before describing in more detail the networked practices of contestation at various levels deployed by GATWU in this particular struggle.

GATWU's Struggle at Avery Dennison from 2017 Till 2019: Background and Chronology of Events

Avery Dennison is a multinational company producing labels, graphic tags and price tags for apparel brands and retailers. In Avery Dennison's Bangalore factory, about 600 workers produce RFID labels and tags for more than 130 international garment brands and retailers. Avery Dennison does not supply brands and retailers directly but acts as a tier two supplier. However, given its quasi-monopoly market position as a producer of RFID labels, Avery Dennison can still be considered a strategic supplier in the garment GPN. Notably, the workforce composition and labour process at Avery Dennison as a label factory differ from the typical workforce composition and labour process organisation in Bangalore garment factories. At Avery Dennison, the workforce is predominantly male and (semi-)skilled, since many tasks involve operating digital design and printing machines. Moreover, compared to wages in the garment manufacturing sector, wages at Avery Dennison were significantly higher than the minimum wage, with blue-collar workers in permanent employment earning an average monthly wage of 25,000 Rupees. Against this backdrop, the company had traditionally relied to a large part on contract labourers, who received much lower wages. Despite the provisions of the Indian Contract Labour Act, which foresees that contract labour can only be used in non-core activities and for a maximum of 240 days per year, Avery Dennison had operated its Bangalore factory in Bangalore with about 70% contract workers up until GATWU intervened in 2017.

By the time GATWU started organising at Avery Dennison, most contract workers had worked at the factory for between two and ten years without a break in service. While most contract workers had initially been hired as unskilled helpers, many had been promoted to skilled positions of digital machine operators and team leaders over the years. At the same time, these contract workers continued to receive the wage rates for helpers, ranging around 7,000 Rupees (approx. 92 US$). As such, wages paid to contract workers at Avery Dennison were significantly below the scheduled minimum wage for the printing industry. Arguing that employment at Avery Dennison does not fall under any of the scheduled employments for which the Government of Karnataka fixes a statutory minimum wage, the Avery Dennison management paid only the minimum wage for so-called 'non-scheduled' employment to contract workers. Hence, while carrying out the same tasks as permanently employed workers, contract workers received significantly lower wages, bonus payments and benefits. Differences in wages and benefits were also particularly salient between contract workers and permanent workers because permanent workers had been organised for years in a factory-level union, the Avery Dennison Workers Union (ADWU). Despite the unfavourable conditions of the collective bargaining agreement negoti- ated by the ADWU compared to industry standards, permanent workers still received total monthly wages of about 25,000 Rupees (approx. 331 US§). Permanent workers' wages were more than three times higher than contract workers' wages. However, the basic wage component defined in the collective bargaining agreement for perma- nent workers was very low, and more than 50% of the total wage consisted of bonus payments. Therefore, social contributions were calculated by the company according to the basic wage of around 10,500 Rupees (approx. 139 US$), meaning significant pension losses for workers.

GATWU's engagement at Avery Dennison in Bangalore began in September 2017, when the factory management dismissed 47 contract workers within one month without further explanation. The collective dismissal instilled fear among contract workers about losing their job and led them to approach GATWU, who, by that time, had built a reputation in Bangalore for its successful networked agency approach in the export-garment industry. GATWU took up the case and, in the first few weeks, concentrated on building a strong membership inside the factory in line with the new strategic factory organising approach. Given contract workers' readiness to unionise, GATWU had organised about 90% of the 300 contract workers who remained at the factory within a few weeks. In October 2017, GATWU handed the management an official charter of demands comprising the following demands: (1) recognition of GATWU as the official representative of contract workers, (2) reinstatement of the dismissed contract workers, (3) payment of adequate sectoral minimum wages for contract workers, (4) provision of equal benefits for permanent and contract workers, and (5) permanent employment for all contract workers who had been working at the factory for more than 240 days as ordered by the Indian Contract Workers Act.

The Avery Dennison management, however, refused to engage in negotiations with GATWU. Instead the management performed two union-busting practices to undermine GATWU's union-building work in the factory. First, the management recurred to a strategy of 'politics of silence' and simply ignored GATWU's requests for dialogue. Even when GATWU filed a complaint at the labour department, the Avery Dennison management did not participate in the tripartite conciliation meetings but instead sent a lawyer as a representative—according to GATWU leaders, a clear sign that the management had no interest in coming to an agreement with the union. Second, the management continued with contract worker lay-offs, targeting workers who had participated in union gate meetings.

The management's refusal to recognise and engage with GATWU as a bargaining partner was followed by a year-long struggle by GATWU, which intertwined practices of contestation at the workplace, state and international levels. These practices included holding protests at the workplace, filing legal complaints and involving transnational union and consumer networks. Through these networked practices of contestation, GATWU was able to pressure employers into implementing several important improvements for contract workers.

First, GATWU achieved permanent employment for 110 contract workers (even though contract workers had to undergo a formal application process, meaning they lost the benefits gained through years of continuous service). Second, the management increased contract workers' wages beyond the mandatory statutory minimum wage increase. When in January 2018, the Government of Karnataka increased minimum wages for non-scheduled employment from 7,000 Rupees (approx. 92 US$) to 12,000 Rupees (approx. 159 US$), the Avery Dennison management raised contract workers' wages to 13,000 Rupees (approx. 172 US$).

Lastly, GATWU achieved several transformations in Avery Dennison's practices of employing contract workers. Until GATWU's intervention, Avery Dennison maintained informal relationships with contract labour agencies. Orders for additional labour supply or terminating workers' services had been made through phone calls rather than in written form, allowing Avery Dennison to deny any responsibility for contract workers. Following GATWU's intervention, Avery Dennison stopped this practice and started giving orders in written form. Moreover, all contract workers carrying out core activities were offered permanent employment in the previously mentioned recruitment process. Contract workers were only employed in non-core activities to ensure compliance with the Indian Contract Labour Act.

GATWU's victories in the contract worker struggle subsequently allowed them to extend its organising and collective bargaining efforts at Avery Dennison to include permanent workers as well. During factory gate meetings, GATWU leaders and activists started to address permanent workers organised in the ADWU to raise awareness about the unfavourable conditions of the collective bargaining agreement negotiated by ADWU. Given GATWU's victory for contract workers, many permanent workers decided to join GATWU, hoping that they could also win additional benefits for permanent workers. By July 2018, GATWU had managed to organise about 70% of permanent workers. GATWU had managed to gain more members than the long-standing factory union ADWU. Following the provisions of Indian labour law, GATWU demanded a secret ballot election to formally establish itself as the majority union and hence as the official partner for collective bargaining. The management, however, ignored GATWU's request to hold elections.

Instead, factory managers employed various union-busting practices to undermine GATWU's unionisation efforts with permanent workers. For example, the Avery Dennison management repeatedly convened staff meetings, warning workers not to engage with GATWU, claiming that the union as an 'outsider' organisation would harm the factory. Against this background, GATWU once more deployed a networked agency strategy combining practices of contestation at the factory, state and international levels to put pressure on management. The pressure placed on brands and the Avery Dennison management through the combination of intertwined practices at all three levels finally culminated in a formal mediation process initiated by the Ethical Trading Initiative (ETI). The ETI is a business-led multi-stakeholder initiative, of which many brands sourcing from Avery Dennison are part. In this mediation process, GATWU and the Avery Dennison management finally agreed to form a joint bargaining committee involving members of GATWU and the ADWU. In the collective bargaining agreement negotiated by this joint committee in December 2019, GATWU achieved various improvements for permanent workers, including inter alia a significant raise in basic wages and annual, wedding and bereavement leave. Figure 7.2 gives an overview of the chronology of events of GATWU's struggle at Avery Dennison.

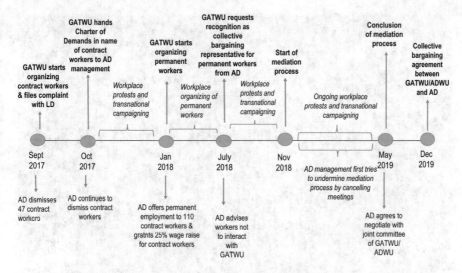

Fig. 7.2 Timeline of events of GATWU's struggle at Avery Dennison Bangalore. *Source* Author

Holding Gate Meetings and Conducting Symbolic Protests at the Workplace Level

Following the strategic factory organising, collective-bargaining approach, the core space of contestation in GATWU's struggle at Avery Dennison was the workplace. At the workplace, GATWU combined several practices of contestation, from holding gate meetings to protesting inside the factory. Gate meetings were constructed as spaces for dialogue and trust building with workers, and consequently for building a strong membership base inside the factory. Accordingly, GATWU leaders and factory activists used gate meetings to distribute leaflets to contract and permanent workers in order to raise workers' awareness of management's wage theft practices affecting both contract and permanent workers. On the other hand, gate meetings were constructed as protest spaces to put pressure on the factory management (see Fig. 7.3). In this vein, workers and GATWU organisers used gate meetings to voice their demands and produce photos and videos that could then be shared with international networks of collaborators.

Protesting practices inside the factory, in turn, included various practices of symbolic protest, including collective canteen boycotts and hunger strikes (see Fig. 7.4), and workers wearing badges that stated: "I belong to GATWU".

These symbolic protest practices at the workplace level contributed to building GATWU's bargaining power vis-à-vis the Avery Dennison management in two ways: First, collective protest practices strengthened GATWU's associational and organisational power because they provided spaces for workers to experience collective action first-hand. Moreover, by actively involving workers in planning these actions, GATWU leaders created spaces for workers to build the strategic capacities needed to lead a labour struggle. Planning of workplace actions happened during regular

Fig. 7.3 Factory gate meeting in front of Avery Dennison factory with workers and GATWU organisers. *Source* ExChains 2019

Fig. 7.4 Worker canteen boycott at Avery Dennison. *Source* ExChains 2019

Sunday meetings with workers, in which GATWU leaders updated workers on every step and new development in the struggle, answered workers' questions, and, together with workers, developed plans for strategic next steps. According to GATWU's president, this form of intensive worker involvement in the planning and decision-making process was a central condition for building associational power in the form of a strong membership base at Avery Dennison and for maintaining this membership base throughout the more than two years of the struggle.

However, GATWU made the conscious decision not to use these associational power resources to strike but instead to conduct only symbolic collective actions inside the factory through the hunger strike and the badge campaign. This decision was based on GATWU's strategic evaluation of the balance of forces: they concluded that the management would most probably counter industrial action with mass lay-offs and potentially even factory closure. Given this risk, GATWU decided to perform symbolic collective action at the factory level. This symbolic action could then be transformed into moral power resources by distributing video and photo footage across their international supporter networks and social media channels. The leverage that GATWU's symbolic protests within the factory space produced vis-à-vis management, therefore, resulted only from intertwining these protests at the workplace level with practices of activating links with international NGO and union networks.

Practices of Contestation at the State Level: Filing Complaints with the Labour Department

Parallel to building and leveraging associational power resources at the workplace, GATWU sought to leverage institutional power resources at the state level by filing repeated complaints with the Labour Department. In total, GATWU filed three complaints with the Labour Department: (1) a complaint about Avery Dennison's illegal use of contract labour, (2) a complaint about non-payment of minimum wages for the printing industry for contract workers and (3) a complaint about the illegal termination of workers engaged in the union. GATWU's practices of filing complaints with the Labour Department, however, also only allowed the union to leverage institutional power resources through interrelations with moral power resources activated through public campaigns with international collaborators.

As explained in Sect. 6.5, the overall pro-business stance of state authorities generally constrains unions' institutional power resources. Therefore, filing a complaint at the labour department is usually not an effective means for unions to ensure the implementation of labour rights, particularly in cases of illegal terminations and union-busting. However, as GATWU leaders explain, filing an official complaint at the labour department can still represent a strategically important practice within a networked agency approach, because it increases the credibility of the union's events report vis-à-vis international collaborators. In the Avery Dennison case, enhancing credibility vis-à-vis international union and consumer networks worked particularly well because the deputy labour commissioner in charge ordered an official inspection, which confirmed Avery Dennison's illicit use of contract labour. Similar to the documentation of workers' hunger strike, the official inspection report by the labour department served GATWU as a tool for leveraging moral power resources through transnational campaigns in collaboration with international union and consumer networks. These networks used the official inspection report as leverage to pressure brands into ensuring that these labour law violations by their supplier would be corrected.

Leveraging Coalitional Power Resources at the International Level

As has been demonstrated, GATWU's practices of contestation at the workplace and at the state level only unfolded their leverage over Avery Dennison when combined with practices of involving international union and consumer networks. These networks used images of workplace protests and the inspection report to leverage moral power over brands through 'naming and shaming' campaigns addressing brands as primary targets. Even though GATWU had, under the new factory organising approach, decided to engage with consumer campaigning networks more selectively, in the struggle at Avery Dennison, GATWU took the strategic decision to involve both union and consumer networks from the onset. Given the quasi-monopoly of Avery Dennison in the garment production network as a supplier of RFID-tags and labels for all major garment retailers and brands, GATWU had decided to draw as many international organisations as possible into the space of contestation to maximise the scope of brands that could be targeted. In that sense, in the Avery Dennison case, GATWU still benefitted from relationships with the WRC and the CCC established during the earlier fire-fighting approach phase. In addition, GATWU was able to activate the connections with the TIE ExChains network, and the League established under the new strategic factory organising and collective-bargaining approach.

To coordinate their actions and to facilitate communication with GATWU, international union and consumer networks established a geographical division of labour with the League and the WRC coordinating actions targeting US brands and the CCC coordinating actions targeting European brands. Union and consumer networks built pressure on brands and on Avery Dennison directly for more than a year through various practices of contestation. The practices included sending letters to brands' central managements, publishing fact finding reports, disseminating workers' demands and calls for solidarity on various social media platforms and creating an online petition that reached more than 80,000 signatures.

In addition to public campaigns, works councils from the TIE ExChains network in Germany exerted in-house pressure on Primark and H&M through two practices. First, works councils created internal awareness among retail workers about the violations at Avery Dennison by talking to colleagues and disseminating information material in workers' social areas. Second, German Primark and H&M works councils used the institutionalised spaces of regular meetings with store and general management

Fig. 7.5 Poster created by Primark works councils in Germany to raise awareness for GATWU's struggle at Avery Dennison. *Source* ExChains 2019

to request the management to ensure that their supplier Avery Dennison respected the rights to Freedom of Association and Collective Bargaining. As mentioned earlier, when deploying these practices, union and consumer networks used documentation of workplace action produced by GATWU to leverage moral power resources vis-à-vis brands. Figure 7.5 illustrates how Primark works councils in Germany that are part of the broader TIE ExChains network, used photos from gate meetings to create posters. These posters were put up in the social rooms of various Primark stores to raise awareness among retail workers and store managers for GATWU's struggle.

As a result of the coordinated and sustained practices of awareness raising and campaigning by worker and consumer networks in the US and Europe, brands finally intervened in the conflict between GATWU and Avery Dennison by initiating a mediation process through the Ethical Trade Initiative. The mediation process brought to the table representatives of GATWU and its national representation, the New Trade Union Initiative, and the Avery Dennison India management. In this mediation process, GATWU finally achieved recognition as official bargaining partner by the management—even though as part of a joint bargaining committee with the already present ADWU. Nevertheless, GATWU's recognition as official bargaining partner and the subsequent signing of the official collective bargaining agreement in December 2019 need to be regarded as a crucial victory. It is the first official bargaining agreement achieved by GATWU or by any local garment union in India.

It is important to note, however, that the historic win for GATWU with the signing of the collective bargaining agreement at Avery Dennison was made favourable by two central conditions for factory organising and collective bargaining that distinguish Avery Dennison from most factories in the Bangalore garment export industry: First, as mentioned above, the workforce composition at Avery Dennison differs significantly from the workforce composition in most tier one garment factories, where 85% of workers are female. Since the workforce at Avery Dennison was predominantly composed of men, it was much easier for GATWU leaders to hold factory gate meetings and regular Sunday meetings. As opposed to many women workers in garment factories, who have domestic responsibilities and often face opposition from their husbands or parents-in-law when they want to attend union meetings, male workers at Avery Dennison were free to join union meetings regularly. Second, as opposed to Bangalore tier one export-garment factories which are usually constructed as tightly controlled, union-free spaces by managers (see Sect. 6.4), at Avery Dennison, there was already a practice of collective bargaining at the workplace level. Hence, workers at Avery Dennison were already familiar with the concepts of unionisation and collective bargaining through first-hand experience at their workplace—an experience that most women workers in typical tier one export-garment factories lack. Both these conditions provided fertile ground for the union to organise, first the contract workers and subsequently permanent workers and allowed GATWU to build a significant membership base in a very short time.

In light of these special conditions, evaluating the significance of the struggle at Avery Dennison for GATWU's broader strategic shift towards a factory organising, collective bargaining approach is important. To which extent did GATWU implement the new strategic priorities of the factory organising, collective bargaining approach in the Avery Dennison struggle? In the Avery Dennison struggle, GATWU achieved to negotiate a collective bargaining agreement for the first time. However, Avery Dennison was not among the five target factories that GATWU had selected in 2017 under their new factory organising, collective bargaining approach. The Avery Dennison case, therefore, sheds light on the tension that GATWU faces under the new strategic factory organising approach between advancing the union-building process in the target factories and reacting to pressing matters and incidents in other factories. GATWU's engagement in the Avery Dennison struggle can hence be considered a case of 'hot shop' organising in the sense that GATWU did not strategically select this factory but rather provided an institutional roof for the wildcat collective organisation among contract workers.

Nevertheless, it can be considered a critical success for GATWU that they achieved to turn this spontaneous collective organisation into a more stable, lasting form of organisation with the formation of GATWU factory unions for contract and permanent workers. However, the Avery Dennison case bound all of GATWU's resources for almost two years. Due to the time-intensive communication and coordination processes with the Avery Dennison management, the labour department and international union and consumer networks in addition to regular meetings with workers, GATWU central leaders had little resources left to advance collective bargaining processes in the originally selected target factories. At the same time, workers leaders in the selected target factories did not yet have the strategic capacities to start collective bargaining processes alone at the respective target factories where GATWU had already achieved a significant level of organisation.

Given the strong anti-union attitude of management in the Bangalore export-garment cluster (see Sect. 6.5), pressuring managers into collective bargaining processes requires additional leverage through international union and consumer networks. Communication with these international collaborators usually takes place in English. However, most GATWU worker leaders do not speak English. Therefore, the capacity to plan and execute networked agency strategies with international supporter networks is still centralised with union leaders.

Hence, the Avery Dennison case reveals how the centralisation of core capacities to construct complex, networked spaces of contestation with GATWU's leadership poses limits for implementing the factory organising, collective bargaining-strategy at a greater scale in the Bangalore export-garment industry. Only after the struggle at Avery Dennison was officially concluded in December 2019 with the signing of the joint collective bargaining agreement between ADWU/GATWU and the Avery Dennison management, the GATWU leadership could dedicate resources again to target factories. Further advancements with building collective bargaining processes in target factories were, however, blocked when with the COVID-19 pandemic, the next pressing issue arrived that required immediate action in terms of ensuring workers' basic labour rights. Hence, particularly in the face of GATWU's limited financial resources, tensions between a 'fire-fighting' strategy directed at addressing immediate labour rights violations and a more long-term oriented strategic organising approach continue to exist in GATWU's day-to-day practice.

7.1.3   Discussion: Lessons from GATWU's Case

What can we learn from GATWU's case regarding which types of relations and interactions can enable local unions in garment producing countries to build sustained bargaining power, and thereby achieve lasting improvements for workers? I propose that there are three important lessons to take away from GATWU's case.

First, regarding the relations that unions construct with workers in spaces of organising, GATWU's case has highlighted the enabling potential but also the limits of community organising strategies that organise workers in their spaces of reproduction around workplace but also around community and household issues (cf. Jenkins 2013). Whereas such a community organising approach can enable unions to circumvent employer control in the workplace and thereby build associational power at the local level, it does not enable unions to build workplace bargaining power. Relations with workers organised through a community organising approach necessarily comprise workers from many different factories, with membership numbers in each factory remaining low. As a result, while workers can be mobilised for public protests at the city level—e.g. to target state actors and exercise political power (Hauf 2017)—low membership numbers in the workplace however do not allow unions to pressure employers into collective bargaining through industrial action. At the same time, GATWU's case has shown that community organising strategies can help prepare the ground for more targeted factory organising, enabling unions to build workplace bargaining power and to engage in collective bargaining.

Regarding local unions' interactions and relations with external actors in spaces of collaboration, GATWU's case has, on the one hand, highlighted the potential enabling effects that external funding from Northern NGOs can have for independent, local unions. Especially in their early stages, independent unions still have a low membership and can, therefore, not fund organisers and offices through membership fees. On the other hand, GATWU's case has also revealed the importance of treating external funding as a short- or, at most, mid-term solution rather than as a long-term arrangement. Union leaders found that the dependence on external funding hampered participatory and democratic internal union structures. Since all strategic activities were carried out exclusively by the unions' full-time organisers, there was little room for building a strong second-rank worker leadership. The case of GATWU therefore corroborates and adds to the findings of past studies pointing out the mixed effects that transnational collaborations with Northern NGOs have on local union-building processes in garment producing countries (see e.g. Fink 2014; Hauf 2017; Zajak 2017).

Against this backdrop, When GATWU decided to cut off funded project ties with NGOs, this opened up new room for a strategic reorientation—not only in the practices through which GATWU constructed spaces of collaboration but also in GATWU's organising practices. New collaborations with grassroots unions and worker networks at the vertical dimension of the garment GPN allowed GATWU to build the necessary strategic planning capacities for developing a networked agency approach. This approach combines strategic organising in selected target factories with the moral and institutional power of workers and civil society organisations in consumer countries to open up room for collective bargaining in target factories. GATWU's case shows that the enabling resources that local unions in garment producing countries can access through transnational spaces of collaboration are not limited to unilateral North–South flows of financial resources and moral power, as highlighted by past research (see e.g. Fink 2014; Zajak et al. 2017). Instead, transnational spaces of collaboration can also encompass a mutual exchange of knowledge and experiences with other labour actors in the Global North *and* the Global South. Hence, this study sheds light on two types of transnational collaborations that have so far remained understudied in literature on labour's networked agency strategies in GPNs: grassroots worker organising along value chains and South-South collaborations between local labour and civil society actors.

Lastly, with regard to relations and interactions with employers, state actors and transnational consumer campaigning networks in spaces of contestation, GATWU's case has shown that active worker participation in spaces of contestation is a central enabling condition for building a unions' lasting associational, organisational and workplace bargaining power. GATWU's minimum wage campaign and struggle at Avery Dennison have illustrated how—in cases where unions do not have sufficient associational power to engage in industrial action—symbolic workplace protests can be an option for unions to foster workers' active participation in spaces of contestation. The collective experiences that workers make by participating in symbolic protests in the workplace contribute to building workers' collective mindset and sense of ownership of the union as a membership-based organisation—two vital enabling conditions for strengthening the unions' associational and organisational power. As a result, labour struggles with active worker participation are more likely to contribute to enhancing a union's bargaining power vis-à-vis employers than labour struggles in which workers do not play an active role. The case of GATWU hence provided further evidence for arguments from labour geographers, who have reiterated that transnational campaigning strategies and linked moral power resources can only reinforce but never substitute workers' associational power on the ground (Kumar 2014, 2019a; Selwyn 2013).

In the next section, I turn to the second-biggest Bangalore garment union: the Garment Labour Union.

7.2   Garment Labour Union (GLU)

In this section, I analyse the strategic approach of GLU, the second-biggest garment union in Bangalore, with around 6000 members at the time the research was conducted. A group of former GATWU members founded GLU in 2012, one year after GATWU had split from Cividep. Whereas in GATWU's executive committee at the time, there were several men, GLU was founded explicitly as a women-led union. To this day, GLU works closely with Cividep and the community organisation 'Munnade'. 'Munnade' was founded together with GLU as a counterpart to the other community organisation, 'Garment Mahila Karmikara Munnade', which continued to work with GATWU.

GLU's strategic approach shows strong continuities with the strategic community organising approach that characterised the first phase of GATWU before the split from Cividep. In recent years, GLU has also sought to increase their foothold inside factories by forming factory committees with a geographical focus on Peenya, an industrial area on the North-Western outskirts of Bangalore. In the following, I will lay out the practices and relationships through which GLU constructs spaces of organising, spaces of collaboration and spaces of contestation. In doing so, I will illustrate the various tensions arising from GLU's attempts to build factory committees while simultaneously continuing with the area-based organising and 'fire-fighting' approach.

7.2.1   Spaces of Organising

As mentioned earlier, GLU's organising practices show strong continuities to the community-based organising approach that informed GATWU's strategy in its early years. At the same time, GLU has, also introduced some practices directed at factory organising and building factory committees over the past five years with the long-term strategic goal of engaging in collective bargaining with employers. In the following, I will first illustrate those practices and strategies that show continuities with the community-based organising approach and after that lay out the practices through which GLU seeks to introduce elements of a factory-centred organising approach.

Continuities in GLU's organising practices with the community organising approach are present mainly in GLU's close collaboration with the community organisation Munnade. Whereas in GATWU's union and community organising work have become rather separate areas of work over time, in GLU, union and community organising continue to be closely intertwined. These close relations also result from personnel overlaps between GLU and Munnade organisers, who are employed by Cividep within the framework of projects sponsored by NGOs from the Global North. These NGOs fund a broad area of GLU's and Munnade's activities, including, for example, family and childcare counselling, psychological counselling for women facing domestic abuse and legal counselling and support for garment workers in case of workplace grievances. GLU's organising work is geographically concentrated in the industrial area of Peenya, representing one of Bangalore's major garment hubs. Here, GLU maintains an office and a worker and community centre where meetings and cultural events take place. In addition, GLU maintains an office on Mysore Road.

Through Munnade, GLU continues to provide social community services for garment workers, including counselling services, savings groups and cultural activities. These community services serve as a space for raising awareness among garment women workers about their rights as workers, citizens and women and to support workers in claiming these rights. As part of their community work, GLU organisers also support garment workers with applying for school stipends provided by the State of Karnataka to children of garment workers. Community work hence continues to be an essential part of GLU's work not merely as a 'pre-union' organising tool but also as a form of addressing women garment workers' needs and problems beyond the workplace. At the same time, GLU organisers stress that community work through Munnade remains a critical practice to gain workers' trust and to familiarise them with the idea of collective organisation. In addition to building relationships with workers in their living areas through community work, GLU has also developed a repertoire of organising practices to reach out to workers outside of their workplaces in areas with a high concentration of garment factories with a particular focus on the industrial zone Peenya. Two central practices for approaching workers outside their workplaces are (1) holding factory gate meetings and (2) holding 'junction meetings'. Where possible, GLU organisers hold factory gate meetings with workers after the end of their daily shifts in front of factories. In these gate meetings, the organisers distribute leaflets and other information on labour rights and the unions' services for workers.

Furthermore, GLU employs creative organising practices such as street theatre. Given that the space in front of factory gates is, however, often tightly controlled by employers, GLU organisers have started to shift their organising activities from factory gates to strategic central junctions, as this GLU organiser explains:

Many factories don't allow us to stand in front of the factory gate. They chase us away. Once, two organisers were even locked up in a factory. They had just been standing outside the factory gate, talking to workers and distributing pamphlets. Then the manager and some supervisors came and they violently dragged them inside the factory. Since then, we don't stand directly in front of the factory any more. [...] There is one junction where workers have to pass by so or so. So now we stand there. (INT5)

Factory gate and junction meetings serve GLU organisers mainly as a first contact point to get in touch with workers and raise awareness among workers for the possibility of receiving support from the union for workplace grievances. Individual workers are then invited to accompany GLU organisers to the union office to register their cases and to talk about potential interventions by GLU. Usually, GLU organisers intervene on behalf of individual workers by writing a letter to the respective factory management. Hence, a large part of GLU organisers' time is spent solving individual workers' problems in the workplace. As a GLU representative explains, the rationale for taking up individual workers' cases is that there continues to be a strong need for awareness building as well as for gaining workers' trust:

Locally here, we are spending a lot of time mainly on spreading awareness among the workers. We end up spending a lot of time on individual cases because it creates a bad impression if you don't support that worker and they will then say: 'The union did not support me'. In general, the situation for organising is very difficult. There is a lot of repression. So, we need to create awareness among workers about their rights to Freedom of Association and about the union. (INT36; translated from Kannada)

Hence, over the past years, GLU has handled a large number of individual workers' cases, such as illegal terminations, non-payment of employers' social contribution, denial of leave or maternity benefits and cases of sexual harassment. According to a GLU representative, for most women, sexual harassment and 'production torture'—i.e. abusive behaviour and excessive work pressure by supervisors—are the main problems that motivate them to seek support from the union. It is hence gender-based and social issues rather than economic issues that motivate women to come to the union in the first place, as GLU's president explains:

You know, 90% of garment workers here in Karnataka are women. So, they join the union because they face harassment and sexual harassment in the factory by their supervisors, and they want it to end. That was also my case. (INT4)

Against this background, GLU seeks to build relationships with workers through practices that address women garment workers not merely as workers but as working *women*. In their community activities and in factory gate and junction meetings, GLU organisers speak to the specific experiences that workers share due to their position as women in the workplace and in the broader society.

Whereas through these organising practices, GLU has achieved a total membership of around 6000 members since their foundation in 2012, their members have traditionally been distributed over a large number of factories. GLU's focus on approaching workers in their living areas and at central street junctions in industrial areas has constrained GLU's capacities for building a strong representation *inside* factories—a crucial precondition for moving from solving individual workers' grievances to collective bargaining. Against this background, in addition to community and area-based organising practices, GLU has, over the past years, adopted various sets of strategic factory organising practices with the ultimate goal of negotiating collective bargaining agreements at the workplace level. These practices include (1) selecting target factories and (2) building and training factory committees. Selecting target factories first involved mapping GLU's membership distribution across factories in Peenya and identifying the four factories with the most members, which were then chosen as target factories.

In the next step, GLU organisers started to build factory committees of 10 workers in each of the four selected target factories. Factory committees are conceptualised as internal union representation at the workplace, which addresses workers' grievances with management independently. Factory committees shall thereby reduce the dependence of workers on GLU full-time organisers. Moreover, factory committee members shall act as organisers at the workplace and seek to expand GLU's membership in the respective factory. Building factory committees involves several practices, including (1) selecting and inviting potential members, (2) holding factory committee meetings and (3) training committee members. When selecting potential factory committee members, GLU organisers apply two criteria: Firstly, they select workers who have shown particular leadership and communication skills in community or gate meetings. Secondly, GLU organisers seek to select workers from the various departments within a specific factory to ensure that committee members can reach out to different groups of workers in the factory. To form the committees, organisers initially invite about 15 to 20 workers from each factory to participate in a meeting, knowing that many workers may stop coming after the first meeting due to time constraints, pressure from their husbands or fear of victimisation, as a GLU representative explains:

> For the factory committee, we first identify the leaders. We will identify the people who can take responsibility and talk to the management. From all sections, e.g. cutting section, packing section etc., we identify people who can take the responsibility, bring them 15–20 of them together, we then train them. Actually, choosing them from the gate meeting is very difficult. We conduct multiple meetings and identify and choose them. We tell them what they can do and check with them if we can form a committee and then select them as a committee member. So, if we select 20 people at this stage, it'll further come down to 10, because 10 will still drop out for various reasons. (INT36, translated from Kannada)

Since workers labouring in Peenya's factories live in geographically dispersed areas within and also outside the city, factory committee meetings usually take place directly after factory gate meetings in the GLU office. Factory committee meetings usually last for about half an hour and encompass lessons on labour law, the union's functioning and the committee members' role and responsibilities. Holding factory committee meetings directly after junction meetings allows GLU organisers to recruit workers directly from these meetings and bring them to the union office for training. However, since in the junction meetings, many workers from various factories come together, organisers report that they find it challenging to build and train a stable group of workers as factory committee members. Whereas about 30 to 40 workers participate in every factory committee meeting, organisers hardly achieve to gather the same 30 to 40 workers every week.

One of GLU's main challenges for building factory committees and workplace bargaining power is building a stable group of workers who can be trained to take over responsibility inside the factory. Moreover, since GLU has not yet reached a critical membership level in target factories, union factory committee members cannot yet act openly as union representatives since they would risk victimisation by the management, as GLU's president explains:

> So right now, we tell the GLU members at the factories: Don't say that you are a union member, you will just act as a worker and you will just speak with the workers, but don't say you are a union member until we get more members in the factory. (INT36, translated from Kannada)

Nevertheless, GLU has managed to train some factory committee members in target factories who currently take up workers' issues and negotiate them with the management, even though not officially in the name of the union. In this vein, GLU's factory committee members in target factories have, for example, led smaller worker protests to successfully redress several law violations and unfair management practices, such as lack of drinking water, late wages or abusive behaviour by supervisors. Whereas the scope of these protests is limited to more minor issues, protests are nevertheless important because they help to increase committee members' and workers' confidence, as testified by this worker and active GLU member:

> Actually, since I have become a union member, I am much more confident also inside the factory. I know now that the supervisor has no right to yell at me and that he must treat me with respect. So, before I was a member of the union I used to just cry silently, when the supervisor scolded me. But now I speak up to him and I tell him: Who gives you the right to speak to me like that? You have no right to speak to me like that. You must treat me with respect! (INT5)

However, over the three years during which research was conducted, GLU had yet to achieve a factory committee and membership base strong enough to engage employers in collective bargaining. As a GLU representative explains, their factory organising approach foresees that the union first needs to organise about 50% of the workforce to submit an official request to the management for collective bargaining. Given the repressive environment for union organising in Bangalore garment factories, GLU needs to be able to ensure worker leaders' protection through collective workplace action, as GLU's president explains:

> As of now, if we write that these are all our committee members, they'll be targeted. [...] So, if the management wants to fire, say, ten of the committee members, then all the workers have to come to their support and tell the company that they will all leave if the committee members leave. [...] So, we want a lot of members and we want them to be aware of the union and its activities. [...] if there are 50% members, if they are strong, we can go to the labour department for collective bargaining. If there are less, we will fail. If all the workers are not aware of this and only leaders and few people support us, it won't be useful. We will fail. (INT36)

Building a strong membership base and a second-rank leadership that can actively organise inside factories hence remains GLU's most important challenge. GLU's challenges in building a strong member and leadership inside factories need to be interpreted in light of the tension between their dominant community and area-based organising practices based on taking up workers' individual grievances, and their attempts to build collective agency structures inside factories. In 2018 alone, GLU and Munnade together took up more than 500 individual cases of domestic or workplace rights violations (FEMNET, nd). Handling such large numbers of individual cases binds significant union resources since each case usually involves several attempts to contact the factory management, file a complaint with the labour department and follow up with individual workers. A GLU representative explains that the union invests these resources since they hope that individual workers who receive active support from the union will become organisers or worker leaders in their respective factories. However, given the ease with which workers can find a job in another garment factory (see Sect. 6.7), many workers regard working in a specific factory as a rather short-term arrangement. Hence, instead of becoming active union members at their respective workplaces, workers, in many cases, leave the job sooner or later, as exemplified in the following case reported by a GLU representative:

> In a unit of [company name], one worker was transferred to another factory unit without his consent. So, I spoke to the management about this and told them that this will have consequences. They had already made all arrangements to transfer him. After I warned them, they cancelled the transfer. But that boy worked only for about 3 months after that and left the job to go back to his hometown where his father was sick. We feel bad when such things happen because we struggle a lot to get their issue solved and we hope that they will take leadership in their factory. But they just use us when they have an issue. And without even telling us, they leave. He could have taken leave and gone to his home-town but he quit the job. After many days of not being able to contact him, we got to know that he has quit the job. Such issues happen sometimes. (INT36, translated from Kannada)

Her statement that workers 'use' the union when they have a problem and then 'leave', reflects the inherent problem to the 'service union' model: the dominant practice through which full-time organisers relate with union members under this model is handling members' individual issues and grievances. As a result, many resources are bound to union activities that do not create or promote spaces for developing workers' 'oppositional consciousness' (Katz 2004) and collective agency. I argue here that the continued focus on solving individual grievances needs to be understood as resulting, at least partly from GLU's continued close collaborations with Cividep and international donor NGOs in the form of project work. To understand the tensions within GLU's practices of constructing spaces of organising, it is therefore essential to scrutinise the practices through which GLU constructs spaces of collaboration.

7.2.2   Spaces of Collaboration

GLU constructs spaces of collaboration through two main sets of practices and relationships: (1) by constructing relationships with local and international NGOs in the context of collaborations for funded project work and (2) by constructing relationships with international consumer campaigning networks through visiting training sessions and filing urgent appeals.

As mentioned before, GLU continues to maintain close relationships with the Bangalore-based NGO Cividep and with other local NGOs that have good networks with NGOs from the Global North. These local NGOs act as an intermediary between GLU and international NGOs funding specific projects implemented by GLU: On the one hand, since GLU full-time organisers speak little English, Cividep facilitates contacts and communication with international NGOs. On the other hand, Cividep also acts as official project partner for foreign NGOs and administers the project funding.

GLU today has projects with several NGOs from the Global North. These collaborations allow the union to tap into coalitional power resources in the form of financial resources. With these financial resources, GLU funds office rent, salaries for full-time organisers, executive and factory committee meetings, training sessions and cultural activities. The focus of these projects is usually on GLU's community and counselling work and on awareness building among garment workers about their rights. As the first union in Bangalore, GLU has begun to engage with migrant workers from the Northern and North Eastern states of India (see Sect. 6.7) in the context of a project funded by several European women and human rights organisations. As part of this project, GLU organisers provide information, counselling and training on labour rights to migrant workers. Moreover, GLU produces a union newsletter in Hindi language. At the same time, the project aims to gather data to produce public reports on the situation of migrant workers in Bangalore. As in other projects, Cividep plays a vital role in the project as an intermediary organisation that administers funds. The space of collaboration constructed around the project is shaped not only by GLU's interests and strategic action frame as a membership-based organisation but also by the institutional logics of the involved NGOs as advocacy organisations.

As a result, the potential for GLU to build associational power resources and lasting bargaining power through this project collaboration has been limited. Since migrant workers represent a growing share of workers in Bangalore export-garment factories, engaging with migrant workers through awareness building may in the mid- and long-term strengthen GLU's associational power, if migrant workers become active members of GLU. This prospect is, however, limited by the fact that many inter-state migrant workers only stay for limited time periods in the city to save a specific amount of money, e.g. for a family member's wedding, and then return to their home villages. Moreover, the goal of the project, as formulated by the funding NGOs, is not primarily to build GLU's associational power but rather to build public awareness for the situation of migrant workers in Bangalore. In the international NGOs' institutional logic, GLU takes on the role of a strategic partner that enables

NGOs to achieve their strategic goals of public awareness raising to address specific problems in the garment industry. This role as a strategic partner for public awareness raising activities is also expressed in this statement by a Cividep representative:

> Sometimes the unions are part of a project. For example, with GLU we have that project with migrant workers from Eastern India. [...] Access to these young workers is difficult. They don't speak the language, they live in hostels. We are publishing a report on this and GLU is doing worker education with these workers. Sometimes we depend on them to get data. Because they are a union, they are able to get better data from workers. (INT8)

Tensions between the institutional logic of NGOs as advocacy organisations and the organisational logic of unions as membership-based organisations also become apparent in the criteria against which international donor NGOs measure GLU's success. Following the advocacy logic of NGOs, scope or outreach are important criteria for funders to legitimise their collaborations with GLU. In this logic, GLU is understood as a multiplicator organisation that should provide services and benefits to as many garment workers as possible. This logic is highlighted on the website of the German NGO FEMNET, who funds GLU's and Munnade's family and legal coun-selling activities. FEMNET writes that GLU and Munnade together have "access to over 25,000 women workers" (FEMNET, nd). The measure against which FEMNET evaluates successful implementation of projects is hence outreach. In this logic, to be successful, GLU organisers need to maximise the number of engagements with women, be it through providing them assistance in the form of family or legal counselling or through providing training.

As a result, how GLU constructs relationships with international NGOs hence influence the practices through which GLU organisers build relationships with workers: to ensure that project funding contracts with NGOs are renewed, GLU organisers need to maximise the number of workers who participate in their training sessions, meetings and counselling activities. However, this need to maximise scope leads to the fact that relationships with individual workers are often limited to inter-actions for a specific training session or to a series of punctual interactions that end when an individual worker's problem has been solved.

The institutional logic of maximising outreach inherent to projects funded by international NGOs, therefore, conflicts with the logic of the union as a membership-based organisation: For the union to build associational and organisational power resources, the *quality* of relationships with workers and union members is more important than maximising the *quantity* of engagements. Consequently, the specific practices through which GLU builds and maintains collaborations with international NGOs in the context of funded projects have mixed effects for GLU's capacities to build lasting bargaining power: On the one hand, through maintaining collaborations for funded project work, GLU is able to acquire financial resources to fund offices and organisers. On the other hand, a large part of organisers' time is invested in building loose ties with workers that do not contribute to building the union's organisational and associational power.

In addition to constructing spaces of collaboration with local and international NGOs around funded projects, GLU constructs spaces of collaboration with inter-national consumer campaigning networks, and primordially with the CCC, through practices of visiting training sessions and filing urgent appeals. Collaborations with international campaigning networks serve to leverage moral power resources when struggling for the correction of labour rights violations. To construct and maintain relations with international consumer networks, GLU organisers regularly visit meetings and training sessions organised by these networks. Topics tackled in these meetings are, for example, labour rights or how to file urgent appeals or complaints. Whereas GLU organisers state that they already possess sufficient knowledge on these issues, they argue that these meetings are still important for networking. Personal interactions during these meetings build trust and understanding and hence allow for a quick contact and response in case of labour rights violations.

In addition to constructing relationships with international consumer campaigning networks, GLU is also an active member of the Asia Floor Wage Alliance, a labour-led international campaigning network involving unions and NGOs from production countries in Asia and from consumer countries in the Global North. The main focus of the Asia Floor Wage Alliance lies in producing reports on and campaigning for a living wage in the Asian garment industry. Members of the alliance meet regularly to discuss strategies for joint public campaigning to put pressure on brands to implement a living wage across their Asian supplier factories. Besides campaigning for living wages in the garment industry, the Asia Floor Wage Alliance also organises multi-stakeholder meetings with brands. GLU organisers have participated in a series of meetings with brands organised by the Asia Floor Wage Alliance on gender violence in Asian garment factories. GLU organisers stress that participating in meetings organised by the Asia Floor Wage campaign has helped them to construct relationships with various brands, whom they can now contact directly when receiving complaints from workers.

It is important to note that the strategic approach prioritised by both international consumer campaigning networks and the Asia Floor Wage Alliance is to hold brands (and not local employers) accountable for ensuring labour rights in the garment industry. This strategic framing of brands as primary agents of change has important consequences for how international campaigning networks construct spaces of contestation: instead of territorially embedded workplaces, international network spaces are constructed as primary arenas of contestation through email exchanges and international social media campaigns. As will be shown in the next section, this shift towards constructing spaces of contestation as dis-embedded network spaces is also characteristic of GLU's strategic approach, which strongly relies on the interventions of brands and international consumer networks.

7.2.3 Spaces of Contestation

GLU constructs spaces of contestation mainly around issues raised to their organisers by individual workers in community or junction meetings. As mentioned before, GLU organisers see it as an essential practice to gain workers' trust to take up all issues brought to them by workers, independent of workers being from a target factory. Therefore, GLU organisers usually collect issues from various factories during gate and junction meetings. Then, organisers group issues from factories belonging to the same company group to raise them with the respective factory and company managers, the labour department, and brands and international consumer organisations. More minor cases, such as late wages, non-payment of legally prescribed bonuses or non-payment of gratuities, can often be resolved by organisers in direct dialogue with the management. Other issues that concern groups of workers or that are not as easy to prove, however, usually require a combination of several practices, such as writing to the management and parallel filing a complaint at the labour department and contacting brands. Organisers report that, in most cases, factory and company managements ignore GLU's initial requests for a meeting. In these cases, GLU contacts brands and international consumer organisations and uses their leverage to get the local garment factory management to meet with GLU organisers. In the remainder of this section, I will illustrate how GLU constructed a space of contestation around a series of labour rights violations that occurred in early 2017 in various factories belonging to the Bangalore export-garment company Gokaldas Exports. I will show how GLU used practices of complaint filing at the labour department and leveraging the influence of brands to engage the factory management in dialogue. At the same time, I will point out the limits of this strategic approach to constructing spaces of contestation for building sustained union bargaining power.

7.2.3.1 Tackling Labour Rights Violations at Gokaldas Exports

In early 2017, GLU organisers were notified by workers about a range of labour rights violations at factories belonging to the company group Gokaldas Exports. The violations included inter alia (1) several practices of overtime wage theft such as extending the regular working hours from eight hours to nine hours per day and giving 'comp-offs' (see Sect. 6.3), (2) sexual harassment and abusive behaviour by supervisors and (3) the illegal dismissal of a union worker activist in one factory. Upon learning about these labour rights violations, GLU organisers immediately took action at various levels: At the company level, GLU organisers contacted the respective factory managers and the central management of Gokaldas Exports. At the state level, GLU organisers filed a complaint with the labour department and the Department of Factories, Boilers, Industrial Safety & Health to leverage institutional power resources. Since the change of working hours was easy to prove for GLU organisers, labour inspectors ordered the factory management to immediately change the regular working hours back to eight hours per day.

Moreover, the management agreed to stop the illegal comp-off practice due to the interventions by GLU and the labour department. However, the Gokaldas management did not take any corrective action regarding the cases of sexual harassment

and illegal termination of a union worker activist. Instead, the Gokaldas management denied that sexual harassment had happened and held that the termination of
the GLU worker activist was justified because he had behaved violently towards
another factory employee. The Gokaldas management not only refused to discuss
these issues with GLU organisers directly but also refused to attend the conciliation
meetings convened by the labour department following GLU's formal complaint.

Given the management's refusal to engage in dialogue with GLU directly or
through the conciliation process in the labour department, GLU decided to take
additional steps and to activate the leverage of brands and consumer campaigning
networks at the international level. Drawing on the relationships constructed with
brands during previous cases and during multi-stakeholder meetings, GLU contacted
all brands sourcing from the respective factories of Gokaldas Exports, including
H&M and GAP. In addition, GLU organisers filed an urgent appeal complaint with
the CCC. Lastly, the organisers also informed the Fair Labour Association, a multi-
stakeholder initiative involving predominantly US-based garment and sportswear
brands and universities and civil society organisations. As an outcome of these
combined practices of directly contacting brands and activating the leverage of
consumer organisations, GLU organisers were invited to meetings with representatives from H&M's and GAP's regional sourcing offices. In these meetings, GLU
organisers convinced both brands to tell the Gokaldas management to meet with
GLU.

As a result, the Gokaldas management finally asked GLU organisers for an
informal meeting to discuss the issues of sexual harassment and the illegal dismissal
of a union activist. In the meeting, the management assured GLU that they would
solve all problems, but neither provided a concrete action plan nor a written statement. When the management had not taken any action after several weeks, GLU
once again wrote emails to brands and consumer organisations. Only after a year
and a half of continued liaising with brands, international consumer organisations
and the Gokaldas management, GLU achieved to engage the Gokaldas management
in serious negotiations and finally won a compensation of 150,000 Rupees (approx.
2,000 US$) for the dismissed worker activist.

Gaining this compensation—amounting to more than a yearly average wage in
the Bangalore garment industry—represented a crucial victory for GLU. However,
this victory did little to strengthen GLU's workplace bargaining power inside the
factory for two reasons. First, since GLU could not reinstate the worker activist,
they lost an important resource for advancing the factory-internal union-building
process. Second, the space of contestation was constructed predominantly through
network practices of appealing to and liaising with brands and consumer campaigning
organisations and did not involve any practices of collective worker protest or action
at the workplace. Hence, even though the struggle continued for more than a year,
workers from the factory did not play an active role in it and therefore had limited
opportunities for developing strategic capacities.

7.2.4   Discussion: Lessons from GLU's Case

Which general implications result from GLU's case for the enabling and constraining effects that different relations of local unions with other actors have for building unions' bargaining power vis-à-vis employers? Three major implications shall be highlighted here: First, with regard to unions' relations and interactions with workers in spaces of organising, the case of GLU illustrates, in particular, the limitations of community organising practices that focus on building loose ties with a large number of workers through punctual interactions in training or counselling sessions. Whereas these organising practices help the union to increase membership numbers, they contribute little to building the union's associational power base since these members can hardly be mobilised for collective action. This fact has been exemplified in the difficulties that GLU organisers face in their attempts to build stable factory committees in selected target factories.

GLU's organising practices are in turn directly shaped by their practices of constructing spaces of collaboration. The union's focus on building loose ties with a large number of workers needs to be understood as shaped by the strategic frameworks of the funded project collaborations GLU maintains with various NGOs from the Global North. Since financial resources flow unilaterally from Northern NGOs to GLU in these collaborations, Northern NGOs have the power to define the collaboration's terms and conditions and strategic goal. NGOs as advocacy organisations, however, tend to have a strategic action frame that differs from the one of unions as membership-based organisations. Whereas NGOs aim to maximise their outreach and prioritise the quantity of interactions, for unions, the quality of interactions is equally important: a large membership on paper is of little use if these members cannot be mobilised for strategic action. The case of GLU hence illustrates, second, that when unions' collaborative relations with external actors are characterised by asymmetrical power relations and incompatible strategic action frames, these collaborations are likely to hamper rather than foster local unions' capacities for building associational and organisational power. In this light, the case of GLU provides essential insights into the structural effects that networks of collaborations constructed by unions themselves with transnational actors have on unions' everyday organisational practices and internal relations (cf. Zajak et al. 2017; see also Fütterer and López Ayala 2018).

Lastly, with regard to the relations constructed by unions with capital actors and allies in spaces of contestation, the findings from the analysis of GLU's agency strategy make a renewed case for arguments from labour scholars that the leverage of moral power through collaborations with transnational consumer networks cannot make up for lack of associational power on the ground (see e.g. Hauf 2017; Zajak 2017). While still working towards building a significant membership base inside factories, GLU has to rely on the leverage of consumer organisations and brands over employers when contesting labour rights—with mixed outcomes. Hence, GLU's experiences also point at the limits of a mere 'up-scaling' strategy that relies exclusively on relations and actions at the transnational level without strategically intertwining them with actions at the workplace and local level.

In the next section, I turn to the third of the three Bangalore garment unions: the Karnataka Garment Workers Union.

7.3 Karnataka Garment Workers Union (KGWU)

KGWU was officially founded in 2009 by former GATWU activists employed at the Bangalore-based NGO FEDINA. FEDINA had also been part of the initial community organising project with garment workers in the early 2000s, funded by Oxfam. During these early years, a geographical division of organising work between FEDINA and Cividep had been established. Whereas Cividep activists had been organising predominantly in workers' living areas along Mysore Road, FEDINA activists had concentrated their organising work along Hosur Road in the areas Bommanahalli and Tavarekere. Like Cividep, FEDINA had started approaching garment workers in their communities through saving groups. According to KGWU's honorary president, FEDINA activists, however, soon felt that workers were ready for unionisation and that the saving groups approach should be abandoned in favour of a more strategic union organising approach. As a result of these internal strategic differences, in 2008, FEDINA activists decided to form a separate union.

At the time research was conducted, KGWU had, according to its own reports, around 3000 members. These are located predominantly in the area along Hosur Road in the East of Bangalore, which has traditionally represented the geographical focus of KGWU's work. Since 2017, KGWU has expanded their organising activities to also include factories along Mysore Road, in the industrial area Peenya and the district of Davanagere, located about 250 km from Bangalore. The expansion of KGWU's organising work has been linked to a collaboration with the Chinese NGO China Labour Bulletin (CLB). In the context of this collaboration, KGWU has undergone a strategic shift from an originally area-based worker organising and reactive fire-fighting approach to a more proactive, strategic factory organising approach.

The strategic evolution of KGWU has hence shown characteristics that can also be found in the trajectories of GATWU and GLU. However, how KGWU has linked changes in their organising strategies with changes in constructing collaborations with external actors differs from the ways in which GATWU and GLU construct spaces of collaboration, as will be illustrated in the following. The remainder of this section illustrates how KGWU's practices of constructing spaces of organising, collaboration and contestation have evolved over the years from a fire-fighting approach to a strategic factory organising, collective bargaining approach. It moreover shows how the decisive driving factor in KGWU's strategic evolution has been a new collaborative relationship that the union has established with the Chinese labour organisation China Labour Bulletin since 2016.

7.3.1   Spaces of Organising

As mentioned earlier, similar to GATWU and GLU, KGWU has undergone a change in how they construct spaces of organising from an area-based organising approach towards a factory-centred organising approach. Until 2015, KGWU followed an area-based organising approach, in which spaces of organising were constructed predominantly *outside* factories through two main organising practices: (1) holding area committees and (2) holding factory committee meetings. Monthly area committee meetings were held in workers' living areas and gathered workers from different factories. These meetings aimed to inform workers about their labour rights, to discuss workplace problems and train workers regarding the role and function of the union. In addition, KGWU held regular factory committee meetings with workers from specific factories, usually at the end of a workday, to discuss factory problems and spread awareness among workers about their rights. When KGWU organisers learned about labour rights violations from workers during these meetings, they would then take up the issue and intervene with the management.

However, dissatisfaction with this area-based organising approach grew among KGWU activists when after almost a decade of organising, the area-based organising approach had still not enabled them to engage in collective bargaining. As their honorary president explains, two factors were hindering KGWU from starting a collective bargaining process in factories in their geographical organising areas: First, in the area along Hosur Road, where KGWU's organising work was (and remains) concentrated, there were a lot of smaller, workshop-like factories. Since in particular, in these smaller factories there were a lot of basic labour rights violations, KGWU activists concentrated their interventions on these smaller factories for a long time. However, due to their small size, these factories could be easily closed down and be reopened in another place under a different name by the management in reaction to worker organising. Hence, the footloose nature of these smaller factories severely constrained KGWU's ability to engage in collective bargaining with the management, as a KGWU representative explains:

> At that time, small factories were violating a lot of laws. And there was no PF [Provident Fund), no ESI [Employees' State Insurance], wages were not paid on time. So, lots of issues, lots of problems were there. And also, those factories were the ones which were closing down often. By the time you started to organise in […] that factory, they would say that the factory was closing down. So, I think we spent a lot of time like that on small factories thinking that small factories workers would be responding easier. (INT45)

Hence, whereas KGWU, in some cases, achieved to win back wages and compensations for workers from small factories that had abruptly closed down, the union, however, never achieved to engage employers from small factories in collective bargaining.

It was not until the year 2015 that KGWU organisers began developing a more systematic, factory-centred organising approach. According to a KGWU representative, KGWU leaders and organisers only then found the external conditions favourable enough to adopt a more strategic factory organising approach. After almost a decade of union organising in the Bangalore export-garment sector, not only by KGWU but also by the two other unions, workers finally became acquainted with the concept, idea and benefits of unionisation and organised collective worker action. Moreover, the market consolidation process in the Bangalore export-garment cluster following the end of the Multi-Fibre-Agreement and the quota system (see also Sect. 5.2.1) had led to the concentration of production within fewer, larger factories, often employing several hundred or even thousands of workers. These larger factories possessed greater financial capacities to withstand economic slumps or industrial action by workers.

Whereas the *external* local conditions provided new opportunities for long-term strategic organising in selected factories, KGWU organisers still lacked the *internal* organisational resources for developing and implementing a strategic factory organising, collective bargaining approach. For KGWU to develop these strategic capacities, a new set of collaborative relations was decisive: In 2016, KGWU started a new collaboration with the Chinese labour rights NGO 'China Labour Bulletin' (CLB). Through this collaboration, KGWU received financial resources to fund full-time organiser positions and union offices in Bommanahalli, Peenya and on Mysore Road. In addition, KGWU received a series of strategy training sessions to develop a new, proactive factory organising and collective bargaining strategy. As a KGWU representative explains, in the meetings with CLB, KGWU leaders and organisers developed the strategic mindset and the planning capacities that enabled them to construct spaces of organising more systematically around selected target factories to build a long-term collective bargaining process:

> So, when we started talking about [how] it could be beneficial to both the factory and the worker if there is [...] a bargaining process, it required a lot of mindset change. [...] For that the [collaboration] with CLB was very crucial [...]. [Now] we are strategising on collective bargaining, whether we can do something in one or two factories at least. That [collaboration] has given us some more focus in our work. Now we identify how the factories are organised, what brand they are manufacturing. (INT30)

As an outcome of the training with the CLB, KGWU has started to concentrate a large part of their resources on organising in two selected target factories. These two factories were selected based on four criteria that KGWU activists developed in training sessuions with the CLB: First, the selected factories are part of large export-garment company groups. As a result, both factories have a relatively stable financial situation, and, according to worker reports, had received stable orders in preceding years. Second, in both factories, there were a lot of labour rights violations and, therefore, ample room for improvement. Third, KGWU organisers have long-standing members in both factories who have participated in prior struggles alongside KGWU. Fourth and last, the selected factories both produce to a large extent for H&M, which KGWU activists saw as an opportunity in two regards: On the one hand, KGWU organisers saw the opportunity to use H&M as an additional leverage on the factory management to prevent or redress management attempts of union busting. On the other hand, KGWU saw H&M's living wage promise[1] as a tool for mobilising workers and engaging them in collective bargaining (FN7).

As part of the target factory selection process, KGWU activists conducted so-called 'factory mappings', which involved gathering information on several aspects of the factory, including the mother company's financial situation, the number of workers and departments, main buyers and present labour rights violations. More-over, factory mappings involved documenting end-consumer prices gathered from price tags and comparing these in relation to workers' wages. This mapping process served as an important organising tool since it allowed KGWU to involve workers in the target factory selection process actively. By gathering and bringing together information, KGWU organisers and workers developed critical strategic and analyt-ical capacities, such as knowledge about the organisation of the factory, the ability to identify potential chokepoints and an enhanced understanding of the profit distri-bution and power relations in the value chain. Many workers stated during a training session with CLB that they got to know and understand the broader organisation and power relations of the factory and the value chain for the first time through the factory mapping process (FN7). According to KGWU's honorary president, this new knowledge and understanding of the power relations and organisation within their own factory also helped to build workers' sense of ownership of the union building and the collective bargaining process:

> Factory mapping helped especially for workers to understand their industry and their position in it. I mean, it is easy to complain and say, you know, these are all the problems. But you need to understand how you tackle it, that the union is not some outside agency which has to come in and correct it for you, but that you also have a role in it. I think the mapping process helped a lot to strengthen that understanding. That they [the workers] have to take responsibility also. As an external agency only, we had no strength really to challenge the management. (INT45)

In the two selected target factories, KGWU's organising work now involves various sets of practices that follow a snowball system. First, an initial core group of worker activists in the factory is asked to gather workers from each department and bring them to the regular target factory meetings. These workers are then prepared and trained to act as worker organisers and bring further workers from their own department to the next meeting. Since organising inside the factory is often not possible at the initial stage of the organising process due to tight management control, KGWU activists also approach workers at the factory gate and in their living areas. To this end, the initial organising process also involves mapping workers' living areas.

Target factory meetings are held regularly and serve as a space for discussing workplace problems building workers' mindset as union representatives and organ-isers inside the factory. As a KGWU representative explains, one of the most important elements in these meetings is building full-time organisers' and worker

---

[1] In 2013, H&M published a document titled 'Roadmap towards a fair living wage in the textile industry' announcing that by 2018 workers at H&M's strategic suppliers should receive a living wage. In subsequent years, this promise, however, disappeared again from H&M's public communication and remains up to date unfulfilled (see also CCC 2018).

activists' strategic capacities to negotiate and reach a strategic compromise with the management:

> And so [when] we started that [factory organising process] in 2017, we [...] took some time to understand the concept and you know what collective bargaining is. [...] When you are always in the mindset of attacking the enemy, often you don't prepare yourself to negotiate or you know strategically compromise that sort of thing. You want to advance. You think you have to win it all at once. (INT45)

Besides serving as spaces to develop workers' strategic mindset, target factory meetings also serve as spaces to discuss any problems in the factory and to develop collective demands in a democratic process. Generally, target factory meetings are open to all workers from the respective factory. However, only union members have the right to vote on strategic decisions. Once a significant number of members from various departments has been reached in a factory, secret ballot elections are conducted to officially form a union factory committee, usually consisting of nine worker representatives. This union factory committee represents the workforce in the collective bargaining process.

As mentioned earlier, KGWU's practices of establishing an international collaboration with the CLB played an important role in enabling KGWU to develop their new factory organising, collective bargaining approach. The following section introduces this collaboration in more detail. It reveals how KGWU's active engagement in constructing the relationship with the CLB was itself the outcome of a strategic shift in KGWU's practices of constructing spaces of collaboration.

7.3.2 Spaces of Collaboration

Collaborations with local and international organisations have played an essential role in KGWU's history and strategic development. As mentioned earlier, KGWU has traditionally maintained close connections with the local NGO FEDINA. At the time the research was conducted, all of KGWU's full-time activists were employed through FEDINA, and many had priorly worked on FEDINA's other projects. Since the 1980s FEDINA has been organising different marginalised groups in South India through local 'Social Action Groups'. These Social Action Groups aim to foster collective grassroots organising and empower marginalised groups such as Dalits, smallholder farmers, landless labourers, informal sector workers, and slum-dwellers (FEDINA, nd).

Whereas during their early years, KGWU leaders and organisers mainly constructed spaces of collaboration at the local level, with the shift towards strategic organising in large supplier factories of major US and EU brands, KGWU leaders began to construct relationships also with international consumer and labour networks. An interviewed KGWU representative recounts that in a first intent to move beyond KGWU's initial fire-fighting approach in the early 2010s, KGWU activists participated in regular meetings of the CCC and the Asia Floor Wage Alliance. The two networks were planning and implementing public media campaigns to pressure H&M into fulfilling their announcement of implementing a living wage at all their strategic suppliers by 2018. To this end, they had initiated a multi-stakeholder process involving unions and labour rights NGOs from several Asian countries and brands to discuss what living wages on the continent should be. As part of this process, KGWU and other involved local unions collected data on living costs, inflation and workers' regular expenses, which were then used to determine a living wage for the Asian garment industry. Findings were then presented to brands at several round table meetings.

The interviewed KGWU representative stresses that participating in the network spaces of collaboration constructed under the Asia Floor Wage Alliance helped KGWU activists to develop strategic resources and capacities. Through discussions in the Alliance, KGWU activists gained an enhanced understanding and knowledge about the structure of the value chain and the tactics of harnessing brands' leverage over suppliers. Nevertheless, KGWU leaders felt that participating in the activities of the Asia Floor Wage Alliance had only very limited effect in terms of producing material improvements for workers on the ground for two reasons: First and foremost, practices of researching workers' living expenses and participating in round table meetings were somewhat disconnected from KGWU's on-the-ground organising work and hence did not help the union to engage employers in collective bargaining. This detachment is exemplified in a statement by a KGWU leader, who says that he experienced the multi-stakeholder process under the Asia Floor Wage Alliance "more [as] a theoretical exercise than [as] a practical unionisation process" (INT30). Second, KGWU leaders found that the strategic approach of the Asia Floor Wage Alliance to prioritise pressure on brands produced only limited material improvements in working conditions on the ground. According to these leaders, brands only intervened and exerted pressure on manufacturers in case of particularly cruel labour rights violations. However, with the industry's structural transformation towards 'organised' production in large tier one garment factories, such cruel labour rights violations were not as prevalent any more as in earlier years. As a result, interventions by brands became a less effective tool to achieve improvements for workers, as a KGWU representative explains:

> While this [multi stakeholder process] was being organised and brands would come and sit [at the roundtable meetings], it would end at that. The only place, to some extent, where brands could help, was when the management or export company was resorting to very cruel violations. If there are very cruel violations then sometimes you can threaten them that you will expose it in Europe or some other country. But most of the export houses, [tier one] manufacturing companies don't resort to very cruel measures: They are within the minimum wage or just above the limit. They provide a crèche, which may be not functioning too well. but they provide a creche. (INT30)

In the face of the limitations of the brand-led strategic approach for improving working conditions favoured by the Asia Floor Wage Alliance, in 2016, KGWU activists decided to shift their focus towards building collaborations that would help them to strengthen their own bargaining power and position vis-à-vis manufacturers. It was then that KGWU's honorary president was provided with a contact of CLB at an international union conference. The official collaboration between CLB and KGWU started in 2017 and involved financial support for three worker centres in Bommana-halli, Peenya and Mysore Road, including three full-time organiser positions in each centre. Moreover, the collaboration comprised a series of training sessions with the explicit objective of developing KGWU's organising and collective bargaining strategy.

As opposed to other project-based collaborations between local unions in the Global South and NGOs in the Global North, in which the funding NGO largely determines the project agenda and activities, the collaboration between CLB and KGWU was set up as a joint and mutual learning process. During the initial training session, CLB's director explained that their rationale for initiating this project was to gain first-hand experience in building a collective bargaining process from scratch—and experience that was difficult for CLB to gain in China, where independent unions are not allowed. KGWU leaders and organisers, in turn, stated as the primary rationale for the collaboration the aim to develop their strategising, organising and negotiating capacities and skills through the collaboration with the CLB (FN7). In all strategic decisions regarding activities to be conducted or steps to be taken in specific labour struggles, KGWU leaders and organisers have the lead and are only supported by CLB with strategic advice.

In summary, KGWU's strategic turn towards developing a proactive, factory-centred collective bargaining approach and the related changes in constructing spaces of organising have also been accompanied by changes in how KGWU constructs spaces of collaboration. Following the realisation that lasting improvements for workers beyond the mere implementation of minimum labour standards can only be brought about through building associational and workplace bargaining power and engaging employers in collective bargaining, KGWU leaders decided to only engage in collaborations that would help the union to develop their collective bargaining strategy. As a result, KGWU now constructs spaces of collaboration with interna-tional organisations, not primarily as spaces for mobilising moral power resources. Instead, KGWU focuses on constructing collaborations as learning spaces where union leaders, organisers and activists can develop their strategic union-building and collective capacities.

The following section describes how KGWU leaders and activists put strategic learnings from the collaboration with CLB into practice. To this end, I zoom in on a struggle for collective bargaining and union recognition that KGWU led between 2017 and 2018 at one of the two target factories called Shahi 8.

7.3.3 Spaces of Contestation

Since the shift to the new factory organising, collective bargaining approach, KGWU has constructed spaces of contestation primarily around collective worker issues that can be linked to the union's collective bargaining strategy. In this line, KGWU has led several struggles at the two selected target factories and in other major Bangalore export-garment factories to address problems affecting a large number of workers. In the remainder of this section, I will zoom in on a struggle for union recognition and collective bargaining that KGWU conducted at the target factory Shahi 8. In this struggle, KGWU activists implemented the pro-active factory mapping and organising strategy developed in strategy meetings with the CLB. Following this strategy, KGWU formed a factory union, developed a charter of demands with workers and handed it over to the management. The management, however, reacted with a violent attack on and suspension of the elected worker leaders. These events were followed by a ten-week-long struggle by KGWU activists for the reinstatement of the suspended worker activists and KGWU's recognition as official collective bargaining partner. As an outcome of this struggle, KGWU signed a memorandum of understanding with the management of Shahi, in which Shahi agreed to reinstate the suspended KGWU worker activists and to respect workers' rights to freedom of association. Moreover, the management agreed to hold monthly meetings with KGWU representatives to discuss any collective worker problems in the factory. Despite falling short of engaging the management in collective wage negotiations, signing the memorandum of understanding still represented a significant victory for KGWU, since it established a formal dialogue structure between the union and the management and de facto secured workers' rights to collective organisation.

In the following, I outline the events that led up to the struggle at Shahi in more detail and provide insights into the various practices and relationships through which KGWU constructed the space of contestation around this struggle. To conclude, I will assess to which extent KGWU was able to use the victory of the memorandum of understanding with Shahi to achieve a lasting shift of capital-labour power relations in the workplace and thereby pave the way for a subsequent collective bargaining agreement.

7.3.3.1   KGWU's Struggle at Shahi 8: Background and Chronology of Events

Shahi 8 is a production unit owned by Shahi Exports, one of India's largest garment exporters. The production unit Shahi 8 employs about 3,000 workers and is located in the West of Bangalore on Magadi Road, slightly on the outskirts of the Bangalore urban area. KGWU had been in contact with workers from the factory since 2011 and handled some individual worker grievances under the area-based organising approach. KGWU organisers started to intensify their organising efforts in the factory after it was selected as a target factory in March 2017. For the rest of the year 2017, KGWU full-time organisers invested significant time in building a membership base in the factory through the strategic factory mapping process (see Sect. 7.3.1). By the end of 2017, KGWU had reached a number of around 140 members, amounting to an organisation rate of about 5% of the total workforce. In January 2018, KGWU held a general body meeting with all members from the factory, during which collective demands were developed and factory worker representatives were elected. The collective demands defined at this meeting were: (1) access to clean drinking water for all workers; (2) reliable and safe bus transportation for workers; (3) a wage increase of 3,000 Rupees per month for all workers.

In the weeks after the general body meeting, KGWU worker representatives collected approximately 700 signatures from Shahi 8 workers in support of the collective Charter of Demands. During this time, factory managers and supervisors began calling workers from different departments for meetings, advising them not to sign any documents from the union. On April 2nd, two worker representatives gave the management a formal letter introducing their collective demands. In the following two days, violent attacks from managers and supervisors on KGWU full-time activists as well as on elected worker representatives took place. When KGWU activists came to the factory on April 3rd to collect worker representatives' signatures on a copy of the charter of demands, the activists were circled by managers and forbidden to leave the factory premises for about three hours. On April 4th, one of the elected worker representatives arrived ten minutes late to work and was stopped by a group of factory managers and supervisors, who attacked him verbally and physically. When other worker representatives and union members came to aid their colleague, another group of workers siding with the management came out of the factory and attacked the unionised workers. In the end, out of the 15 unionised workers who had been attacked, five had to be treated in hospital. The management, in turn, framed the attack on the unionised workers as a clash between two groups of workers and suspended all 15 unionised workers under the pretence of having instigated violence in the factory.

In the following twelve weeks, KGWU conducted a struggle for the reinstatement of the KGWU worker representatives and the right to freedom of association and collective bargaining. In this struggle, KGWU leaders and organisers constructed and intertwined different sets of relationships with various actors—including labour department officers, police officers, international labour rights NGOs, consumer networks, and brands. The struggle finally led to the signing of the memorandum of understanding between the Shahi management and KGWU on the June 25th 2018, in which the Shahi management agreed to reinstate all suspended workers and to respect workers' rights to Freedom of Association and Collective Bargaining. In this line, the management also agreed to hold monthly meetings with KGWU to discuss issues in the factory and to conduct free and secret elections to the mandatory workplace committees. Figure 7.6 provides a graphic representation of the timeline of events.

In the following, I will lay out the different practices and relationships through which KGWU constructed the space of contestation around the Shahi 8 case and examine to which extent KGWU activists and workers were able to develop strategic capacities and to activate different sets of power resources within the various relationships constructed at multiple levels.

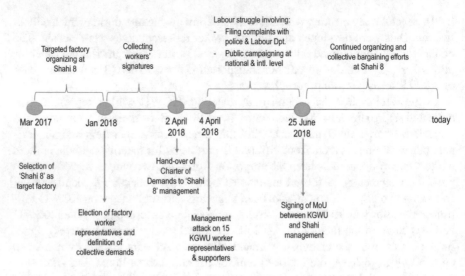

Fig. 7.6 Timeline of events of KGWU's struggle at Shahi 8 in 2018. *Source* Author based on interview data

7.3.3.2   Filing Complaints with the Labour Department and the Police at the State Level

As a first measure, immediately after the attack on the unionised workers, KGWU activists filed complaints at the police department and the labour department against the management for attacking KGWU worker representatives. In both cases, KGWU's practices of constructing relations with higher officials at the state level were decisive for leveraging institutional power vis-à-vis management. In the case of the police complaint, KGWU activists found that on the day of the attack, the management had already filed a criminal case at the local police station against KGWU's worker representatives for instigating violence inside the factory. As a consequence, the police officers at the local station refused to register KGWU's complaint against the management. Since KGWU activists had experienced this type of situation in the past, they had already developed the practice of contacting the police division higher in rank. Having established this contact in previous situations allowed KGWU activists to quickly file their complaints without having to inquire about the right person to contact. By filing the police complaint against the managers involved in the attack, KGWU ensured that their version of events was officially recorded, providing a counterweight to the complaint made by the management.

Moreover, to increase the weight of their complaint and maximise the chance that the police would take action, KGWU activated additional institutional and moral power resources by registering their complaint under the Prevention of Atrocities Scheduled Castes and Scheduled Tribes Act of 1989. The Indian Prevention of Atrocities Act states that any kind of discrimination or violence against members of 'scheduled castes', presents a criminal offense of particular gravity. Therefore, complaints filed under this Act are usually given priority by the police and state authorities. Since three of the attacked KGWU worker representatives belonged to the Dalit caste—a caste formerly considered as 'untouchables', which is now registered as a 'scheduled caste'—KGWU filed the complaint against the management attack under the Prevention of Atrocities Act. This strategy proved successful since several prominent newspapers with national coverage, including the Deccan Herald, published articles on the management attack on workers at Shahi 8.[2]

As mentioned before, in addition to the complaint at the police department, KGWU filed a complaint at the labour department against the violation of the right to freedom of association and against the illegal suspension of the KGWU worker activists. However, as explained in Sect. 6.5, filing a complaint at the labour department is usually not a very effective way for unions to appeal violations of freedom of association: Usually, the management's word stands against the union's word. As a result, disputes around violations of the right to freedom of association are usually referred to the court for adjudication, where they may be pending for several years. In the Shahi 8 case, KGWU organisers, therefore, combined filing the complaint at the labour department with making an emergency call directly to the Bangalore deputy labour commissioner while the attack at the Shahi 8 factory was happening. Due to the gravity of the attack, the deputy labour commissioner immediately sent a labour inspector to the factory to conduct an independent inquiry. In his report, the labour inspector confirmed that a management attack on KGWU worker leaders had occured. Again, here KGWU organisers benefitted from having already established a relationship with the deputy labour commissioner in previous interactions, allowing them to act quickly.

KGWU's practice of constructing strategic relationships with high-level state officials also played an essential role in the subsequent conciliation process at the labour department. When the Shahi management blocked the conciliation meetings and the assistant labour commissioner chairing the meetings remained passive about it, KGWU organisers personally approached the labour commissioner to ask for support. Creating attention for the case at the highest level within the state labour department led the assistant labour commissioner to finally take on a more active stance during the meetings and to demand the management to make concessions, as a KGWU representative explains: "We used that [the personal contact with labour commissioner] just to keep the labour department a little more in our favour. Normally they easily collude with the management" (INT48). As a result of this more active role by the assistant labour commissioner and combined pressure on the Shahi management from various public campaigns, KGWU finally achieved a settlement in the conciliation process. This settlement included the reinstatement of the suspended workers and the signing of the memorandum of understanding.

---

[2] The article in the Deccan Herald published on 5 April 2018 under the title "HR Staff Booked over Beating Workers for Joining Union" was, however, removed from the newspaper's website on 7 April after pressure from Shahi and replaced by an article portraying the assault as a clash between workers (for more details see WRC 2018).

It is important to note that the settlement was achieved by KGWU not only by leveraging institutional power resources through filing complaints at the police and labour department. It was through the combination of institutional power resources with various other coalitional power resources that KGWU was able to push the Shahi management into signing the memorandum of understanding. The relationships and practices through which KGWU could leverage these power resources will be laid out in the next section.

7.3.3.3   Seizing Coalitional Power Resources from Relationships with International Labour Rights Organisations

In terms of leveraging coalitional power resources in the Shahi 8 case, two types of relationships were of particular importance for KGWU: (1) the relationship with the CLB which provided KGWU with resources in form of strategic advice and (2) a new relationship with the WRC, which helped KGWU to put pressure on brands and thereby to activate their leverage over local manufacturers.

The collaboration with the CLB proved important in the Shahi 8 case since regular counselling with the CLB helped KGWU develop their strategic actions and plan the next step. In this context, the CLB also encouraged and trained KGWU organisers to develop a public social media campaign targeting the Shahi management. With the support of CLB, KGWU organisers, who had not engaged with social media before set up a Twitter campaign that generated considerable public attention for the Shahi 8 case in India. According to KGWU leaders, the campaign contributed significantly to make the Shahi management sign the memorandum of understanding. The success of the campaign was linked to KGWU's strategy of seizing the public attention of the wedding of Anand Ahuja, owner of Shahi Exports, to Bollywood star Sonam Kapoor in May 2018—one month after the attack on KGWU worker representatives at Shahi 8. Since Sonam Kapoor had publicly endorsed feminist positions in the past, KGWU took the opportunity to address her publicly on Twitter and ask her to take a stand for the rights of women workers at her fiancé's company. Whereas the idea of starting a social media campaign had come from the CLB, KGWU used their knowledge of territorially embedded social relations to leverage moral power resources vis-à-vis the Shahi management.

In addition to setting up the public campaign at the national level, KGWU also constructed relationships with international consumer organisations to leverage pressure from brands on the Shahi management. To this end, KGWU activists engaged with the WRC. According to KGWU's honorary president, the decision to engage with the WRC was motivated by the fact that by letting the WRC handle contacts with brands, KGWU would be able to concentrate their resources on pushing the conciliation process at the labour department forward and on keeping the suspended KGWU activists engaged in the union's activities. Hence, KGWU had established a division of labour with the WRC, in which KGWU activists provided detailed information about ongoing events and the WRC, in turn, handled communication with brands. In addition, WRC put pressure on brands by publishing an independent investigation report laying out the details of the management attack on KGWU worker activists (WRC 2018). The wide attention in international and Indian press and social media that the report received led brands to finally put pressure on the Shahi management to sign the memorandum of understanding with KGWU.

As shown, the combination of different types of coalitional relationships helped KGWU to develop the strategic capacities and power resources that led to the signing of the memorandum of understanding with the Shahi management. Strategic advice from the CLB enabled KGWU leaders and organisers to leverage moral power over Shahi through the Twitter campaign at the national level. The collaboration with the WRC, in turn, allowed KGWU to leverage indirect pressure on the Shahi management from brands at the international level as a secondary tactic.

7.3.3.4 From the MoU to Collective Bargaining at the Workplace Level?

It is important to note that the strategic practices undertaken by KGWU in the Shahi 8 struggle constructed spaces of contestation mainly outside the factory. Direct confrontations with the management took place either at the labour department or in network media spaces but not at the workplace. This shift from the workplace to other places and network spaces as central arenas of contestation was due to two reasons: First, after the suspension of the KGWU worker representatives, the management had installed an atmosphere of fear in the factory, threatening to dismiss anyone who would talk to KGWU. Second, due to the upcoming Karnataka state elections in May 2018, the police did not grant KGWU permission to hold public protests outside the factory. As a result, workers from Shahi 8 participated only marginally in the space of contestation. KGWU's involvement with workers was limited to informing workers about the ongoing developments in the struggle through home visits.

As a result, after signing the memorandum of understanding and the return of the 15 workers to the factory, KGWU faced severe difficulties in turning this victory into sustained workplace bargaining power. In a secret ballot election for worker representatives conducted by the management as part of the memorandum of understanding, KGWU worker representatives could not win the majority vote and hence failed to be elected as worker representatives. KGWU leaders attribute this failure to the following reasons: First, the elections had taken place shortly after worker activists' reinstatement, giving KGWU limited time and opportunities to use the victory of the memorandum of understanding to organise workers. Second, the immediate heat of the struggle in response to the attack had already worn off among worker leaders. Moreover, KGWU had lost members in the factory during the months following the attack, characterised by limited worker involvement. Hence, KGWU activists and worker leaders had to start over with the organising and collective bargaining process, as this union coordinator explains:

> It was very difficult initially. We thought after they were reinstated we would increase the numbers. That was what all of us were expecting but which didn't happen. But now slowly

the membership is increasing. I think 40, 50 they have enrolled beyond that 140 members we had when we handed over the charter of demands. (INT48)

As a consequence of these difficulties, KGWU had in March 2019 not yet been able to negotiate any issues beyond workers' day-to-day grievances in the monthly meetings with the Shahi 8 management. KGWU's honorary president reports that they have tried to use the monthly meetings with the Shahi management to try to negotiate higher wages, but without success:

> Wage is a topic which has been difficult to address. As soon as you raise it, they [the management] reject it. So, they are not even willing to talk too much because they say the industry is going through a very bad period. (INT48)

Against this backdrop, the outcome of KGWU's struggle at Shahi 8 exemplifies once more that coalitional and moral power resources can only reinforce but never substitute associational workplace bargaining power. Given that KGWU's membership base at Shahi 8 was only around 5% of the total workforce, the Shahi management did not feel pressured to recognise KGWU as a collective bargaining partner. At the same time, with the official commitment of the Shahi 8 management to recognise workers' rights to freedom of association and collective bargaining in the memorandum of understanding, KGWU had paved the way for an open organising process in the factory. However, organising the majority of workers in the factory is a long process requiring sustained organising efforts. Efforts by KGWU to engage the Shahi management in collective wage negotiations are, therefore, still ongoing.

7.3.4   Discussion: Lessons from KGWU's Case

Which lessons can we draw from KGWU's case with regard to which types of interactions and relations enable unions in garment producing countries to build lasting bargaining power vis-à-vis employers? I suggest that we can learn three important lessons from KGWU's case: First, regarding interactions and relations with workers in spaces of organising, KGWU's case highlights both the new opportunities and the challenges for organising in larger tier one supplier factories. Past studies have primarily stressed the new opportunities for collective bargaining for unions in garment producing countries resulting from the emergence of new, strategic large tier one suppliers with enhanced financial capacities, which are able to endure through prolonged periods of worker strikes (see e.g. Kumar 2019a, b). KGWU's experiences partially confirm this argument since the increasing concentration of production in the Bangalore export-garment cluster in large factories represented one central enabling condition in KGWU's strategic turn towards a factory organising, collective bargaining approach. At the same time, however, the union faced significant challenges when implementing this strategic approach in a large tier one garment factory with about 3,000 workers. The large number of workers combined with tight manager and supervisor control inside the factory made it difficult for KGWU to reach a significant membership level in the workplace, forcing the union to initiate a collective bargaining process with an organisation rate of only about 5% of the total workforce. As a result, KGWU was not able to counter the management's repression after the handover of the charter of demands with collective industrial action. Therefore, KGWU could push through only a part of the demands, excluding demands for higher wages. KGWU's experiences, therefore, indicate that past studies stressing exclusively the enabling aspects of the emergence of large tier one suppliers for unions' collective bargaining strategies might have been overly optimistic (cf. Kumar 2019b).

With regard to interactions and relations with external actors in spaces of collaboration, KGWU's case shows that collaborations with other labour actors, which are characterised by balanced power relations and a shared strategic action frame can have enabling effects on unions' capacities to build associational, organisational and workplace bargaining power. As opposed to collaborations with Northern NGOs or consumer networks that have been in the focus of previous studies (see e.g. Anner 2015; Fink 2014; Merk 2009), KGWU's collaboration with the China Labour Bulletin illustrates a different type of transnational collaboration that is constructed primarily as a learning space, in which KGWU leaders, organisers and activists can develop strategic capacities. KGWU's collaboration further illustrates that unions in garment producing countries may also construct transnational collaborations as South-South co-operations—as opposed to the predominant focus in past studies on North–South labour-consumer co-operations in the garment GPN (see e.g. Hauf 2017; Zajak et al. 2017).

Lastly, with regard to the practices and interactions with employers, brands and allies in spaces of contestation, KGWU's struggle at Shahi 8-has once more highlighted that when spaces of contestation are constructed primarily outside of the workplace, and without the active participation of workers, victories in a specific labour struggle may not easily be translated into workplace bargaining power. In the case of KGWU's struggle for collective bargaining at Shahi 8, after almost three months of public campaigning that did not involve workers in the factory, KGWU leaders and organisers had to start building the membership in the factory almost from scratch again. Therefore, despite the memorandum of understanding granting KGWU the rights to collective organisation and bargaining, the union could not use these rights immediately. Hence, KGWU's experience highlights the importance of constructing spaces of collaboration through networked relationships that also involve workers. Only under these circumstances do workers get a chance to have first-hand experiences of collective action and strengthen their collective mind-set and strategic capacities—both central conditions for building unions' associational and organisational power resources and thereby unions' bargaining power vis-à-vis employers (cf. Lévesque and Murray 2010).

The following section summarises and synthesises the central findings and lessons from the case studies of all three Bangalore garment unions.

7.4   Interim Conclusion: Networked Labour Agency and Lessons for Building Sustained Union Power 'at the Bottom' of GPN

In this chapter, I have analysed the agency strategies of three local garment unions that are active in the Bangalore export-garment cluster: the Garment and Textile Workers Union (GATWU), the Garment Labour Union (GLU) and the Karnataka Garment Workers Union (KGWU). Specifically, I have examined the extent to which the agency strategy followed by each union enabled unions to build sustained bargaining power vis-à-vis employers and thereby achieve lasting improvements for workers. To this end, I have conceptualised unions' agency strategies as emerging at the intersection of three spaces of labour agency constituted through intertwining sets of practices and relationships: (1) spaces of organising constituted through unions' practices of building relationships with workers as (potential) union members; (2) spaces of collaboration constituted through unions' practices of constructing solidary or collaborative relationships with external organisations such as other unions, NGOs or consumer organisations; and (3) spaces of contestation constructed by unions around specific labour struggles through practices of targeting employers and lead firms on the one hand, and through strategically activating coalitional power resources from external actors on the other hand.

Based on the empirical analysis, I propose that we can distinguish between two basic strategic approaches, which can be observed as part of the three case study unions' historical evolution: (1) a strategic approach combining practices of organising workers at the community level, of collaborating with international donor NGOs and consumer networks, and of tackling basic labour rights violations; and (2) a strategic approach combining practices of organising workers in selected target factories, of maintaining solidary relations with international labour organisations and networks, and of negotiating collective bargaining agreements with employers.

The first strategic approach, which I call the *community organising, fire-fighting approach*, has been prevalent among all three garment unions in the first decade of union agency in the Bangalore export-garment cluster from 2005 until 2015: Given that 85% of workers in the Bangalore export-garment industry are women and first-time industrial workers, in their early years, the main aim of unions was to familiarise workers with the concepts of unionisation and collective action and to win workers trust. To this end, unions constructed spaces of organising primarily through community organising practices that addressed garment workers not only in their identities as wage labourers but also in their identities as women, mothers, wives and community members. In this line, not the factory but the community represented the central physical space of organising. Community organising practices were linked to workplace action mainly through full-time union organisers' practices of taking up individual workers' grievances and intervening with the management.

Consequently, spaces of contestation were constructed by unions in their early years, mainly around individual labour rights violations. Given the lack of a strong membership base inside the factory, full-time union organisers relied primarily

on interventions by state authorities to pressure management into taking corrective action. Hence, constructing and maintaining relationships with international
consumer networks were central practices of constructing spaces of collaboration
under the community organising, fire-fighting approach. In addition, unions relied
on maintaining close collaborations with local and international NGOs to acquire
financial resources through funded projects, allowing unions to pay for organisers'
salaries, office spaces and community organising activities.

Under the community organising, fire-fighting approach, all three local garment
unions achieved essential improvements for workers by stopping various large-scale
labour rights violations, such as non-payment of minimum wages. These improvements in working conditions were also facilitated by the consolidation of the garment
industry after the end of the quota regime in 2005. As a result, many smaller factories
closed, and production was increasingly concentrated in larger, tier one supplier factories integrated into brands' social auditing regimes. Nevertheless, workers' reality
was still characterised by below-subsistence wages, high work pressure and abusive
behaviour by supervisors. At the same time, unions found that with a community organising, fire-fighting strategic approach, they could not gain concessions
concerning wages or other benefits that would incur additional costs on employers.
Due to the focus on community organising, union members were distributed over a
large number of factories. As a result, unions were not able to deploy industrial action
to pressure employers into collective bargaining and thereby achieve improvements
for workers beyond minimum labour standards.

Against this background, over the past five to seven years, all unions have undergone a strategic reorientation process that involved implementing elements of a
second strategic agency approach that I call the *strategic factory organising, collective bargaining approach*. Under this approach, unions have started to shift their
organising activities from the community to the factory as the primary physical
space for organising. Unions now construct spaces of organising through practices of selecting target factories, forming factory committees and training workers
leaders, who should then lead the organising and collective bargaining process in the
respective factories. For at least two of the three unions, KGWU and GATWU,
restructuring spaces of organising was linked also to restructuring the practices
through which they construct spaces of collaboration. Instead of investing time
and personnel resources into maintaining relations with donor NGOs from the
Global North and with consumer organisations, these two unions now concentrate
on building collaborations with international labour organisations and networks.

This shift was also motivated by the GATWU's and KGWU's realisation that
maintaining close collaborations with international donor NGOs and campaigning
networks constrained their capacities for building a strong union base on the ground.
In the context of these collaborations, significant time and personnel resources were
spent on performing research and documentation activities for funders—resources
that could, in turn, not be invested in organising activities. Against this background,
both GATWU and KGWU started constructing spaces of collaboration through
building close relationships with labour organisations and networks that focussed
on joint strategy development. Hence, the types of coalitional power resources that

GATWU and KGWU sought to build and access through these collaborations were not financial resources and 'borrowed' moral power, but rather strategic knowledge and solidarity based on shared experiences.

The strategic knowledge and capacities developed through collaborations with international labour organisations and networks enabled GATWU and KGWU to construct spaces of contestation that allowed the unions to advance collective bargaining processes in target factories. KGWU and GATWU now limit the deployment of networked agency strategies involving consumer organisations and brands to labour rights violations linked to workplace organising and collective bargaining campaigns in specific factories.

Following this strategic organising, collective bargaining approach, GATWU achieved to sign a collective bargaining agreement with the Avery Dennison management for workers at the company's Bangalore plant—the first collective bargaining agreement signed by a local garment union in India. KGWU signed a memorandum of understanding with India's largest garment export company, Shahi Exports. In this memorandum, the management officially commits to respecting workers' rights to freedom of association and collective bargaining at its Bangalore-based plant Shahi 8.

Albeit generally, a strategic shift from the community organising, fire-fighting approach to a factory organising, collective bargaining approach can be observed among Bangalore export-garment unions, this shift has also been characterised by various tensions. These tensions can be observed most strikingly in GLU's practices, which show the strongest path dependency on the community organising, fire-fighting approach. On the one hand, GLU has introduced elements of a strategic factory organising approach into their practices of constructing spaces of organising by selecting four target factories and trying to establish factory union committees in these factories. On the other hand, GLU continues to invest significant personnel and time resources into community or area-based organising practices, such as providing family and psychological counselling and gathering workers from large numbers of factories in junction meetings. Target factory committee members are recruited through these practices. As a result, GLU organisers find it difficult to recruit a stable group of workers from each target factory for committee meetings. GLU's strategic focus on community and area-based organising practices, in turn, needs to be understood as inextricably linked to GLU's continued practices of constructing spaces of collaboration around funded projects with international NGOs. However, the institutional logic of these funded project collaborations focused on maximising outreach through loose ties with workers clashes with the institutional logic of a strategic factory organising approach. The latter, in contrast, requires unions to focus resources on building strong, long-term ties with selected groups of workers. As a result of these tensions, GLU's practices of constructing spaces of organising also remain rooted predominantly in the fire-fighting approach. As a result, by the time the research was conducted, GLU had not yet achieved the necessary membership strength in any factory for engaging in collective bargaining with employers.

Tensions between a strategic factory organising, collective bargaining approach and practices of fire-fighting are present in the agency approaches of GATWU and KGWU as well. GATWU's most significant challenge lies in working with limited financial and personnel resources after ending collaborations with international NGOs for funded projects. Whereas ending these collaborations has enabled GATWU to concentrate all their resources on organising work, the fact that GATWU has no more full-time, paid organisers also limits the unions' abilities to conduct several struggles at the same time. During the extended struggle at Avery Dennison, which went on for almost two and a half years, union leaders' resources were bound, and collective bargaining processes in other factories were put on hold. At the same time, worker leaders have taken on a more active role in negotiating workers' everyday grievances and problems with management independently.

In the case of KGWU, tensions in shifting from a community organising approach to a strategic-factory organising, collective bargaining approach were, in turn, manifested in the unions' difficulties to organise a majority of workers in the Shahi 8 factory despite employing strategic factory-centred organising practices. Even though KGWU organisers employed a snowballing organising model to systematically organise workers from different departments at Shahi 8, training sessions and discussions with these workers took place outside of the factory due to a prevalent anti-union climate inside the factory. For the same reason, during the struggle for collective bargaining at Shahi 8, KGWU was not able to mobilise workplace protests and instead had to rely on leveraging institutional power and moral power resources by filing complaints at the police and the labour department and through conducting public 'naming and shaming' campaigns with support from the China Labour Bulletin and the Worker Rights Consortium. Nevertheless, with the signing of the memorandum of understanding, KGWU has achieved to open the factory as a safe space for organising and can now continue their organising activities *inside* the factory. KGWU's experiences show that the shift from a community organising, fire-fighting approach to a strategic factory organising collective bargaining approach needs to be understood as a long-term process rather than as a radical break.

In summary, which general lessons can we draw from the analyses of GATWU's, GLU's and KGWU's agency strategies regarding the enabling conditions for building sustained union bargaining power in garment producing countries? I propose that we can draw three main lessons. First, the analysis has shown that the practices through which unions construct spaces of organising matter: Under repressive employer and state regimes, direct workplace organising may not always be an option. Moreover, especially in sectors with a high share of first-generation industrial and migrant workers, unions may need to win workers' trust and familiarise workers with the concepts of collective organisation and unionisation. Hence, to get a foothold among workers, it can be beneficial for unions to develop organising strategies that address workers not primarily as wage labourers but rather as women or community members and that build on the collective experiences of workers in these contexts (cf. Jenkins 2013; Doutch 2021). At the same time, the analysis has shown that community organising approaches cannot replace but merely pave the way for more focused workplace organising strategies that actively seek to develop workers' strategic capacities for collective action. In the end, to achieve concessions from employers, unions' strongest leverage that unions have over employers is their associational power that can be deployed for industrial action at the workplace (cf. Kumar 2019a).

Second, the analysis has shown that the types of collaborations that unions construct with external actors matter: Collaborations with external actors can enable unions to build associational and organisational power resources, but they can also constrain unions' capacities to do so. As we have seen in the analysis, in particular relationships of local unions with international donor NGOs for funded project work can have a rather constraining effect with regard to building unions' associational and organisational power resources when asymmetrical power relations characterise these relations. Asymmetric power relations are present when central planning and decision-making capacities are centralised with NGOs, while local unions depend on funding from these NGOs for daily organising activities. In these cases, union leaders and full-time organisers are accountable primarily to NGOs as funders instead of being primarily accountable to the union's members. This external accountability, in turn, poses constraints for building participatory and democratic union relations, which are, however, crucial for building organisational and associational power resources (see Lévesque and Murray 2010). Moreover, when unions depend on financial resources tied to specific projects led by international NGOs, the project framework will likely shape the practices through which unions construct relationships with workers. Since NGO-funded projects tend to rely on an institutional logic of maximising outreach, unions are required to maximise the number of worker engagements, e.g. through training or counselling activities. However, this requirement contrasts with unions' need to build closer ties with smaller groups of workers who can then act as worker leaders in their respective factories. Therefore, to ensure that collaborations with external actors strengthen unions' associational and organisational power, local unions must retain strategic decision-making competences. In this regard, collaborations with other labour organisations and unions can be particularly fruitful since, in these collaborations, unions are more likely to develop strategic capacities (see also Fütterer and López Ayala 2018).

Third, the analysis has also shown that how unions construct spaces of contestation matters: when unions construct spaces of contestation primarily as network spaces that are detached from workers' territorially embedded everyday spaces, workers have little opportunity to be part of the struggle and hence to develop strategic capacities and to make first-hand experiences of collective organisation. This is the case, for example, when unions rely on filing complaints with state authorities and transnational consumer campaigns as *primary* power resources in a struggle. Whereas unions often use such an approach to constructing spaces of contestation to compensate for low associational and organisational power resources, exclusively relying on institutional and moral power resources does little to help unions to build the power resources they are lacking.

To build unions' associational and organisational power resources it is therefore of strategic importance for unions to involve workers into spaces of contestation, if not through industrial action, then through other forms of symbolic collective action. On the other hand, the analysis has shown that transnational consumer campaigns can have enabling effects for building sustained local union bargaining power when unions deploy them as *secondary* power resources to support struggles for collective organisation and bargaining (see also Kumar 2014). In these cases, moral power resources leveraged through public campaigns can reinforce and strengthen local unions' organising and union-building efforts and thereby help to shift the capital labour-power balance in favour of workers.

The next chapter concludes this study with a summary of the most important empirical results and this study's empirical and theoretical contributions.

References

Anner M (2015) Social downgrading and worker resistance in apparel global value chains. In: Newsome K, Taylor P, Bair J, Rainnie A (eds) Putting labour in its place: labour process analysis and global value chains. Palgrave Macmillan, London and New York, pp 152–170

Clean Clothes Campaign (2018) Lost and found: H&M's living wage roadmap. https://turnaroun dhm.org/static/background-hm-roadmap-0f39b2ebc3330eead84a71f1b5b8a8d4.pdf. Accessed 28 Aug 2021

Doutch M (2021) A gendered labour geography perspective on the Cambodian garment workers' general strike of 2013/2014. Globalizations 18:1406–1419. https://doi.org/10.1080/14747731. 2021.1877007

ExChains (2015) ExChains network: strategy for strengthening the negotiation power of South Asian Garment Trade Unions. http://www.exchains.org/exchains_newsletters/2015/ExChains_ strategy_eng_screen.pdf. Accessed 8 Jan 2022

FEDINA (nd) Introduction. https://fedina.in/introduction-2. Accessed 31 Dec 2021

FEMNET (nd) Munnade and GLU (Garment Labour Union), India. https://femnet.de/en/what-we-do/dialog-networks/partner-organisations/1594-munnade-india.html. Accessed 31 Dec 2021

Fink E (2014) Trade unions, NGOs and transnationalization: experiences from the ready-made garment sector in Bangladesh. ASIEN 130:42–59

Fütterer M, López Ayala T (2018) Challenges for organizing along the garment value chain. Experiences from the union network TIE ExChains. https://www.rosalux.de/en/publication/id/39369/challenges-for-organizing-along-the-garment-value-chain/. Accessed 5 Apr 2022

Hauf F (2017) Paradoxes of transnational labour rights campaigns: the case of play fair in Indonesia. Dev Chang 48:987–1006. https://doi.org/10.1111/dech.12321

International Union League for Brand Responsibility (2021) Who we are. https://www.union-lea gue.org/who. Accessed 8 Jan 2022

Jenkins J (2013) Organizing 'spaces of hope': union formation by indian garment workers. Br J Ind Relat 51:623–643. https://doi.org/10.1111/j.1467-8543.2012.00917.x

Katz C (2004) Growing up global: economic restructuring and children's everyday lives. University of Minnesota Press, Minneapolis

Kumar A (2014) Interwoven threads: building a labour countermovement in Bangalore's export-oriented garment industry. City 18:789–807. https://doi.org/10.1080/13604813.2014.962894

Kumar A (2019a) A race from the bottom? Lessons from a workers' struggle at a Bangalore warehouse. Compet Chang 23:346–377. https://doi.org/10.1177/1024529418815640

Kumar A (2019b) Oligopolistic suppliers, symbiotic value chains and workers' bargaining power: labour contestation in South China at an ascendant global footwear firm. Global Netw 19:394–422. https://doi.org/10.1111/glob.12236

Lévesque C, Murray G (2010) Understanding union power: resources and capabilities for renewing union capacity. Transf Eur Rev Labour Res 16:333–350. https://doi.org/10.1177/102425891037 3867

Merk J (2009) Jumping scale and bridging space in the era of corporate social responsibility: cross-border labour struggles in the global garment industry. Third World Q 30:599–615. https://doi.org/10.1080/01436590902742354

Selwyn B (2013) Social upgrading and labour in global production networks: a critique and an alternative conception. Compet Chang 17:75–90. https://doi.org/10.1179/1024529412Z.00000000026

Worker Rights Consortium (2015a) Compensation secured for worker whose child died in care of factory nursery (India). https://www.workersrights.org/communications-to-affiliates/compensation-secured-for-worker-whose-child-died-in-care-of-factory-nursery-india/. Accessed 31 Dec 2021

Worker Rights Consortium (2015b) Safety risks at factory-based childcare centers. Gokaldas Exports Ltd., Bangalore, India. https://www.workersrights.org/wp-content/uploads/2019/12/WRC-Gokaldas-Report-062215.pdf. Accessed 31 Dec 2021

Worker Rights Consortium (2018) Worker Rights Consortium Assessment. Shahi Exports Pvt. Ltd., Bangalore India. Findings and Recommendations. https://www.workersrights.org/wp-content/uploads/2018/06/WRC-Assessment-re-Shahi-India-6.20.18.pdf. Accessed 31 Dec 2021

Zajak S (2017) International allies, institutional layering and power in the making of labour in Bangladesh. Dev Chang 48:1007–1030. https://doi.org/10.1111/dech.12327

Zajak S, Egels-Zandén N, Piper N (2017) Networks of labour activism: collective action across Asia and beyond. An introduction to the debate. Dev Chang 48:899–921. https://doi.org/10.1111/dech.12336

Part VI
Conclusion

Chapter 8
Conclusion: Lessons for Building Union Power in Garment Producing Countries and Benefits of a Relational Approach for Analysing Labour Control and Labour Agency in GPNs

Abstract This chapter summarises central findings in light of the posed research questions and discusses the empirical and conceptual contributions of this book. In terms of empirical contributions, the book highlights the central role of local worker organisations in improving working conditions in the garment industry while simultaneously revealing the complex, networked labour control structures that constrain the terrain for labour agency in garment producing countries. Against this background, unions need to develop networked agency strategies that employ coalitional and moral power resources from international consumer and labour organisations to open up space for workplace organising and collective bargaining. Conceptually, the relational approach for studying labour control and labour agency in GPNs developed in this book contributes to reinvigorating a relational understanding of labour dynamics in GPNs as constituted through power-laden, networked relationships at the vertical and horizontal dimension of the GPN. Thereby the book addresses a gap in past scalar analyses, which have not sufficiently explored the specific links between network dynamics and territorial outcomes for labour at specific nodes of a GPN. The chapter concludes with final reflections on challenges and strategies for improving working conditions in the global garment industry and directions for further research.

Keywords Global production networks · Garment industry · Working conditions · Labour control · Labour agency · Practice-oriented approaches · Relational approaches

This study has set out to explore the conditions that enable and constrain the capacities of local unions in garment producing countries to build sustained bargaining power vis-à-vis capital and state actors and thereby bringing about lasting improvements for workers. To this end, I have integrated theoretical concepts from research on labour control and labour agency in GPNs with the analytical perspective of relational and practice-oriented approaches within economic geography. Building on the concepts of the labour control regime and of labour's networked agency as central heuristics, this study has been guided by two central research questions:

© The Author(s) 2023
T. López, *Labour Control and Union Agency in Global Production Networks*, Economic Geography, https://doi.org/10.1007/978-3-031-27387-2_8

1. How do labour control regimes at specific nodes of the garment GPN—constituted through place-specific articulations of processual labour control relations at the horizontal and vertical dimension of the GPN with localised labour processes—shape and constrain the terrain for the agency of workers and unions in garment producing countries?
2. Which relationships and routinised interactions allow unionists and workers in garment producing countries to develop strategic capacities and power resources that enable them to shift the capital-labour power balance in favour of workers lastingly?

In this chapter, I first summarise the central findings of this study in relation to each research question before highlighting the main empirical and theoretical contributions. The chapter concludes by reflecting on the challenges and pathways for improving working conditions in the global garment industry and indicating directions for further research.

8.1 Answering the Research Questions: Summary of Central Findings

8.1.1 Labour Control Regime in the Bangalore Export-garment Cluster and Constraints for Union Agency

To answer the first research question, I have developed a practice-oriented, relational approach for analysing labour control regimes in GPNs. I have conceptualised labour control regimes at specific nodes of a GPN as emerging from place-specific articulations of various 'network' and territorially embedded processual relations of labour control with localised labour processes (see Sect. 3.2). In this context, I have identified six processual relations of labour control that fulfil exploiting or disciplining functions and/or contribute to securing the broader conditions for capital accumulation at specific nodes of a GPN. These processual relations of labour control are: (1) *sourcing relations* that link global lead firms within a GPN with local suppliers at the various nodes of the GPN; (2) *wage relations* linking workers, employers and state actors within a specific region, state or country; (3) *workplace relations* constituted through the interactions between workers and management in a specific site of production; (4) *industrial relations* constituted through territorially embedded relationships between employers and their organisations, workers and their organisations, and the state in a specific state, sector or country; (5) *employment relations*, i.e. the relationship between employers and workers, in which workers sell their labour power to an employer; and (6) *labour market relations* linking employers, workers and (potentially) third-party actors such as contract labour or recruiting agencies and training organisations. Through the lens of this relational, practice-oriented heuristic framework of labour control regimes in GPNs, I have then explored how the labour control regime in the Bangalore export-garment cluster constrains and shapes the agency strategies of three local garment unions (see Chap. 6).

The results of the empirical analysis highlight that major challenges for the agency of unions result from the complex intersections and interdependencies between the above-mentioned six processual relations of labour control. Due to the intricate intertwinings of more localised and spatially more extensive processual relations, employers' exploitation and disciplining practices are often shaped or enabled by the practices of other actors located at more or less geographical distance. For example, the practices through which Bangalore export-garment manufacturers construct labour process, workplace, wage and industrial relations are *shaped* directly by retailers' predatory purchasing practices at the vertical dimension of the GPN: To remain profitable in the face of retailers' practices of squeezing prices, placing irregular orders and demanding shorter lead times, Bangalore garment manufacturers exercise tight control over workers' productivity, rely on wage theft practices to keep production costs low, engage in practices of 'hiring and firing' to flexibilise employment relations, and perform union-busting practices to mitigate collective worker organisation. These practices are, in turn, *enabled* through pro business state practices, undermining unions' capacities to contest illegal exploiting and disciplining practices through legal channels. Moreover, the exploiting and disciplining practices deployed by Bangalore garment manufacturers intersect with and deliberately exploit broader social power asymmetries along the lines of age, gender and geographical origin. For example, manufacturers deliberately recruit predominantly young women from rural areas within Karnataka and increasingly also from North-East India since these workers are less likely to resist exploitation practices.

In a nutshell, the empirical analysis has highlighted three ways in which the labour control regime in the Bangalore export-garment cluster constrains the terrain for the agency of the three local case study unions. First, unions' opportunities for leveraging *structural power resources* are constrained by the specific practices through which retailers construct sourcing relations, and through which manufacturers organise the labour process. By constructing spatially asymmetric relationships with suppliers, retailers are able to play manufacturers in different locations off against each other and pressurise suppliers into offering lower prices (Sect. 6.2). Consequently, garment manufacturers in the Bangalore export-garment cluster also continuously seek novel strategies to keep wages low. Most export-garment companies operating in Bangalore maintain regional factory networks, enabling employers to pass price pressures on to workers. Employers argue that due to retailers' price pressures, they cannot increase wages and would be forced to shift production to lower-wage locations in India if workers attempted collective bargaining (see Sect. 6.3). Spatial asymmetries between transnational retailers and local manufacturers are hence reproduced at a smaller scale in the relations between manufacturers and workers. As a result, workers' and unions' capacities to exercise workplace bargaining power vis-à-vis employers through production stoppages are limited. Workers constantly face the risk that either manufacturers or retailers shift production orders to another factory. Besides limited workplace bargaining power, workers and unions in the Bangalore export-garment cluster also possess limited marketplace bargaining power. Since manufacturers organise the labour process in an assembly line system, with more complex operations being increasingly automated, the majority of tasks in Bangalore garment factories can be carried out by unskilled or semi-skilled workers (see Sect. 6.1). This rather low-skilled work profile, in turn, enables managements to balance local labour shortages by recruiting migrant workers from rural areas that have undergone a three-month training in one of the many private and public training centres set up under the Integrated Skill Development Scheme (see Sect. 6.7). As a result, workers possess limited workplace *and* marketplace bargaining power, which in turn constrains unions' abilities to implement proactive collective bargaining strategies that could achieve improvements for workers beyond the legally prescribed minimum standards.

Second, limitations for unions' capacities to exercise power over employers through industrial and workplace action also result from the various constraints that the labour control regime places on unions' opportunities for *building associational power*. These constraints result, on the one hand, from Bangalore export-garment manufacturers' disciplining practices directed at preventing collective worker organisation. These disciplining practices include constructing factories as tightly controlled spaces, reproducing gendered structures of domination on the shop floor (Sect. 6.4), and a range of union-busting practices directed at discouraging workers from engaging with unions (Sect. 6.5). On the other hand, challenges for unions to build and leverage associational power resources also result from the spatial restructuring from Bangalore garment manufacturers' practices of expanding the labour market frontier: In the face of an increasing shortage of unskilled labour in the Bangalore urban area, garment manufacturers are moving factories to the outskirts of the city or neighbouring rural areas within the State of Karnataka and are also increasingly recruiting workers from villages near Bangalore. Consequently, there is an increasing spatial separation between workplaces and workers' living areas, with many workers being transported to the factory in company buses from distances of up to 80 kms (Sect. 6.7). As a result, unions face severe challenges for organising workers outside of the factory after their shifts or in their communities— two organising strategies that unions have for a long time relied on in the face of the tight management control inside factories.

Third and last, pro-state business practices constrain workers' and unions' *institutional power resources*. India has traditionally possessed strong labour legislation offering workers various means to challenge illegal exploiting and disciplining practices such as requesting labour inspections, filing complaints with the labour department or filing a lawsuit in the labour court. With the general shift towards neoliberal politics in India's post-liberalisation area, these traditional sources of institutional power have, however, been dwindling. In the context of a general political climate that prioritises the creation of a business-enabling environment over the implementation of labour rights, many labour officers refrain from taking an active stance for workers in industrial dispute settlement mechanisms, thereby paving the way for an increasing employer dominance in industrial relations (Sect. 6.5). Moreover, due to the chronic understaffing of labour courts, processes are often dragged on for several years until a ruling is made. As a result, unions' capacities to use institution-alised dispute settlement or legal mechanisms as institutional power resources for challenging employers' illegal exploiting and disciplining mechanisms are severely constrained.

Despite these constraints for the agency of unions on the Bangalore export-garment cluster, the empirical analysis has also shown that workers and unions have, nevertheless, achieved to stop or transform specific practices of exploitation or disci-plining through networked agency strategies. These strategies simultaneously target multiple actors through combined actions at various levels. GATWU, for example, achieved to stop employer and state practices of delaying minimum revisions by targeting retailers, the state and employers at the same time through combined work-place action, local public protests and transnational consumer campaigns. Unions have also used pressure from transnational worker and consumer campaigning networks to achieve interventions by retailers and, thereby, to stop employers' illegal union-busting strategies. In doing so, unions have opened up spaces for organising and collective dialogue in selected factories. In these factories, unions were able to stop particularly harsh exploiting practices such as verbal abuse of workers or wage theft through 'giving comp-offs'. Hence, the analysis has also showcased that the labour control regime as an institutionalised framework for capital accumula-tion is not only unilaterally imposed on workers and unions by state and capital actors. Instead, workers and unions also challenge and transform and thereby co-shape the practices and relations that constitute the labour control regime through their everyday actions and struggles.

In the next section, I turn towards the second research question and summarise insights into the different ways in which unions in the Bangalore garment unions construct networked agency strategies. Specifically, I recapitulate the most impor-tant findings regarding which relationships and routinised interactions have allowed workers and unionists in the Bangalore export-garment cluster to develop strategic capacities and power resources.

8.1.2 Building Unions' Strategic Capacities and Power Resources in Relational Spaces of Labour Agency

To answer the second research question addressing the potential of different agency strategies for enabling unionists and workers to build strategic capacities and power resources, I have developed a relational heuristic framework for analysing union agency (see Sect. 3.3). Building on the concepts of 'Networks of Labour Activism' and worker and union power resources, I have developed a heuristic framework for analysing unions' strategic approaches through the lens of three intersecting 'spaces of labour agency'. These spaces are: (1) *spaces of organising* constituted through unions' practices of building relationships with workers as (potential) union members; (2) *spaces of collaboration* constituted through unions' collaborative relationships with external organisations such as other unions, NGOs or consumer organisations; and (3) *spaces of contestation* constructed by unions around specific labour struggles through practices of targeting employers, lead firms and (in some cases) state actors, and through practices of 'drawing' allies such as other workers, consumer or labour rights groups into struggles. Spaces of labour agency are hence relational networks of processual relations and routinised interactions, within which workers and unions can potentially develop the strategic capacities and power resources that ultimately enable unions to build lasting bargaining power vis-à-vis capital and state actors.

The empirical analysis of the three Bangalore-based garment unions has highlighted that unions construct spaces of organising, collaboration and contestation through different routinised interactions and relationships—with varying implications for developing workers' and unionists' strategic capacities and power resources (see Chap. 7). In the analysis, I have identified two stylised strategic agency approaches characterised by different practices of constructing spaces of organising, collaboration and contestation. I have labelled these two approaches as (1) the 'community organising, fire-fighting' approach and (2) the 'strategic factory organising, collective bargaining' approach. In the following, I will summarise the main practices and relations that characterise each approach as well as the potential and limits of each approach for building sustained union bargaining power.

Under the *community organising, fire-fighting approach*, unions construct spaces of organising mainly at the community level. Relationships with workers are not constructed primarily around workplace issues but also around issues of workers' everyday life in the household and the community through organising saving groups and area committees, or providing counselling for family problems. To this end, unions usually collaborate closely with local community organisations. Community organising strategies were prevalent especially in the early phase of the three garment unions, which emerged as independent unions from an NGO-led support project for garment workers. In this context, the community organising approach responded to the specific composition of the workforce in the Bangalore export-garment cluster. Being mostly female, first-generation industrial workers from rural areas, many garment workers in the cluster were unfamiliar with the concept of unionisation and unions as collective organisations. This lack of awareness, combined with tight employer control in the workplace and patriarchal power structures, made traditional workplace organising unviable. Against this backdrop, building relationships with workers through community organising strategies enabled unions to gain workers' trust, foster collective experiences and thereby build associational power at the local level. However, this organising strategy also had limits in building bargaining power vis-à-vis employers. While organising workers in their communities allowed unions to mobilise relatively large numbers of workers for punctual public protests, it did not enable unions to build a strong membership base inside specific factories. A strong membership at the factory level is, however, a precondition for the exercise of workplace bargaining power through industrial action.

Unions' early community organising strategies need to be understood as inter-related with the practices of constructing spaces of collaboration that characterised unions' early stages. In their early years, unions constructed collaborations with external actors, mainly around projects funded by NGOs from the Global North. As mentioned before, the three unions emerged from an NGO-led garment worker support project funded by Oxfam International, through which the unions' activities and full-time organisers were paid. As for community organising strategies, the empirical analysis has highlighted mixed effects of project collaborations with Northern NGOs for unions' capacity to build sustained bargaining power. On the one hand, funding from NGOs provided unions with the financial means to sustain their organising and community work when unions were unable to sustain themselves through membership fees. On the other hand, collaborations with NGOs for funded projects had constraining effects on unions' capacities to engage in strategic factory organising and foster internal union democracy. Constraints for strategic factory organising resulted from the primary objective of these NGO-funded projects, which was usually not to strengthen union-building processes but instead to provide aid for garment workers. In this light, the project's success was not measured against its contributions to strengthening a unions' bargaining power but instead against the number of workers reached through the project. This 'institutional logic' of maximising outreach (c.f. Egels-Zandén et al. 2015) hence requires unions to create loose ties with a large number of workers rather than strong ties with a smaller number of workers from specific factories. Moreover, funded project collaborations with Northern NGOs tended to bind significant personnel resources for documenting and research tasks—resources that, in turn, could not be invested in union building and worker organising.

The focus on building relations with many workers at the community level and the resulting lack of a strong associational power base inside factories also influenced how unions constructed spaces of contestation under the community organising, fire-fighting approach. As the term 'fire-fighting' suggests, unions mainly constructed spaces of contestation in a reactive manner by addressing labour rights violations reported to the union by individual workers. In the face of the constraints for unions to exercise leverage on employers through industrial action, unions relied primarily on the moral power of transnational consumer campaigns to harness the leverage of retailers over garment manufacturers and thereby achieve corrections of labour rights violations in specific factories. However, as the empirical analysis has shown, relying on the borrowed moral power of transnational consumer campaigning networks had limited potential in building lasting union bargaining power vis-à-vis employers. The scope of issues that could be addressed through transnational consumer campaigning strategies was limited to reacting to particularly harsh labour rights violations. Therefore, transnational consumer campaigns alone did not enable unions to pressure employers into collective bargaining processes and achieve improvements for workers beyond the mere implementation of minimum labour standards.

In the face of the limitations of the community organising, fire-fighting approach for building union bargaining power vis-à-vis employers, the three Bangalore garment unions have—to varying extents—implemented a different strategic agency approach over the past years. This approach prioritises the factory as space for organising and seeks to foster collective bargaining with employers. Under this *strategic factory organising, collective bargaining* approach, unions have started to concentrate their organising activities on selected target factories to build strong ties with a core group of workers from each factory. These workers then act as worker representatives and organisers *inside* the factory. Complementary to this shift in spaces of organising from the community to the workplace, two of the three Bangalore garment unions have constructed new spaces of collaboration. Instead of focussing their resources on constructing alliances with transnational consumer networks and NGOs, these unions are now building collaborations with labour organisations and networks. In this context, the empirical analysis has exemplified that such collaborations between labour actors can represent important spaces for union organisers and workers to gain strategic knowledge—for example, about the value chain structure—and to develop strategic capacities. These capacities include, for example, strategy development capacities for designing networked agency strategies that use transnational solidarity only as an instrument to open up spaces for organising and collective bargaining in target factories.

As a result, with the shift to a factory organising, collective bargaining approach, unions have shifted the practices through which they construct spaces of contestation as well. With increased associational power inside factories and enhanced capacities to engage in collective action, the workplace has become the central arena in struggles for collective bargaining agreements. Transnational campaigning strategies and labour solidarity, in turn, are employed by unions increasingly only as a secondary source of leverage. Following such a networked agency approach, the Bangalore garment union GATWU signed in 2019 the first collective bargaining agreement in the history of the union. As opposed to previous victories under the 'fire-fighting' approach, with this bargaining agreement, Avery Dennison not only committed to refrain from union-busting practices but also to grant workers benefits and wages beyond the legally prescribed minimum (see Sect. 7.1.2).

However, it is important to note that in the everyday practices of all three garment unions, up to date, tensions exist between practices rooted in the community organising, fire-fighting approach and unions' declared strategic goals of factory organising and collective bargaining. The two presented stylised models of strategic union agency approaches, hence, need to be understood as opposite ends of a spectrum, with Bangalore garment unions currently finding themselves somewhere in between. Nevertheless, the findings from the empirical analysis have provided important insights into the potentials and limitations of the two agency strategies for building sustained union bargaining power.

Together with the insights from the analysis of the labour control regime and resulting constraints for union agency, this study, hence, makes several critical empirical contributions to debates in labour geography and GPN analysis about how to build sustained union power in garment producing countries. These empirical contributions will be discussed in the next section.

8.2 Empirical Contributions of This Study: Lessons for Building Local Union Power in Garment Producing Countries

Which lessons can we learn from the analysis presented in this study for building local union power and improving working conditions in the global garment industry? Four essential teachings shall be pointed out here that are valuable for local unions in garment producing countries as well as for labour rights and consumer organisations in the Global North concerned with improving labour conditions in the garment industry.

First, the empirical analysis has highlighted that a significant challenge for improving working conditions in the global garment industry results from the highly complex structural frameworks for labour exploitation and capital accumulation at the various nodes of the garment GPN. These structural frameworks—designated in this study as labour control regimes—are constructed and reproduced through the intertwined practices of a multitude of actors located in more or less distant places, who all seek to extract and appropriate surplus value from living labour (c.f. Cumbers et al. 2008). As a result, to challenge institutionalised frameworks for labour exploitation, local unions need to develop networked agency strategies, which allow unions to target capital and state actors simultaneously at multiple levels. This lesson contradicts the arguments made by earlier studies in labour geography that different types of labour control regimes in garment producing countries—e.g. market, state, employer control regimes—are conducive to different agency strategies by local unions, e.g. wildcat strikes, engaging in multi-stakeholder initiatives and transnational labour organising (see e.g. Anner 2015a). Instead, the findings from this study highlight how local unions—in the face of complex, networked labour control regimes—need to intertwine all of the aforementioned strategies to contest and transform structural relations of exploitation (see also Tufts 2007; Wills 2002). In this context, this study has exemplarily illustrated how the local garment union GATWU has achieved stopping employer and state practices of undermining and circumventing statutory minimum wage revisions through a networked minimum wage campaign (see Sect. 7.1.1). Strategic actions within the campaign comprised (1) conducting symbolic protests in the workplace, (2) holding public protests at the local level, (3) filing a lawsuit to contest illegal state practices of withdrawing already issued minimum wage notifications and (4) harnessing the leverage of lead firms over manufacturers through transnational consumer campaigns. It was only through the combination of all these actions that GATWU was able to stop the interrelated set of employer and state practices that had prevented the implementation of mandatory minimum wage increases. Hence, this study makes a case for a heightened sensitivity towards the networked character of capital and state-produced labour control structures. It has shown that no scale can be singled out as particularly dominant within the local labour control regime in the Bangalore export-garment cluster. Consequently, unions must develop agency approaches that combine and strategically intertwine actions at *various* scales.

Second, the study has illustrated that in countries or regions where the garment industry is characterised by a highly feminised workforce or by a high share of migrant, first-generation industrial workers, traditional workplace organising strategies focusing exclusively on economic demands may not be conducive. In these contexts, it can be helpful for unions to construct initial relationships with workers through organising practices that address workers not only as wage workers but also as women, mothers, daughters, migrants and community members (see also Doutch 2021; Jenkins 2013). Such organising practices can help to raise workers' awareness of intersecting lines of structural exploitation along categories of class, gender and geographical origin and thereby to foster workers' collective mindset— an essential precondition for building associational and organisational power (c.f. Lévesque and Murray 2010). Therefore, the findings from this study align with the arguments of prior work by labour geographers and researchers stressing the potential of 'community' or 'social movement' unionism approaches for building union power in countries of the Global South (see e.g. Moody 1997; Nowak 2017). In contrast to prior work, this study, however, also highlights the risks that come with organising strategies that prioritise the community as space for organising at the expense of more targeted workplace organising strategies: While community organising strategies can enable unions to build associational power at the local level, they do not enable unions to build a strong membership base inside the factory. However, such a membership base inside the factory is necessary for unions to be able to exercise workplace bargaining power through industrial action. As a result, unions that rely exclusively on a community organising strategy are likely to limit their scope of action to correcting individual labour rights violations since they do not possess the necessary workplace bargaining power to engage in proactive collective bargaining. Therefore, to achieve improvements for workers beyond the implementation of basic minimum labour standards, community organising practices need to be combined with a targeted workplace organising strategy.

Third, the results of this study call for heightened sensitivity to the mixed effects that collaborations between local unions in garment producing countries and NGOs in consumer countries from the Global North have on local unions' capacities for building sustained bargaining power vis-à-vis employers. This lesson is particularly relevant since the global garment industry has seen a proliferation of such North-South collaborations since the 1990s with the rise of transnational anti-sweatshop movements and multi-stakeholder initiatives (Zajak et al. 2017: 914; see also Anner 2015b; Esbenshade 2004; Fütterer and López Ayala 2018; Hauf 2017; Merk 2009). At the same time, this study shows that while such collaborations can help local unions to access and leverage different types of coalitional power resources (e.g. financial resources from funded projects, moral power resources from consumer campaigns), North-South collaborations can also hamper unions' abilities to develop organisational and associational power resources. Financial flows from Northern NGOs to local unions in garment producing countries are often linked to strict accountability regimes requiring unions to document all activities and expenses. As a result, union organisers invest significant time into activities that do not directly contribute to the union-building process. Moreover, when unions sustain themselves primarily through external funding and not through membership fees, the union leadership becomes primarily accountable to external donors instead of being primarily account-able to the union members. This shift from internal accountability to external account-ability of the union leadership, in turn, hampers internal union democracy. This study has hence provided further evidence for arguments from past studies that external funding for local unions can be a double-edged sword (see e.g. Banse 2016; Fink 2014).

Similarly, this study has pointed out potential hampering effects for building local unions' associational and organisational power resources linked to local unions' engagement with transnational consumer campaigning networks. Especially when unions rely exclusively on moral power resources from consumer campaigns and on the leverage of retailers to compensate for a lack of associational power in the workplace, this strategy may create a path dependency that limits unions' capaci-ties to engage in workplace organising and collective bargaining. In transnational campaigns, NGOs from the Global North take on strategic planning and decision-making capacities while the role of unions and workers is reduced to providing information about and documentation of labour rights violations. As a result, neither unionists nor workers develop the strategic capacities that are necessary to build the unions' associational and organisational power resources in the long term. However, as this study and others have pointed out, victories achieved through transnational campaigning strategies tend to be temporary and limited to correcting particularly harsh labour rights violations (see e.g. Anner 2015b; López and Fütterer 2019; Ross 2006). Therefore, building associational and organisational power resources is central for unions to lastingly shift the capital-labour power balance in favour of workers and achieve lasting and comprehensive improvements for workers (see also Kumar 2014, 2019; Oka 2016).

As a result of the third lesson, this study has, fourth, highlighted the importance of transnational collaborations and union strategies that prioritise the development of unionists' and workers' strategic capacities and collective experiences as a vehicle for building sustained union bargaining power. In this regard, this study has, on the one hand, showcased the potential of transnational collaborations with other labour organ-isations that—as opposed to collaborations with consumer organisations—prioritise union building and collective bargaining processes as a long-term goal. In doing so, these collaborations can provide network spaces for mutual learning, planning and strategy development as well as for the exercise of transnational solidarity to combat union-busting practices, and thereby open up wiggle room for local union building and collective bargaining in garment producing countries (see also Lohmeyer et al. 2018; López and Fütterer 2022). On the other hand, this study has highlighted the importance of unions in garment producing countries to actively involve workers in planning and decision-making processes in specific labour struggles and to foster collective experiences of resistance. Such experiences are, in turn, essential to build the associational and organisational power resources that will allow a union to shift the power balance between employers and unions in the long term. In this context, this study has shown that in cases where unions do not have sufficient associa-tional power at the factory level to engage in industrial action, symbolic protests in the workplace can be an instrument for fostering workers' active participation and collective resistance experience in a specific labour struggle. Developing workers' strategic capacities is particularly important to mitigate professionalisation processes that lead to the increasing centralisation of decision-making and planning competencies on full-time union staff (see e.g. Choudry and Kapoor 2013; Fink 2014; Fütterer and López Ayala 2018). Such a centralisation of strategic capacities weakens unions since they hamper the development of a strong second-rank leadership in the workplace, which can serve as a nucleus for organising and which can negotiate everyday problems with the management independently.

In summary, this study has highlighted that to achieve lasting improvements for workers, unions not only need to construct collaborations with external actors in ways that foster unionists' and workers' strategic capacities but also need to construct internal union relations in horizontal and democratic ways. In local contexts characterised by a highly feminised workforce and strong patriarchal social relations—as in many Asian garment producing countries—fostering horizontal and democratic internal union relations may, therefore, also require unions to actively combat internal gendered power asymmetries by fostering women leadership (see also Doutch 2021; Evans 2017).

8.3   Theoretical Contributions of This Study: Producing New Insights Through a Relational Analytical Perspective

Besides offering important empirical findings regarding the challenges and strategies for building sustained union bargaining power and improving working conditions in garment-producing countries, this study has made several theoretical contributions to current debates within economic and labour geography. Specifically, the relational analytical approach presented in this book contributes to advancing theoretical concepts and debates within three strands of research in economic and labour geography: (1) research on labour in GPNs; (2) GPN analysis more generally and (3) practice-oriented research in economic geography.

8.3.1   Contributions to Research on Labour Control and Labour Agency in GPNs

Most importantly, by introducing a relational approach as an alternative to dominant scalar approaches, this book centrally advances the theoretical discussion of labour control and labour agency in GPNs. As illustrated in the literature review (Chapter 2), the dominance of scalar heuristics has limited past studies' capacity to recognise the deeply relational nature of the multi-scalar 'labour control architectures' underpinning GPNs and of workers' multi-scalar agency strategies in GPNs. The relational analytical approach developed in this book achieves to overcome these limitations by shifting the analytical focus from pre-defined scales to networks of relationships as a central heuristic.

As a result, the relational analytical approach developed here firstly, achieves to overcome a crucial limitation of past studies on labour control regimes in GPNs regarding their ability to grasp the socio-spatial relations underpinning specific local labour control regimes. These studies have tended to presuppose a universal, hierarchical nested scalar order as characteristic of all labour control regimes in GPNs. This presupposition has limited past studies' capacity to map the empirically existing socio-spatial relations that constitute labour control regimes. In contrast, the relational heuristic framework for studying labour control in GPNs introduced in this book leaves analytical space for carving out the individual and place-specific socio-spatialities of labour control regimes at different nodes of a GPN. Instead of seeking to fit dynamics and relations of labour control into pre given scalar categories, the proposed relational framework takes empirically existing practices, relations and their interrelations as an analytical point of departure and maps their spatial extensions and characteristics. As a result, the here-developed relational approach is more sensitive to the different ways, in which geographically more delimited and spatially more unbounded processes and relationships of labour control shape and enable each other. It, therefore, provides an apt tool for addressing recent calls from labour geographers who have called on researchers to pay increased attention to the "mix of geographically distant and proximate relationships across different scales" (Wickramasingha and Coe 2021) that characterise labour control regimes in GPNs.

Second, the relational approach developed in this study sheds light on another aspect that has remained understudied in past research on labour control regimes, namely the dialectical relationships between labour control and labour agency (see e.g. Hastings and MacKinnon 2017; Wickramasingha and Coe 2021). Past studies working with scalar heuristics have tended to conceptualise labour control regimes as structural contexts at various levels that are unilaterally imposed on local workers (see e.g. Pattenden 2016; Smith et al. 2018). In contrast to this dominant 'top-down' conceptualisation, the here-developed relational approach stresses that labour control regimes as structural contexts are constructed through practices and relationships that are situated in space and time and can be challenged and transformed by workers and unions. Consequently, the here-developed relational, practice-oriented analytical approach can shed light on the 'small transformations' (Latham 2002) of labour control practices and relations achieved by workers and unions through strategies of reworking (Cumbers et al. 2010). Even though such strategies may not challenge hegemonic power and capitalist exploitative relations per se, they may still recalibrate local power relations and thereby redistribute resources in favour of workers. Therefore, the relational, practice-oriented framework developed in this book allows producing analyses that are sensitive to how labour control regimes as institutional frameworks for capital accumulation are produced and continuously transformed in a "dialectical process of interaction" between capital, state and labour actors (Hastings and MacKinnon 2017: 104).

Third, the relational approach to union agency developed in this book can generate enhanced insights into how different processes and relationships of labour control at various stages enable and shape each other and how unions can strategically intertwine actions at various levels. Thereby, the relational approach presented here has expanded the scope of past studies of labour agency in GPNs, which have tended to adopt a one-sided focus on the 'up-scaling' of local labour struggles to the international level while neglecting other scales of agency (see e.g. Anner 2015b; Merk 2009). By giving visibility to how unions deploy strategic actions at *various* levels and how these interplay in building unions' bargaining power, the here-introduced relational approach facilitates understanding labour's networked agency "as constituted by interdependent scales of action that are not nested in a hierarchy privileging one scale over another" (Tufts 2007: 2387). Such an understanding heightens researchers' sensitivity towards the structural effects that collaborations with external actors at various levels may have on internal union relations and practices (see also Zajak et al. 2017). As has been shown in the previous section, only by analysing different scales of action as interrelated can researchers evaluate which types of union collaborations have enabling effects and which types of collaborations have constraining effects for fostering workers' and unionists' strategic capacities.

Fourth and last, the relational approach to labour agency developed in this book allows tackling a blind spot in past research on union agency in GPNs. In past research, internal union relations and their intersections with broader social relations have largely remained a black box (see also Cumbers 2015). By conceptualising internal union relations as a vital dimension of union agency, the relational approach to union agency developed in this book opens this black box. Thereby, it allows for a critical analysis of unions' everyday practices not only with regard to constructing alliances with external actors but also with regard to constructing internal relations between union leadership and members as well as between union members and non-members. As a result, the here-developed relational approach also sharpens our understanding of the intersections of internal union relations with other social relations, such as relations of gender or geographical provenience and the power structures enshrined in these relations. It can, therefore, refine our understanding of the embeddedness of labour agency not only within the structural-relational formations of capital and the state but also within broader, place-specific socio-cultural relations (Coe and Jordhus-Lier 2010; Doutch 2021; Hastings 2016).

8.3.2   Contributions to GPN Analysis

Beyond providing new insights into the dynamics of labour control and labour agency in GPNs, the relational approach developed in this study contributes to reviving and reinvigorating a relational perspective within GPN analysis more broadly. As laid out in Sect. 2.1, early work with the GPN approach was underpinned by a profoundly relational understanding of the global economy as constituted through networked vertical and horizontal relations of production, exchange and consumption (c.f. Dicken et al. 2001). However, in the further evolution of GPN analysis, there has only been sporadic engagement with this incipient relational analytical perspective (Cumbers 2015). As a result, GPN scholars have recently voiced critique towards many (self-attributed) GPN-studies for focussing exclusively on territorial dynamics at the horizontal dimension without systematically exploring their interconnections with "the configuration and operation of the global production network in question" (Coe and Yeung 2019: 788; see also Yeung 2020).

Against this background, the relational approach developed in this book firstly provides an innovative analytical framework for producing empirically grounded demonstrations of the "causal links between […] network dynamics and territorial outcomes" (Coe and Yeung 2019: 778). The focus of this study has been on analysing the conditions and role of labour within GPNs. Nevertheless, the here-developed conceptual approach of analysing place-specific territorial outcomes within GPNs through the lens of interwoven network and territorial relations laden with power can enrich other research areas in GPN analysis as well, such as research at the intersection of GPN analysis and political ecology (see e.g. Bridge and Bradshaw 2017; Dorn and Huber 2020; Irarrázaval and Bustos-Gallardo 2019) or research on the development effects of integrating specific places into GPNs (see e.g. McGrath 2018; Tessmann 2018; Vicol et al. 2019).

Second, the relational approach developed in this book also contributes to a more nuanced understanding of "the relational, networked and institutional qualities of how power is generated and ultimately exercised" in GPNs (Hess 2008: 456; see also Arnold and Hess 2017; Raj-Reichert 2020). Whereas the GPN framework has traditionally conceptualised power predominantly as a static resource held by specific actors within a GPN (Henderson et al. 2002), this study has highlighted that power in GPNs is profoundly dynamic, relational and networked. Power in GPNs is relational, since it only becomes effective in shaping material outcomes when actors exercise it in relation to other actors. Moreover, power within GPNs has a networked character when actors strategically direct flows of power within networks of relationships to exercise leverage over other actors. This study has illustrated networked power on the example of Bangalore-based unions' strategies of leveraging the influence of geographically distant lead firms, consumer and worker groups from the Global North to shift the local power balance with employers. By foregrounding the relationships through which power flows and is exercised, the here-developed relational approach enhances our understanding of the complex power flows within GPNs that "reach and stretch across distances where the lives of others far away are shaped by those nearby and vice versa" (Raj-Reichert 2020: 654).

8.3.3   Contributions to Practice-Oriented Research in Economic Geography

Last but not least, the relational analytical approach to labour control regimes and labour agency developed in this book also contributes to advancing practice-oriented research in economic geography: It provides a novel conceptual tool for theorising the links between micro-scale practices and macro-scale social and economic phenomena. As practice-oriented economic geographers have reiterated: the value but also the central challenge for practice-oriented research lies in demonstrating how "higher order phenomena", such as institutions or class structures, "are enacted, reproduced, and/or transformed through the everyday actions embedded within them" (Jones and Murphy 2010: 372; see also Everts 2016; Wiemann et al. 2019). In this vein, by emphasising the links between the manifold labour control practices and relations that constitute the structural context for worker and union agency in GPNs, the here-developed analysis shows "how context, structures, and individual agency or action come together in the doing of economic and industrial activities" (Jones and Murphy 2010: 3050).

8.4   Final Reflections and Directions for Further Research

This book highlights the challenges for improving working conditions in the global garment industry, which result from uneven power relations between multinational retailers and local manufacturers on the one hand, and between manufacturers and workers, on the other. Despite an increasing consolidation of supplier networks and the emergence of large tier one suppliers over the past 15 years, spatial asymmetries between retailers and suppliers persist. Retailers continue to maintain large and geographically dispersed networks of suppliers and to establish new sourcing relations with manufacturers in ever lower-wage countries, such as Myanmar and Ethiopia. Garment manufacturers, in contrast, are usually forced to concentrate the largest share of their business on a few key buyers due to variations in buyers' technical and social standards and requirements. As a result, especially large retailers like H&M, Inditex or G.A.P are still able to 'squeeze' manufacturers by demanding lower prices, shorter lead times and enhanced flexibility from their suppliers.

Manufacturers pass on the pressures for reducing costs while increasing productivity to workers through a complex web of disciplining and exploiting practices. Whereas strong state control and regulation in garment producing countries could help to mitigate worker exploitation, such a regulatory role of the state conflicts with the aim of governments in many garment producing countries to boost economic development through providing a business enabling environment for capital. Due to its ability to generate employment for the unskilled, rural population and to attract foreign investments, the export-garment industry enjoys a special status in many industrialising countries. As a result, not only in India but also in other garment producing countries, the state and public institutions do not provide a counterweight to the dominance of employers over labour. Instead, they frequently tolerate or actively support illegal employer practices of exploiting and disciplining, such as wage theft or union busting (see e.g. Anner 2022; Hossain 2019; Wickramasingha and Coe 2021).

In light of these complex entanglements of intersecting relationships of domination at the vertical, 'network' dimension of the garment GPN and at the horizontal dimension, i.e. within individual garment producing countries, it becomes evident that 'soft' regulation attempts through codes of conducts and ethical trading initiatives alone can only have a limited effect with regard to improving working conditions. Without strong pressure from labour and consumer organisations, capital and state actors have little incentive for effectively putting the social standards defined in the context of such initiatives into practice. At the same time, this study has highlighted once more that interventions of consumer organisations without the presence of strong local unions can, at best, contribute to correcting and mitigating particularly cruel violations of workers' rights. In contrast, to bring about lasting improvements of garment workers beyond minimum labour standards, strong labour movements are needed that can shift the capital-labour power balance in favour of workers. Shifting the capital-labour power balance in garment producing countries is particularly important to improve workers' wages, which still remain below subsistence levels in most Asian countries. Where statutory minimum wages have been increased in past years, these raises have usually been the result of sustained worker campaigning and strike action, for example, in Bangladesh (Wickramasingha and Coe 2021), Cambodia (Lawreniuk and Parsons 2018) and India (see Sect. 7.1). At the same time, in the face of the highly feminised and informalised nature of work in the garment industry, unions also need to tackle internal patriarchal structures of domination and develop innovative organising approaches that address the struggles of women and informal workers beyond the sphere of production (Doutch 2021; Evans 2017).

In this light, the scope of this and other studies on labour organising in garment production countries consists of a relatively narrow focus on the struggles of workers labouring in tier one garment factories acting as direct suppliers for transnational retailers. Less attention has been paid so far to the challenges and strategies for organising workers in the subcontracted tier two and three segments of the Asian export-garment industry, where work is frequently carried out in the form of piece-based, informal homework production arrangements (see e.g. Mezzadri 2016; Neve 2014). As a result, the labour process in these lower tiers tends to be characterised by a higher level of spatial segmentation and a dilution of the employee-employer relationship, with workers being formally self-employed and relationships with factories often being mediated through a chain of intermediaries. These organisational characteristics bring along distinct challenges for collective organisation, representation and bargaining. At the same time, home-based, subcontracted workers represent the weakest link in the garment value chain, since they fall through the cracks of both social auditing regimes and state regulation. Work in the subcontracted tier two and three segments of retailers' supplier networks is, therefore, often characterised by an even higher level of precarity and insecurity than in retailers' direct supplier factories.

Against this backdrop, a stronger engagement of researchers and unions with workers in informal settings and the conditions that constrain and foster these workers' collective agency is needed. The relational, practice-oriented approach for analysing labour control and labour agency in GPN developed in this book can provide a conceptual starting point for such an engagement by labour geographers and other researchers.

References

Anner M (2015a) Labor control regimes and worker resistance in global supply chains. Labor Hist 56:292–307. https://doi.org/10.1080/0023656X.2015.1042771

Anner M (2015b) Social downgrading and worker resistance in apparel global value chains. In: Newsome K, Taylor P, Bair J, Rainnie A (eds) Putting labour in its place: labour process analysis and global value chains. Palgrave Macmillan, London and New York, pp 152–170

Anner M (2022) National labour control regimes and worker resistance in global production networks. In: Baglioni E, Campling L, Coe NM, Smith A (eds) Labour regimes and global production. Agenda Publishing, Newcastle upon Tyne, pp 191–208

Arnold D, Hess M (2017) Governmentalizing Gramsci: topologies of power and passive revolution in Cambodia's garment production network. Environ Plann A 49:2183–2202. https://doi.org/10.1177/0308518X17725074

Banse F (2016) Geld für Gewerkschaften. Über die Intentionen und Wirkungen gewerkschaftlicher Förderung. PERIPHERIE – Politik. Ökonomie, Kultur 36:289–306. https://doi.org/10.3224/peripherie.v36i142-143.24681

Bridge G, Bradshaw M (2017) Making a global gas market: territoriality and production networks in liquefied natural gas. Econ Geogr 93:215–240. https://doi.org/10.1080/00130095.2017.1283212

Choudry A, Kapoor D (2013) Introduction: NGOization: complicity, contradictions and prospects. In: Choudry A (ed) NGOization: complicity, contradictions and prospects, 1st edn. Zed Books, London, pp 1–23

Coe NM, Jordhus-Lier DC (2010) Re-embedding the agency of labour. In: Bergene AC, Endresen SB, Knutsen HM (eds) Missing links in labour geography. Ashgate Publishing, Farnham, pp 29–42

Coe NM, Yeung HW-C (2019) Global production networks: mapping recent conceptual developments. J Econ Geogr 19:775–801. https://doi.org/10.1093/jeg/lbz018

Cumbers A (2015) Understanding labour's agency under globalization; embedding GPNs within an open political economy. In: Newsome K, Taylor P, Bair J, Rainnie A (eds) Putting labour in its place: labour process analysis and global value chains. Palgrave Macmillan, London and New York, pp 135–151

Cumbers A, Helms G, Swanson K (2010) Class, agency and resistance in the old industrial city. Antipode 42:46–73. https://doi.org/10.1111/j.1467-8330.2009.00731.x

Cumbers A, Nativel C, Routledge P (2008) Labour agency and union positionalities in global production networks. J Econ Geogr 8:369–387. https://doi.org/10.1093/jeg/lbn008

de Neve G (2014) Entrapped entrepreneurship: labour contractors in the South Indian garment industry. Mod Asian Stud 48:1302–1333. https://doi.org/10.1017/S0026749X13000747

Dicken P, Kelly PF, Olds K, Yeung HW-C (2001) Chains and networks, territories and scales: towards a relational framework for analysing the global economy. Glob Netw 1:89–112

Dorn FM, Huber C (2020) Global production networks and natural resource extraction: adding a political ecology perspective. Geogr Helv 75:183–193. https://doi.org/10.5194/gh-75-183-2020

Doutch M (2021) A gendered labour geography perspective on the Cambodian garment workers' general strike of 2013/2014. Globalizations 18:1406–1419. https://doi.org/10.1080/14747731. 2021.1877007

Egels-Zandén N, Lindberg K, Hyllman P (2015) Multiple institutional logics in union-NGO relations: private labor regulation in the Swedish clean clothes campaign. Bus Ethics 24:347–360. https://doi.org/10.1111/beer.12091

Esbenshade JL (2004) Monitoring sweatshops: workers, consumers, and the global apparel industry. Temple University Press, Philadelphia

Evans A (2017) Patriarchal unions = weaker unions? Industrial relations in the Asian garment industry. Third World Q 7:1–20. https://doi.org/10.1080/01436597.2017.1294981

Everts J (2016) Connecting sites: practice theory and large phenomena. Geogr Z 104:50–67

Fink E (2014) Trade unions, NGOs and transnationalization: experiences from the ready-made garment sector in Bangladesh. Asien 130:42 59

Fütterer M, López Ayala T (2018) Challenges for organizing along the garment value chain. Experiences fom the union network TIE ExChains. https://www.rosalux.de/en/publication/id/39369/ challenges-for-organizing-along-the-garment-value-chain/. Accessed on 5 Apr 2022

Hastings T (2016) Moral matters: de-romanticising worker agency and charting future directions for labour geography. Geogr Compass 10:307–318. https://doi.org/10.1111/gec3.12272

Hastings T, MacKinnon D (2017) Re-embedding agency at the workplace scale: workers and labour control in Glasgow call centres. Environ Plann A 49:104–120. https://doi.org/10.1177/030851 8X16663206

Hauf F (2017) Paradoxes of transnational labour rights campaigns: the case of play fair in Indonesia. Dev Change 48:987–1006. https://doi.org/10.1111/dech.12321

Henderson J, Dicken P, Hess M, Coe N, Yeung HW-C (2002) Global production networks and the analysis of economic development. Rev Int Polit Econ 9:436–464. https://doi.org/10.1080/096 92290210150842

Hess M (2008) Governance, value chains and networks: an afterword. Econ Soc 37:452–459. https:// doi.org/10.1080/03085140802172722

Hossain N (2019) Rana Plaza, disaster politics, and the empowerment of women garment workers in Bangladesh. Contemp South Asia 27:516–530. https://doi.org/10.1080/09584935.2019.1683719

Irarrázaval F, Bustos-Gallardo B (2019) Global salmon networks: unpacking ecological contradictions at the production stage. Econ Geogr 95:159–178. https://doi.org/10.1080/00130095.2018. 1506700

Jenkins J (2013) Organizing 'spaces of hope': union formation by Indian garment workers. Br J Ind Relat 51:623–643. https://doi.org/10.1111/j.1467-8543.2012.00917.x

Jones A, Murphy JT (2010) Practice and economic geography. Geogr Compass 4:303–319. https:// doi.org/10.1111/j.1749-8198.2009.00315.x

Kumar A (2014) Interwoven threads: building a labour countermovement in Bangalore's export-oriented garment industry. City 18:789–807. https://doi.org/10.1080/13604813.2014.962894

Kumar A (2019) A race from the bottom? Lessons from a workers' struggle at a Bangalore warehouse. Compet Chang 23:346–377. https://doi.org/10.1177/1024529418815640

Lawreniuk S, Parsons L (2018) For a few dollars more: towards a translocal mobilities of labour activism in Cambodia. Geoforum 92:26–35. https://doi.org/10.1016/j.geoforum.2018.03.020

Latham A (2002) Retheorizing the scale of globalization: topologies, actor-networks, and cosmopolitanism. In: Herod A, Wright MW (eds) Geographies of power: placing scale. Blackwell, Malden, MA, pp 115–144

Lévesque C, Murray G (2010) Understanding union power: resources and capabilities for renewing union capacity. Transfer: Eur Rev Labour Res 16:333–350. https://doi.org/10.1177/102425891 0373867

Lohmeyer N, Schüßler E, Helfen M (2018) Can solidarity be organized "from below" in global supply chains? The case of ExChains. Indust Bezieh 25:400–424

López T, Fütterer M (2019) Herausforderungen und Strategien für den Aufbau gewerkschaftlicher Verhandlungsmacht in der Bekleidungswertschöpfungskette: Erfahrungen aus dem TIE-ExChains-Netzwerk. In: Ludwig C, Simon H, Wagner A (eds) Bedingungen und Strategien gewerkschaftlichen Handelns im flexiblen Kapitalismus. Westfälisches Dampfboot, Münster, pp 175–191

López T, Fütterer M (2022) Warenketten – aktuelle Konflikte und transnationale Solidarität. Zeitschrift Marxistische Erneuerung 33:85–96

McGrath S (2018) Dis/articulations and the interrogation of development in GPN research. Progr Human Geogr 42:509–528. https://doi.org/10.1177/0309132517700981

Merk J (2009) Jumping scale and bridging space in the era of corporate social responsibility: cross-border labour struggles in the global garment industry. Third World Q 30:599–615. https://doi.org/10.1080/01436590902742354

Mezzadri A (2016) Class, gender and the sweatshop: on the nexus between labour commodification and exploitation. Third World Q 37:1877–1900. https://doi.org/10.1080/01436597.2016.1180239

Moody K (1997) Workers in a lean world: unions in the international economy. Haymarket series. Verso, London, New York

Nowak J (2017) Mass strikes in India and Brazil as the terrain for a new social movement unionism. Dev Change 48:965–986. https://doi.org/10.1111/dech.12320

Oka C (2016) Improving working conditions in garment supply chains: the role of unions in Cambodia. Br J Ind Relat 54:647–672. https://doi.org/10.1111/bjir.12118

Pattenden J (2016) Working at the margins of global production networks: local labour control regimes and rural-based labourers in South India. Third World Q 37:1809–1833. https://doi.org/10.1080/01436597.2016.1191939

Raj-Reichert G (2020) The powers of a social auditor in a global production network: the case of Verité and the exposure of forced labour in the electronics industry. J Econ Geogr 20:653–678. https://doi.org/10.1093/jeg/lbz030

Ross RJS (2006) A tale of two factories: successful resistance to sweatshops and the limits of firefighting. Labor Stud J 30:65–85. https://doi.org/10.1177/0160449X0603000404

Smith A, Barbu M, Campling L, Harrison J, Richardson B (2018) Labor regimes, global production networks, and European Union trade policy: labor standards and export production in the Moldovan clothing industry. Econ Geogr 94:550–574. https://doi.org/10.1080/00130095.2018.1434410

Tessmann J (2018) Governance and upgrading in South-South value chains: evidence from the cashew industries in India and Ivory Coast. Global Netw 18:264–284. https://doi.org/10.1111/glob.12165

Tufts S (2007) Emerging labour strategies in Toronto's hotel sector: toward a spatial circuit of union renewal. Environ Plan A 39:2383–2404. https://doi.org/10.1068/a38195

Vicol M, Fold N, Pritchard B, Neilson J (2019) Global production networks, regional development trajectories and smallholder livelihoods in the Global South. J Econ Geogr 19:973–993. https://doi.org/10.1093/jeg/lby065

Wickramasingha S, Coe N (2021) Conceptualizing labor regimes in global production networks: uneven outcomes across the Bangladeshi and Sri Lankan apparel industries. Econ Geogr 98:68–90. https://doi.org/10.1080/00130095.2021.1987879

Wiemann J, Schäfer S, Faller F (2019) Praxistheorie in der Wirtschaftsgeographie. In: Schäfer S, Everts J (eds) Handbuch Praktiken und Raum: Humangeographie nach dem Practice Turn, 1st edn. transcript, Bielefeld, pp 299–316

Wills J (2002) Bargaining for the space to organize in the global economy: a review of the Accor-IUF trade union rights agreement. Rev Int Political Econ 9:675–700. https://doi.org/10.1080/0969229022000021853

Yeung HW-C (2020) The trouble with global production networks. Environ Plan A 53:428–438

Zajak S, Egels-Zandén N, Piper N (2017) Networks of labour activism: collective action across Asia and beyond. an introduction to the debate. Dev Change 48:899–921. https://doi.org/10.1111/dech.12336

Annex I
List of Interviews

Code	Interview partners	Place	Date	Duration	Category
INT1	HR Manager of garment export factory	Bangalore, Karnataka, India	09.03.2017	30 min	Factory manager
INT2	Regional Secretary, National Garment Industry Association I	Bangalore, Karnataka, India	09.03.2017	30 min	Industry association
INT3	Social Compliance Manager/Production Manager of garment factory	Bangalore, Karnataka, India	10.03.2017	30 min	Factory manager
INT4	Garment Labour Union (GLU) leaders	Bangalore, Karnataka, India	11.03.2017	180 min	Local garment union
INT5	Garment workers	Bangalore, Karnataka, India	11.03.2017	90 min	Garment workers
INT6	Garment and Textile Workers Union (GATWU) leaders	Bangalore, Karnataka, India	13.03.2017	60 min	Local garment union
INT7	Garment workers	Bangalore, Karnataka, India	13.03.2017	30 min	Garment workers
INT8	General Secretary, Cividep	Bangalore, Karnataka, India	13.03.2017	90 min	Local labour rights NGO
INT9	Social Compliance Manager of export-garment factory	Bangalore, Karnataka, India	14.03.2017	90 min	Factory manager

(continued)

T. López, *Labour Control and Union Agency in Global Production Networks*,
Economic Geography, https://doi.org/10.1007/978-3-031-27387-2

(continued)

Code	Interview partners	Place	Date	Duration	Category
INT10	Leaders, Garment and Textile Workers Union (GATWU)	Bangalore, Karnataka, India	16.03.2017	90 min	Local garment union
INT11	Country Representative, Fair Wear Foundation	Bangalore, Karnataka, India	16.03.2017	75 min	International labour rights NGO
INT12	Research Head, Centre for Workers Management	Bangalore, Karnataka, India	20.03.2017	60 min	Representative labour rights NGO
INT13	General Secretary of the Karnataka State Committee, Centre of Indian Trade Unions (CITU)	Bangalore, Karnataka, India	21.03.2017	30 min	Local garment union
INT14	Labour researcher, Institute for Social and Economic Change	Noida, New Capital Region, India	22.03.2017	30 min	Labour rights researcher
INT15	Labour researcher, V.V. Giri National Labour Institute	Noida, New Capital Region, India	27.03.2017	60 min	Labour rights researcher
INT16	Leader, New Trade Union Initiative (NTUI)	New Delhi, National Capital Territory of Delhi, India	28.03.2017	80 min	Labour rights researcher
INT17	Country Representative, Worker Rights Consortium	New Delhi, National Capital Territory of Delhi, India	29.03.2017	180 min	International labour rights NGO
INT18	Asia Coordinator, TIE Global Union Network	New Delhi, National Capital Territory of Delhi, India	29.03.2017	30 min	International union network
INT19	South Asia Regional Secretary, IndustriAll Global Union	New Delhi, National Capital Territory of Delhi, India	30.03.2017	70 min	Representative global union
INT20	Representative, Indian National Trade Union Congress (INTUC)	New Delhi, National Capital Territory of Delhi, India	31.03.2017	40 min	Representative national union federation

(continued)

(continued)

Code	Interview partners	Place	Date	Duration	Category
INT21	Senior Coordinator, Nari Shakti Manch	Gurugram, New Capital Region, India	31.03.2017	80 min	Local labour rights NGO
INT22	Labour researcher, Jawaharlal Nehru University	New Delhi, National Capital Territory of Delhi, India	02.04.2017	30 min	Labour researcher
INT23	Labour researcher, Jawaharlal Nehru University	New Delhi, National Capital Territory of Delhi, India	02.04.2017	30 min	Labour researcher
INT24	Garment workers	Bangalore, Karnataka, India	09.04.2017	30 min	Garment workers
INT25	Organisers, Garment Mahila Karmikara Munnade	Bangalore, Karnataka, India	09.04.2017	30 min	Local labour rights NGO
INT26	Leaders, Garment and Textile Workers Union (GATWU)	Bangalore, Karnataka, India	12.04.2017	30 min	Local garment union
INT27	Regional Director, National Garment Industry Association II	Bangalore, Karnataka, India	15.09.2017	105 min	Industry association
INT28	HR Manager, export-garment factory	Bangalore, Karnataka, India	15.09.2017	30 min	Factory manager
INT29	HR Manager, export-garment factory	Bangalore, Karnataka, India	18.09.2017	60 min	Factory manager
INT30	Leaders, Karnataka Garment Workers Union (KGWU)	Bangalore, Karnataka, India	18.09.2017	110 min	Local garment union
INT31	Leaders, Garment and Textile Workers Union (GATWU)	Bangalore, Karnataka, India	20.09.2017	120 min	Local garment union
INT32	Representative, Karnataka State Textile Infrastructure Development Corporation	Bangalore, Karnataka, India	22.09.2017	50 min	State department

(continued)

(continued)

Code	Interview partners	Place	Date	Duration	Category
INT33	Union leader, Garment Labour Union (GLU)	Bangalore, Karnataka, India	23.09.2017	60 min	Local garment union
INT34	Labour researcher, National Law School of India University	Bangalore, Karnataka, India	23.09.2017	70 min	Labour researcher
INT35	Organiser, Karnataka Garment Workers Union (KGWU)	Bangalore, Karnataka, India	25.09.2017	120 min	Local garment union
INT36	Leader, Garment Labour Union (GLU)	Bangalore, Karnataka, India	26.09.2017	120 min	Local garment union
INT37	Researcher, Centre for Workers Management	Bangalore, Karnataka, India	27.09.2017	90 min	Local labour rights NGO
INT38	Representatives, Karnataka State Government Department of Handlooms and Textiles	Bangalore, Karnataka, India	28.09.2017	40 min	State department
INT39	Organisers, Garment Mahila Karmikara Munnade	Bangalore, Karnataka, India	28.09.2017	120 min	Local labour rights NGO
INT40	Asia Coordinator, TIE Global Union Network	New Delhi, National Capital Territory of Delhi, India	09.10.2017	120 min	Global union network
INT41	International Coordinator, Asia Floor Wage Alliance	New Delhi, National Capital Territory of Delhi, India	10.10.2017	80 min	Global union network
INT42	Advisor, National Garment Industry Association II	New Delhi, National Capital Territory of Delhi, India	11.10.2017	30 min	Industry association
INT43	Representative, Apparel Training and Design Centre	New Delhi, National Capital Territory of Delhi, India	12.10.2017	30 min	Industry association

(continued)

(continued)

Code	Interview partners	Place	Date	Duration	Category
INT44	Director, Garment Sourcing Consultancy	Bangalore, Karnataka, India	25.02.2019	60 min	Sourcing company
INT45	Leaders, Karnataka Garment Workers Union (KGWU)	Bangalore, Karnataka, India	27.02.2019	60 min	Local garment union
INT46	Leaders, Garment and Textile Workers Union (GATWU)	Bangalore, Karnataka, India	04.03.2019	180 min	Local garment union
INT47	Leader, Garment Labour Union (GLU)	Bangalore, Karnataka, India	06.03.2019	60 min	Local garment union
INT48	Leaders, Karnataka Garment Workers Union (KGWU)	Bangalore, Karnataka, India	07.03.2019	120 min	Local garment union
INT49	Representative, Bangalore Council, All India Trade Union Congress (AITUC)	Bangalore, Karnataka, India	08.03.2019	90 min	Local garment union
INT50	Representative, Bangalore Council, All India Trade Union Congress (AITUC)	Bangalore, Karnataka, India	12.03.2019	120 min	Local garment union
INT51	Representative, Bangalore Council, All India Trade Union Congress (AITUC)	Bangalore, Karnataka, India	12.03.2019	30 min	Local garment union
INT52	Labour lawyer	Bangalore, Karnataka, India	13.03.2019	90 min	Labour lawyer
INT53	Director, Garment Sourcing Consultancy	Via Zoom	07.04.2021	120 min	Garment sourcing company

Annex II
List of Field Notes

Code	Event	Date	Place
FN1	International union meeting with representatives of Tie, GATWU, ver.di, Karmikara Mahila Munnade, WRC, CWM, NTUI	22.10.2016	Bangalore
FN2	Excursion to garment factory hub along Mysore Road with GATWU union activists	23.10.2016	Bangalore
FN3	Meeting with workers and GATWU union activists at warehouse of garment manufacturer	23.10.2016	Ramanagara
FN4	Meeting of ver.di delegation with garment workers	23.10.2016	Madduur
FN5	Visit of garment factory with ver.di delegation	24.10.2016	Bangalore
FN6	GATWU workers leaders training	19.03.2017	Bangalore
FN7	KGWU union training with China Labour Bulletin	10–11.04.2017	Bangalore
FN8	GATWU union meeting with Avery Dennison workers	24.09.2017	Bangalore
FN9	GATWU training for Avery Dennison contract workers	24.02.2019	Bangalore
FN10	Tie ExChains international network meeting	16–18.03.2019	New Delhi

© The Editor(s) (if applicable) and The Author(s) 2023
T. López, *Labour Control and Union Agency in Global Production Networks*,
Economic Geography, https://doi.org/10.1007/978-3-031-27387-2

Index

© The Editor(s) (if applicable) and The Author(s) 2023
T. López, *Labour Control and Union Agency in Global Production Networks*,
Economic Geography, https://doi.org/10.1007/978-3-031-27387-2

Printed in the United States
by Baker & Taylor Publisher Services